THE ARTIFICIAL INTELLIGENCE INFRASTRUCTURE WORKSHOP

Build your own highly scalable and robust data storage systems that can support a variety of cutting-edge AI applications

Chinmay Arankalle, Gareth Dwyer, Bas Geerdink, Kunal Gera, Kevin Liao, and Anand N.S.

THE ARTIFICIAL INTELLIGENCE INFRASTRUCTURE WORKSHOP

Authors: Chinmay Arankalle, Gareth Dwyer, Bas Geerdink, Kunal Gera, Kevin Liao, and Anand N.S.

Reviewers: Brent Broadnax, John Wesley Doyle, Tim Hoolihan, Rochit Jain, Sasikanth Kotti, Asheesh Mehta, Arunkumar Nair, Madhav Pandya, Ashish Patel, Shovon Sengupta, and Ashish Tulsankar

Managing Editor: Ashish James

Acquisitions Editors: Manuraj Nair, Royluis Rodrigues, Kunal Sawant, Anindya Sil, Archie Vankar, and Karan Wadekar

Production Editor: Shantanu Zagade

Editorial Board: Megan Carlisle, Samuel Christa, Mahesh Dhyani, Heather Gopsill, Manasa Kumar, Alex Mazonowicz, Monesh Mirpuri, Bridget Neale, Dominic Pereira, Shiny Poojary, Abhishek Rane, Brendan Rodrigues, Erol Staveley, Ankita Thakur, Nitesh Thakur, and Jonathan Wray

First published: August 2020

Production reference: 2020321

ISBN: 978-1-80020-984-8
Published by Packt Publishing Ltd.
Livery Place, 35 Livery Street
Birmingham B3 2PB, UK

WHY LEARN WITH A PACKT WORKSHOP?

LEARN BY DOING

Packt Workshops are built around the idea that the best way to learn something new is by getting hands-on experience. We know that learning a language or technology isn't just an academic pursuit. It's a journey towards the effective use of a new tool—whether that's to kickstart your career, automate repetitive tasks, or just build some cool stuff.

That's why Workshops are designed to get you writing code from the very beginning. You'll start fairly small—learning how to implement some basic functionality—but once you've completed that, you'll have the confidence and understanding to move onto something slightly more advanced.

As you work through each chapter, you'll build your understanding in a coherent, logical way, adding new skills to your toolkit and working on increasingly complex and challenging problems.

CONTEXT IS KEY

All new concepts are introduced in the context of realistic use-cases, and then demonstrated practically with guided exercises. At the end of each chapter, you'll find an activity that challenges you to draw together what you've learned and apply your new skills to solve a problem or build something new.

We believe this is the most effective way of building your understanding and confidence. Experiencing real applications of the code will help you get used to the syntax and see how the tools and techniques are applied in real projects.

BUILD REAL-WORLD UNDERSTANDING

Of course, you do need some theory. But unlike many tutorials, which force you to wade through pages and pages of dry technical explanations and assume too much prior knowledge, Workshops only tell you what you actually need to know to be able to get started making things. Explanations are clear, simple, and to-the-point. So you don't need to worry about how everything works under the hood; you can just get on and use it.

Written by industry professionals, you'll see how concepts are relevant to real-world work, helping to get you beyond "Hello, world!" and build relevant, productive skills. Whether you're studying web development, data science, or a core programming language, you'll start to think like a problem solver and build your understanding and confidence through contextual, targeted practice.

ENJOY THE JOURNEY

Learning something new is a journey from where you are now to where you want to be, and this Workshop is just a vehicle to get you there. We hope that you find it to be a productive and enjoyable learning experience.

Packt has a wide range of different Workshops available, covering the following topic areas:

- Programming languages

- Web development

- Data science, machine learning, and artificial intelligence

- Containers

Once you've worked your way through this Workshop, why not continue your journey with another? You can find the full range online at http://packt.live/2MNkuyl.

If you could leave us a review while you're there, that would be great. We value all feedback. It helps us to continually improve and make better books for our readers, and also helps prospective customers make an informed decision about their purchase.

Thank you,
The Packt Workshop Team

Table of Contents

Preface i

Chapter 1: Data Storage Fundamentals   1

Introduction 2

Problems Solved by Machine Learning 3

    Image Processing – Detecting Cancer in Mammograms
with Computer Vision 3

    Text and Language Processing – Google Translate 4

    Audio Processing – Automatically Generated Subtitles 6

    Time Series Analysis 7

Optimizing the Storing and Processing of Data
for Machine Learning Problems 7

Diving into Text Classification 8

    Looking at TF-IDF Vectorization 9

Looking at Terminology in Text Classification Tasks 11

    Exercise 1.01: Training a Machine Learning Model
to Identify Clickbait Headlines 12

Designing for Scale – Choosing the Right Architecture
and Hardware 21

    Optimizing Hardware – Processing Power,
Volatile Memory, and Persistent Storage 21

    Optimizing Volatile Memory 23

    Optimizing Persistent Storage 24

    Optimizing Cloud Costs – Spot Instances and Reserved Instances 25

Using Vectorized Operations to Analyze Data Fast 26

    Exercise 1.02: Applying Vectorized Operations to Entire Matrices 27

    Activity 1.01: Creating a Text Classifier for Movie Reviews 33

Summary 36

Chapter 2: Artificial Intelligence Storage Requirements 39

Introduction 40

Storage Requirements 41

    The Three Stages of Digital Data 43

Data Layers 44

    From Data Warehouse to Data Lake 45

    Exercise 2.01: Designing a Layered Architecture for an AI System 47

    Requirements per Infrastructure Layer 49

Raw Data 50

    Security 50

        Basic Protection 50

        The AIC Rating 51

        Role-Based Access 52

        Encryption 53

    Exercise 2.02: Defining the Security Requirements
for Storing Raw Data 54

    Scalability 55

    Time Travel 56

    Retention 57

    Metadata and Lineage 59

Historical Data 60

    Security 60

    Scalability 61

    Availability 61

    Exercise 2.03: Analyzing the Availability of a Data Store 62

        Availability Consequences..63

    Time Travel 65

    Locality of Data 67

    Metadata and Lineage 67

Streaming Data 68

    Security 68

    Performance 69

    Availability 69

    Retention 70

    Exercise 2.04: Setting the Requirements for Data Retention 71

Analytics Data 72

    Performance 72

    Cost-Efficiency 73

    Quality 73

Model Development and Training 74

    Security 75

    Availability 75

    Retention 75

    Activity 2.01: Requirements Engineering
    for a Data-Driven Application 76

Summary 77

Chapter 3: Data Preparation

Introduction 80

ETL 80

Data Processing Techniques 82

    Exercise 3.01: Creating a Simple ETL Bash Script 83

    Traditional ETL with Dedicated Tooling 93

    Distributed, Parallel Processing with Apache Spark 94

    Exercise 3.02: Building an ETL Job Using Spark 95

    Activity 3.01: Using PySpark for a Simple ETL Job
    to Find Netflix Shows for All Ages 102

    Source to Raw: Importing Data from Source Systems 104

    Raw to Historical: Cleaning Data 105

    Raw to Historical: Modeling Data 106

    Historical to Analytics: Filtering and Aggregating Data 107

    Historical to Analytics: Flattening Data 107

    Analytics to Model: Feature Engineering 107

    Analytics to Model: Splitting Data 109

Streaming Data 110

      Windows 110

      Event Time 112

      Late Events and Watermarks 113

    Exercise 3.03: Streaming Data Processing with Spark 114

    Activity 3.02: Counting the Words in a Twitter Data Stream
    to Determine the Trending Topics 123

Summary 125

Chapter 4: The Ethics of AI Data Storage 127

Introduction 128

    Case Study 1: Cambridge Analytica 130

    Summary and Takeaways 135

    Case Study 2: Amazon's AI Recruiting Tool 136

    Imbalanced Training Sets 136

    Summary and Takeaways 139

    Case Study 3: COMPAS Software 140

    Summary and Takeaways 142

    Finding Built-In Bias in Machine Learning Models 143

    Exercise 4.01: Observing Prejudices and Biases
    in Word Embeddings 145

    Exercise 4.02: Testing Our Sentiment Classifier
    on Movie Reviews 151

    Activity 4.01: Finding More Latent Prejudices 158

Summary 160

Chapter 5: Data Stores: SQL and NoSQL Databases 163

Introduction 164

Database Components 165

SQL Databases 166

MySQL 167

    Advantages of MySQL 167

    Disadvantages of MySQL 167

    Query Language 167

        Terminology 168

        Data Definition Language (DDL) 168

Data Manipulation Language (DML)..169

Data Control Language (DCL) 170

Transaction Control Language (TCL) 171

Data Retrieval...172

SQL Constraints 176

Exercise 5.01: Building a Relational Database
for the FashionMart Store 179

Data Modeling 186

Normalization 187

Dimensional Data Modeling...190

Performance Tuning and Best Practices 193

Activity 5.01: Managing the Inventory of an E-Commerce
Website Using a MySQL Query 194

NoSQL Databases 198

Need for NoSQL 199

Consistency Availability Partitioning (CAP) Theorem 200

MongoDB 201

Advantages of MongoDB 201

Disadvantages of MongoDB 202

Query Language 202

Terminology 202

Exercise 5.02: Managing the Inventory of an E-Commerce
Website Using a MongoDB Query 210

Data Modeling 217

Lack of Joins..217

Joins 219

Performance Tuning and Best Practices 221

Activity 5.02: Data Model to Capture User Information 221

Cassandra 226

Advantages of Cassandra 226

Disadvantages of Cassandra 227

Dealing with Denormalizations in Cassandra 227

Query Language 228

Terminology 228

Exercise 5.03: Managing Visitors of an E-Commerce
Site Using Cassandra 231

Data Modeling 237

Column Family Design..238

Distributing Data Evenly across Clusters........................239

Considering Write-Heavy Scenarios 239

Performance Tuning and Best Practices 240

Activity 5.03: Managing Customer Feedback Using Cassandra 240

Exploring the Collective Knowledge of Databases 242

Summary 245

Chapter 6: Big Data File Formats   247

Introduction 248

Common Input Files 248

CSV – Comma-Separated Values 249

JSON – JavaScript Object Notation 249

Choosing the Right Format for Your Data 250

Orientation – Row-Based or Column-Based 251

Row-Based 251

Column-Based 252

Partitions 253

Schema Evolution 254

Compression 254

Introduction to File Formats 255

Parquet 255

Exercise 6.01: Converting CSV and JSON Files
into the Parquet Format 260

Avro 266

Exercise 6.02: Converting CSV and JSON Files
into the Avro Format 267

ORC 274

Exercise 6.03: Converting CSV and JSON Files
into the ORC Format 276

Query Performance 282

Activity 6.01: Selecting an Appropriate Big Data File
Format for Game Logs 284

Summary 285

Chapter 7: Introduction to Analytics Engine (Spark) for Big Data   287

Introduction 288

Apache Spark 289

Fundamentals and Terminology 290

How Does Spark Work? 294

Apache Spark and Databricks 295

Exercise 7.01: Creating Your Databricks Notebook 296

Understanding Various Spark Transformations 304

Exercise 7.02: Applying Spark Transformations
to Analyze the Temperature in California 306

Understanding Various Spark Actions 311

Spark Pipeline 312

Exercise 7.03: Applying Spark Actions to the Gettysburg Address 313

Activity 7.01: Exploring and Processing a Movie
Locations Database Using Transformations and Actions 319

Best Practices 321

Summary 322

Chapter 8: Data System Design Examples   325

Introduction 326

The Importance of System Design 327

Components to Consider in System Design 328

Features 328

Hardware 329

Data 329

Architecture 330

Security 330

Scaling 331

Examining a Pipeline Design for an AI System 331

Reproducibility – How Pipelines Can Help Us Keep
Track of Each Component 334

Exercise 8.01: Designing an Automatic Trading System 334

Making a Pipeline System Highly Available 342

Exercise 8.02: Adding Queues to a System to Make
It Highly Available 344

Activity 8.01: Building the Complete System
with Pipelines and Queues 348

Summary 350

Chapter 9: Workflow Management for AI 353

Introduction 354

Creating Your Data Pipeline 355

Exercise 9.01: Implementing a Linear Pipeline
to Get the Top 10 Trending Videos 357

Exercise 9.02: Creating a Nonlinear Pipeline
to Get the Daily Top 10 Trending Video Categories 363

Challenges in Managing Processes in the Real World 372

Automation 372

Failure Handling 373

Retry Mechanism 373

Exercise 9.03: Creating a Multi-Stage Data Pipeline 375

Automating a Data Pipeline 382

Exercise 9.04: Automating a Multi-Stage Data Pipeline
Using a Bash Script 382

Automating Asynchronous Data Pipelines 385

Exercise 9.05: Automating an Asynchronous Data Pipeline ... 388

Workflow Management with Airflow 392

Exercise 9.06: Creating a DAG for Our Data Pipeline
Using Airflow 394

Activity 9.01: Creating a DAG in Airflow to Calculate
the Ratio of Likes-Dislikes for Each Category 405

Summary 408

Chapter 10: Introduction to Data Storage on Cloud Services (AWS)   411

Introduction 412

Interacting with Cloud Storage 413

Exercise 10.01: Uploading a File to an AWS S3 Bucket
Using AWS CLI 416

Exercise 10.02: Copying Data from One Bucket
to Another Bucket 421

Exercise 10.03: Downloading Data from Your S3 Bucket 423

Exercise 10.04: Creating a Pipeline Using AWS SDK Boto3
and Uploading the Result to S3 425

Getting Started with Cloud Relational Databases 430

Exercise 10.05: Creating an AWS RDS Instance via
the AWS Console 431

Exercise 10.06: Accessing and Managing the AWS RDS Instance 442

Introduction to NoSQL Data Stores on the Cloud 450

Key-Value Data Stores 452

Document Data Stores 452

Columnar Data Store 453

Graph Data Store 454

Data in Document Format 455

Activity 10.01: Transforming a Table Schema into
Document Format and Uploading It to Cloud Storage 456

Summary 458

Chapter 11: Building an Artificial Intelligence Algorithm 461

Introduction 462

Machine Learning Algorithms 462

Model Training 463

    Closed-Form Solution 463

    Non-Closed-Form Solutions 464

Gradient Descent 464

    Exercise 11.01: Implementing a Gradient Descent
    Algorithm in NumPy 467

Getting Started with PyTorch 478

    Exercise 11.02: Gradient Descent with PyTorch 481

Mini-Batch SGD with PyTorch 488

    Exercise 11.03: Implementing Mini-Batch SGD with PyTorch 492

    Building a Reinforcement Learning Algorithm to Play a Game 500

    Exercise 11.04: Implementing a Deep Q-Learning Algorithm
    in PyTorch to Solve the Classic Cart Pole Problem 506

    Activity 11.01: Implementing a Double Deep Q-Learning
    Algorithm to Solve the Cart Pole Problem 513

Summary 516

Chapter 12: Productionizing Your AI Applications 519

Introduction 520

pickle and Flask 521

    Exercise 12.01: Creating a Machine Learning Model API
    with pickle and Flask That Predicts Survivors of the Titanic 522

    Activity 12.01: Predicting the Class of a Passenger
    on the Titanic 536

Deploying Models to Production 537

Docker 538

Kubernetes 539

Exercise 12.02: Deploying a Dockerized Machine
Learning API to a Kubernetes Cluster 541

Activity 12.02: Deploying a Machine Learning Model to
a Kubernetes Cluster to Predict the Class of Titanic Passengers 555

Model Execution in Streaming Data Applications 557

PMML 558

Apache Flink 559

Exercise 12.03: Exporting a Model to PMML and Loading it
in the Flink Stream Processing Engine for Real-time Execution 559

Activity 12.03: Predicting the Class of Titanic Passengers
in Real Time 572

Summary 575

Appendix   579

Index   681

PREFACE

ABOUT THE BOOK

Social networking sites see an average of 350 million uploads daily - a quantity impossible for humans to scan and analyze. Only AI can do this job at the required speed, and to leverage an AI application at its full potential, you need an efficient and scalable data storage pipeline. *The Artificial Intelligence Infrastructure Workshop* will teach you how to build and manage one.

The Artificial Intelligence Infrastructure Workshop begins taking you through some real-world applications of AI. You'll explore the layers of a data lake and get to grips with security, scalability, and maintainability. With the help of hands-on exercises, you'll learn how to define the requirements for AI applications in your organization. This AI book will show you how to select a database for your system and run common queries on databases such as MySQL, MongoDB, and Cassandra. You'll also design your own AI trading system to get a feel of the pipeline-based architecture. As you learn to implement a deep Q-learning algorithm to play the CartPole game, you'll gain hands-on experience with PyTorch. Finally, you'll explore ways to run machine learning models in production as part of an AI application.

By the end of the book, you'll have learned how to build and deploy your own AI software at scale, using various tools, API frameworks and serialization methods.

AUDIENCE

Our goal at Packt is to help you be successful, in whatever it is you choose to do. *The Artificial Intelligence Infrastructure Workshop* is the ideal tutorial for experienced Python developers and software architects who are getting started with data storage for AI. It is assumed that you understand and have used various filesystems, file formats, databases, and storage solutions for digital data and that you also have some experience in the field of big data.

Familiarity with basic Natural Language Processing (NLP) concepts would also be beneficial. Pick up a Workshop today and let Packt help you develop skills that stick with you for life.

ABOUT THE CHAPTERS

Chapter 1, Data Storage Fundamentals, introduces you to some real-world scenarios where AI plays a role. You'll take a close look at different kinds of hardware that are used to optimize AI systems, and gain some hands-on experience by building a basic text classification model.

Chapter 2, Artificial Intelligence Storage Requirements, contains a comprehensive overview of the requirements for an AI solution. We will explore the typical layers of a data lake and discuss topics such as security, scalability, and maintainability. The exercises will give you some practice with these subjects and prepare you for setting the requirements for AI applications in your organization.

Chapter 3, Data Preparation, is a hands-on chapter that lets you practice several forms of data processing, with the end goal of evaluating a machine learning model. You'll learn a few common ways to clean data and do feature engineering.

Chapter 4, The Ethics of AI Data Storage, shows you how AI systems can be used in malicious ways, intentionally or through ethical oversights. You'll walk through some examples of AI systems that have brought harm to people or society as a whole, and you'll build your own prejudiced AI so that you can learn how to avoid some common pitfalls.

Chapter 5, Data Stores: SQL and NoSQL Databases, teaches you about SQL and NoSQL databases and the decision-making process for choosing the ideal database. You will learn how to execute common queries on databases such as MySQL, MongoDB, and Cassandra. The chapter will give you hands-on experience with different databases.

Chapter 6, Big Data File Formats, explores popular file formats for big data environments and shows you how to choose between them. You will walk through various code snippets to get started with understanding different file types and executing queries to measure performance.

Chapter 7, Introduction to Analytics Engines (Spark) for Big Data, is about Spark, **Resilient Distributed Datasets (RDDs)**, and Spark transformations and actions. We will use Databricks, a popular integrated cloud-based Spark analytics engine, for our hands-on learning.

Chapter 8, Data System Design Examples, takes a step back from the implementation details to look at the bigger picture of data system design. You'll see why this birds-eye view is often important for collaboration, optimization, and scaling. You'll also design your own AI trading system to get a feel for a pipeline-based architecture.

Chapter 9, Workflow Management for AI, takes you through the journey of prototyping your AI data pipeline, automating the pipeline from end to end, and finally managing your pipelines using the Airflow workflow management system.

Chapter 10, Introduction to Data Storage on Cloud Services (AWS), introduces you to data storage solutions on **Amazon Web Services** (**AWS**). You will become familiar with using the AWS CLI and SDK to perform common data tasks. You will implement a data pipeline that integrates with AWS S3.

Chapter 11, Building an Artificial Intelligence Algorithm, gives you hands-on experience with PyTorch in building deep learning training algorithms. You will implement a deep Q-learning algorithm to play the CartPole game.

Chapter 12, Productionizing Your AI Applications, is the final chapter of this book, where we'll explore a few ways to run machine learning models in production as part of an AI application. You'll get hands-on experience with API frameworks (such as Flask), serialization methods (such as Pickle and PMML), container solutions (such as Docker and Kubernetes), and running models as part of a stream processing application with Flink. After this, you'll be ready to build and deploy your own AI software.

CONVENTIONS

Code words in the text, database table names, folder names, filenames, file extensions, pathnames, dummy URLs, user input, and Twitter handles are shown as follows: "Import the **pandas** module and read the downloaded data using the **read_csv** method."

Words that you see on the screen, for example, in menus or dialog boxes, also appear in the text like this: "Create a cluster by clicking the **Clusters** button on the left pane."

A block of code is set as follows:

```
if __name__ == "__main__":
    import os
    import sys
    from os import path
    # read data from cache
    try:
        df_vids = pd.read_csv('./tmp/data_vids.csv')
        df_cats = pd.read_csv('./tmp/data_cats.csv')
```

New terms, abbreviations, and important words are shown like this: "There are many types of encryption possible. **Advanced Encryption Standard** (**AES**) and **Rivest-Shamir-Adleman** (**RSA**) are two examples of popular asymmetric encryption algorithms."

Long code snippets are truncated and the corresponding names of the code files on GitHub are placed at the top of the truncated code. The permalinks to the entire code are placed below the code snippet. It should look as follows:

filter_data.py

```
28      # filter
29      df_filtered = filter_by_date(df_data, date)
30      # cache
31      dir_cache = './tmp'
32      try:
33          df_filtered.to_csv(os.path.join(dir_cache,
            'data_vids.csv'), index=False)
34      except FileNotFoundError:
35          os.mkdir(dir_cache)
36          df_filtered.to_csv(os.path.join(dir_cache,
            'data_vids.csv'), index=False)
37      print('[ data pipeline ] finish filter data')
```

The complete code for this step is available at https://packt.live/2DhTEK3.

CODE PRESENTATION

Lines of code that span multiple lines are split using a backslash (\ ). When the code is executed, Python will ignore the backslash, and treat the code on the next line as a direct continuation of the current line.

For example:

```
history = model.fit(X, y, epochs=100, batch_size=5, verbose=1, \
                    validation_split=0.2, shuffle=False)
```

Comments are added into code to help explain specific bits of logic. Single-line comments are denoted using the # symbol, as follows:

```
# Print the sizes of the dataset
print("Number of Examples in the Dataset = ", X.shape[0])
print("Number of Features for each example = ", X.shape[1])
```

Multi-line comments are enclosed by triple quotes, as shown below:

```
"""
Define a seed for the random number generator to ensure the
result will be reproducible
"""
seed = 1
np.random.seed(seed)
random.set_seed(seed)
```

SETTING UP YOUR ENVIRONMENT

Before we explore the book in detail, we need to set up specific software and tools. In the following section, we shall see how to do that.

INSTALLING ANACONDA

Anaconda is a tool that is used to manage your Python environment. After installation, Anaconda takes care of package management by enabling multiple environments on your machine.

Follow the installation instructions at https://docs.anaconda.com/anaconda/install/. Make sure to download the Python 3.7 version (from https://python.org).

> #### NOTE
>
> Anaconda for Windows comes with a useful tool called Anaconda Prompt. This is a command-line interface, like Terminal for Linux and macOS, with all the necessary libraries installed. We will use Anaconda Prompt in some exercises and activities, so anyone who is using Windows should make sure that this works.

INSTALLING SCIKIT-LEARN

Scikit-learn is a simple but powerful machine learning library, providing a common API for many machine learning algorithms. It also interfaces with Python's other scientific libraries to provide a common interface to efficient data structures, and some additional tools to help with data preparation and reporting.

To install scikit-learn, follow the instructions at https://scikit-learn.org/stable/install.
html#install-official-release.

INSTALLING GAWK

The GAWK tool is used for parsing comma-separated files. All documentation can be
found at https://www.gnu.org/software/gawk/manual/gawk.html, with detailed installation
instructions at https://www.gnu.org/software/gawk/manual/gawk.html#Installation.

Short versions of the installation instructions for the binary distributions are
as follows:

- Windows:

    Download the **gawk** package from http://gnuwin32.sourceforge.net/downlinks/gawk.
    php and install it.

- Linux:

    Install **gawk** with your package manager. For example, for Debian/Ubuntu-based
    distributions, use this:

    ```
    sudo apt-get install gawk
    ```

- macOS

    Install **gawk** with the Homebrew package manager:

    ```
    brew install gawk
    ```

INSTALLING APACHE SPARK

You can download Apache Spark by visiting the following link: https://spark.apache.org/
downloads.html. You can find the installation steps further down in this Preface in *Step
8* of the section titled *Installing Apache Spark and Scala*. On Windows, you may refer
to the following link for detailed instructions: https://medium.com/big-data-engineering/
how-to-install-apache-spark-2-x-in-your-pc-e2047246ffc3.

INSTALLING PYSPARK

PySpark is the Python interface that is used to communicate with a Spark
environment. It is distributed as an Anaconda package and comes with a local
Spark cluster.

Make sure that Java 8, 9, or 10 is installed on your system. Java 11 is not supported by Spark yet. You can install Java by following the instructions at https://docs.oracle.com/javase/10/install/overview-jdk-10-and-jre-10-installation.htm. Also, make sure that your **JAVA_HOME** environment variable is set; if not, do so by pointing it to your local **java** folder – for example, **export JAVA_HOME=/usr/lib/jvm/java-1.8.0-openjd-amd64**:

```
(base) bas@Hedgehog:~$ export JAVA_HOME=/usr/lib/jvm/java-1.8.0-openjdk-amd64
(base) bas@Hedgehog:~$ echo $JAVA_HOME
/usr/lib/jvm/java-1.8.0-openjdk-amd64
(base) bas@Hedgehog:~$
```

Figure 0.1: Setting the JAVA_HOME variable

To install PySpark, type one of the following commands in your Terminal or Anaconda Prompt:

```
pip install pyspark
```

Or:

```
conda install -c conda-forge pyspark
```

After installation, verify that Spark is running by executing the **pyspark** command in Terminal or Anaconda Prompt. You'll see output similar to the following:

Figure 0.2: Running PySpark

INSTALLING TWEEPY

To connect to Twitter, we'll use the Tweepy library. Follow one of the following installation instructions to install it:

- If you're going for a **pip** installation, go to http://docs.tweepy.org/en/latest/install.html.

- If you're going for a **conda** installation, go to https://anaconda.org/conda-forge/tweepy.

INSTALLING SPACY

spaCy is a powerful **Natural Language Processing** (**NLP**) library for Python. It has pre-trained models built-in, which makes it very easy to get up and running with. It has a focus on speed and efficiency.

To install spaCy, follow the instructions at https://spacy.io/usage.

You will also have to install the pre-trained statistical models spaCy provides by following the instructions at https://spacy.io/models/en.

In the examples, we use **v2.2.5** of the **en_core_web_lg** model, which can be installed by running the following command:

```
python -m spacy download en_core_web_lg
```

INSTALLING MYSQL

MySQL is easy to install, and there are various ways to do so. It's part of the LAMP stack and it can be downloaded separately as well:

1. The 32-bit or 64-bit version of MySQL Server and the MySQL Client can be downloaded from https://dev.mysql.com/downloads/installer/. Choose the type of server according to your requirements if you are using a UI tool for installation.

Alternatively, if you are using Linux, you can use any package-management utility to install **mysql-server**, as in either of the following snippets:

```
sudo apt-get install mysql-server
```

Or:

```
sudo yum install mysql-server
```

> **NOTE**
>
> For further information, refer to the MySQL documentation at the following link:https://dev.mysql.com/doc/refman/8.0/en/.

2. On Linux, after installation, the security script needs to be executed for the default security options, such as remote root logins and sample users. This can be done using the following command:

```
sudo mysql_secure_installation
```

3. The MySQL Client can then be accessed and a database can be created:

```
sudo mysql
mysql> create database <database_name>;
mysql> use <database_name>;
```

The preceding code will create a database for later use.

macOS Setup

MySQL can be installed on macOS using Homebrew (https://brew.sh/) and can be used via Homebrew services:

```
brew update
brew install mysql@5.7
```

This will install the latest version of MySQL. You can then start it as follows:

1. Install Homebrew services first:

```
brew tap homebrew/services
```

2. Start the **mysql** service:

```
brew services start mysql@5.7
```

3. Check whether the **mysql** service has been loaded:

```
brew services list
```

4. Force-link the appropriate version of MySQL:

```
brew link mysql@5.7 --force
```

5. Verify the installed MySQL instance:

```
mysql -V
```

INSTALLING MONGODB

MongoDB is supported by macOS, Linux, and Windows. In this section, we will take a look at how to set up single-node and multi-node clusters of this database.

macOS

1. We can use Homebrew, a package manager for macOS (https://brew.sh/). Let's update it if it is already installed:

```
$ brew update
```

2. You need to **tap** the **mongodb** package first and then install it:

```
$ brew tap mongodb/brew
$ brew install mongodb-community@4.2
```

Here, we are installing version **4.2** – we could choose **3.4** or **3.6** as well, but we will use **4.2**.

3. MongoDB can be configured with this:

```
/usr/local/etc/mongod.conf
```

In the **dbPath** variable, you can specify the path where data should be stored. Logs are stored in **/usr/local/var/log/mongodb**. Data will be stored in BSON format in **/usr/local/var/mongodb**.

4. You can start your MongoDB server with either of the following commands:

```
$ services start mongodb-community@4.2
```

Or:

```
$ mongo --config /usr/local/etc/mongod.conf
```

5. Access the **mongo** shell with this command:

```
$ mongo
```

Linux (Ubuntu)

1. Install **gnupg**:

```
$ sudo apt-get install gnupg
```

2. Install the public key for the package management system:

```
$ wget -qO - https://www.mongodb.org/static/pgp/server-4.2.asc | sudo
apt-key add -
```

3. Create a **list** file for MongoDB:

```
$ echo "deb [ arch=amd64 ] https://repo.mongodb.org/apt/ubuntu
bionic/mongodb-org/4.2 multiverse" | sudo tee /etc/apt/sources.
list.d/mongodb-org-4.2.list
$ sudo apt-get update
```

4. Install the MongoDB package:

```
$ sudo apt-get install -y mongodb-org
```

5. Start the MongoDB service:

```
$ sudo service mongod start
```

6. Access the **mongo** shell:

```
$ mongo
```

Windows

1. Download and install MongoDB from https://www.mongodb.com/download-center/
community?jmp=docs.

2. While installing, select **Install MongoDB as service** and specify the
service name as **MongoDB**. Also, specify the **data** and **log** directories.

3. Once installed, set a path to the MongoDB directory in the **Path**
environment variable.

4. You can access the **mongo** shell as follows:

```
>mongo.exe
```

INSTALLING CASSANDRA

To install the latest Cassandra setup, you can use the Homebrew package manager (https://brew.sh/). Use the following steps to set up a single-node instance.

macOS

1. Make sure that you have Java installed and that the **JAVA_HOME** path is set.

2. Install Cassandra's latest stable version (we are using **3.11.4**) with the following command:

    ```
    $ brew update
    $ brew install cassandra
    ```

3. You can run the Cassandra service using one of two ways:

    ```
    $ cassandra
    ```

    Or:

    ```
    $ brew services start cassandra
    ```

    This will start the Cassandra single-node test cluster.

4. Access the command line for CQL with the following command:

    ```
    $ cqlsh
    ```

    You should get the following output:

    ```
    Connected to Test Cluster at 127.0.0.1:9042.
    [cqlsh 5.0.1 | Cassandra 3.11.4 | CQL spec 3.4.4 | Native protocol v4]
    Use HELP for help.
    cqlsh>
    ```

    Figure 0.3: Running Cassandra

5. Find the configuration for Cassandra at the following location:

    ```
    $ cd /usr/local/etc/cassandra
    ```

    Here, there are three files that you need to worry about:

    *   **README.txt**: This file has information about the other files. We would suggest reading this first.

    *   **cassandra.yaml**: This is the main Cassandra configuration file.

    *   **logback.xml**: This is the logback configuration file for the Cassandra server.

6. You can configure a multi-node cluster with Cassandra. To do so, these steps should be followed for all the nodes in a cluster. First, install Cassandra single-node servers on all the nodes.

7. Go to the **cassandra.yaml** file and edit the following information:

   - Add the current machine's IP address under **listen_address** and **rpc_address**. Here, **listen_address** specifies your current server's IP, and **rpc_address** is for remote procedure calls. Add each server's IP addresses as a comma-separated list under **-seeds**. This is where you list all the components of your cluster.

   - Set **auto_bootstrap** to **false**. If this value is not in the configuration file, you can explicitly add it yourself. This setting is important if you want your new nodes to point to the location of your data correctly.

   Once done, start the Cassandra service on each of your nodes.

8. A tool called **nodetool** will be installed with your Cassandra setup. Using this, you can see the health of your cluster as well as many other things:

   ```
   $ nodetool status
   ```

   You should get the following output:

   ```
   Datacenter: datacenter1
   =======================
   Status=Up/Down
   |/ State=Normal/Leaving/Joining/Moving
   --  Address    Load       Tokens       Owns (effective)  Host ID                               Rack
   UN  127.0.0.1  172.59 KiB  256          100.0%            94df9b54-7be9-4e5b-b6f5-a0236b4ffb40  rack1
   ```

   Figure 0.4: nodetool status

   This is an important step. You need to open ports for all the nodes to communicate with each other. Open port **7000** to transfer data over the network and port **9042** to connect from the CLI to the required servers in the cluster.

Linux

To install Cassandra on Linux, you need to perform the following steps:

1. Download and extract the Cassandra binary files with the following commands:

   ```
   $ wget http://supergsego.com/apache/cassandra/3.11.5/apache-
   cassandra-3.11.5-bin.tar.gz
   $ tar zxvf apache-cassandra-3.11.5-bin.tar.gz
   ```

2. Export the **CASSANDRA_HOME** environment variable and point it to the extracted location.

3. As per the **cassandra.yml** file, the configured directories (**/var/lib/cassandra** and **/var/log/cassandra**) need to be created and given full permissions using the following command:

```
$ chmod 777 /var/lib/cassandra
$ chmod 777 /var/log/cassandra
```

4. You can add the Cassandra home path in the **$PATH** variable so that you can start Cassandra with the following command:

```
$ cassandra
$ cqlsh
```

5. To run Cassandra in a multi-node cluster, just follow *Step 6* onward in the *macOS* installation section.

Windows

1. Download and extract the Cassandra setup from this link: http://cassandra.apache.org/download/.

2. Add the path to the Cassandra folder in **Environment variables**.

3. Run the Cassandra cluster with the following command:

```
> cassandra
```

4. Start the Cassandra CLI using this (command is not compatible with Python 3):

```
> cqlsh
```

5. You can follow *Step 6* onward in the *macOS* installation section to run a multi-node cluster.

INSTALLING APACHE SPARK AND SCALA

Before installing Spark, Java is needed for your system:

Linux Installation Steps

1. Check the current version of Java on your machine. The following command will verify the version of Java that is installed on your system:

```
$java -version
```

If Java is already installed on your system, you will see the following output:

```
java version «1.7.0_71»
Java(TM) SE Runtime Environment (build 1.7.0_71-b13)
Java HotSpot(TM) Client VM (build 25.0-b02, mixed mode)
```

You have to install Java if it is not installed on your system.

2. Now you need to ensure that Scala is installed on your system.

   Installing the Scala programming language is mandatory before installing Spark as it is important for Spark's implementation. The following command will verify the version of Scala used on your system:

```
$scala -version
```

   If Scala is already installed on your system, you will see the following response on the screen:

```
Scala code runner version 2.11.12 -- Copyright 2002-2013, LAMP/EPFL
```

   If you don't have Scala, then you have to install it on your system. Let's see how to install Scala.

3. First, download Scala for your operating system from the following link:

   https://www.scala-lang.org/download/2.11.12.html

   You need to download the latest version of Scala. Here, you will see the **scala-2.11.12** version being used. After downloading, you will be able to find the Scala tar file in the **Downloads** folder.

4. Extract the Scala tar file using the following command:

```
$ tar xvf scala-2.11.12.tgz
```

5. Move the Scala software files to the **/usr/local/scala** directory using the following commands:

```
$ su -
Password:
# cd /home/Hadoop/Downloads/
# mv scala-2.11.12 /usr/local/scala
# exit
```

6. Set **PATH** for Scala using the following command:

```
$ export PATH = $PATH:/usr/local/scala/bin
```

7. Now verify the installation of Scala by checking the version:

```
$scala -version
```

If your Scala installation is successful, then you will get the following output:

```
Scala code runner version 2.11.12 — Copyright 2002-2013, LAMP/EPFL
```

Scala installation for macOS

Scala can be installed via Homebrew on macOS:

```
brew update
brew install scala
```

Once you are ready with Java and Scala on your system, we can download Apache Spark.

Linux Installation Steps

1. Download Apache Spark (https://spark.apache.org/downloads.html) by using the following command:

```
spark-2.4.5-bin-hadoop2.6 version
```

After this, you can find a Spark tar file in the **Downloads** folder.

2. Extract the Spark tar file using the following command:

```
$ tar xvf spark-2.4.5-bin-hadoop2.6.tgz
```

3. Move the Spark software files to the directory using the following commands:

```
/usr/local/spark
$ su -
Password:
# cd /home/Hadoop/Downloads/
# mv spark-2.4.5-bin-hadoop2.6 /usr/local/spark
# exit
```

4. Now configure the environment for Spark. You need to add the following path to the ~/.**bashrc** file, which will add the location where the Spark software files are located to the **PATH** variable:

```
export PATH = $PATH:/usr/local/spark/bin
```

5. Use the following command for sourcing the ~/.**bashrc** file:

```
$ source ~/.bashrc
```

With this, you have successfully installed Apache Spark on your system. Now you need to verify it.

6. Verify the installation of Spark on your system using the following command:

```
$spark-shell
```

The command will display the Spark Shell application version. If Spark has been installed successfully, then you will get the following output:

```
Spark assembly has been built with Hive, including Datanucleus jars
on classpath
Using Spark's default log4j profile: org/apache/spark/log4j-defaults.
properties
15/06/04 15:25:22 INFO SecurityManager: Changing view acls to: hadoop
15/06/04 15:25:22 INFO SecurityManager: Changing modify acls to:
hadoop
15/06/04 15:25:22 INFO SecurityManager: SecurityManager:
authentication disabled;
ui acls disabled; users with view permissions: Set(hadoop); users
with modify permissions: Set(hadoop)
15/06/04 15:25:22 INFO HttpServer: Starting HTTP Server
15/06/04 15:25:23 INFO Utils: Successfully started service naming
'HTTP class server' on port 43292.
Welcome to the Spark World!
```

INSTALLING AIRFLOW

Airflow is a platform created by the community to programmatically author, schedule, and monitor workflows. Before installing Airflow, you will need to have Python and **pip** on your machine. If you have installed Anaconda already, then you will already have Python and **pip** on your machine. To install Airflow and get started with it, please go to this page: https://airflow.readthedocs.io/en/stable/start.html

We have used Airflow version **1.10.7** with Python version **3.7.1** in *Chapter 9, Workflow Management for AI*.

INSTALLING AWS

We will use AWS as the primary cloud provider for our exercises. AWS offers reliable, scalable, and inexpensive cloud computing services. It's free to join, and you pay only for what you use. We will use their free tier service for our exercises. However, if you use its other tiers, you will be charged.

We will be using the AWS S3 and RDS services. S3 is free for up to 5 GB of storage. RDS is free for up to 750 hours within 12 months of a new registration.

For more information about the free tier, please refer to https://aws.amazon.com/free/.

REGISTERING YOUR AWS ACCOUNT

To use AWS services, we need to register a new account. Please follow the instructions here: https://aws.amazon.com/premiumsupport/knowledge-center/create-and-activate-aws-account/.

CREATING AN IAM ROLE FOR PROGRAMMATIC AWS ACCESS

Directly accessing or managing AWS services via the AWS console isn't the best practice. By default, you will be using your root account, which has access to all resources, to interact with AWS services if you don't set up an IAM user. When we create an IAM user, we limit privileges to certain resources. If an account is hacked and IAM user credentials are leaked, the damage caused by the data loss will be limited:

1. Go to the AWS website: https://aws.amazon.com/. Log in to your console using your root account:

Figure 0.5: AWS website

2. Use your account ID or account email to sign in:

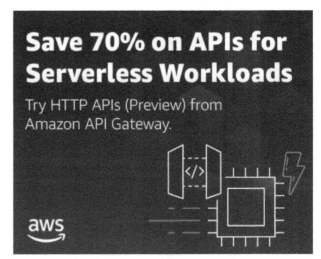

Figure 0.6: AWS sign in

3. After you sign in in to the console, search **IAM** in the **Find Services** search bar:

Figure 0.7: AWS Management Console

4. After you click on **IAM**, you will be directed to the **Identity and Access Management (IAM)** console, and you will need to click on **Add user** to create an IAM user:

Figure 0.8: AWS IAM dashboard

5. There are five steps that you need to go through when you create a new IAM user. The first step is to configure **User name** and **Access type**. We want this user to have programmatic access to our services:

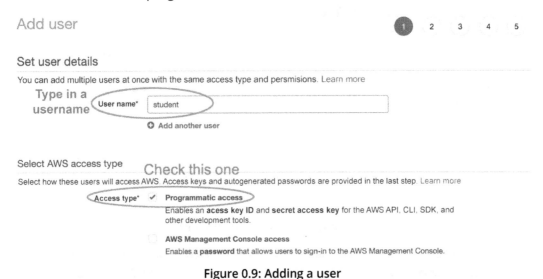

Figure 0.9: Adding a user

6. Next, we will need to set permissions for this IAM user. We can either create our own permissions or just choose from the existing policies. For now, let's give this user **Administrator Access**, which has most of the privileges for most services:

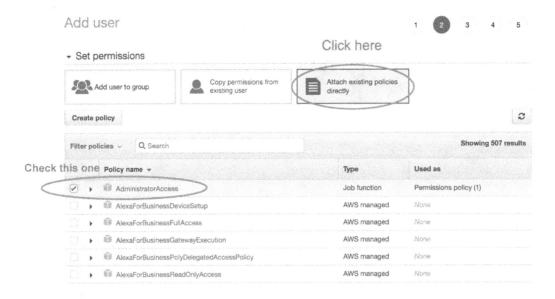

Figure 0.10: Checking AdministratorAccess

7. After we set permissions for this IAM user, the next step is to add tags for this user. This step is optional and is useful for organizing IAM users when you have too many of them:

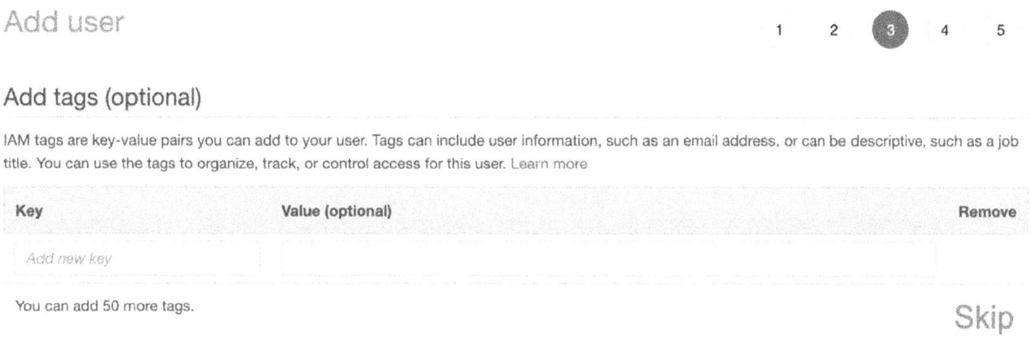

Add user

1   2   **3**   4   5

Add tags (optional)

IAM tags are key-value pairs you can add to your user. Tags can include user information, such as an email address, or can be descriptive, such as a job title. You can use the tags to organize, track, or control access for this user. Learn more

Key	Value (optional)	Remove
Add new key		

You can add 50 more tags.

Skip

Figure 0.11: Adding tags

8. The next step is to verify that all the information and settings are correct:

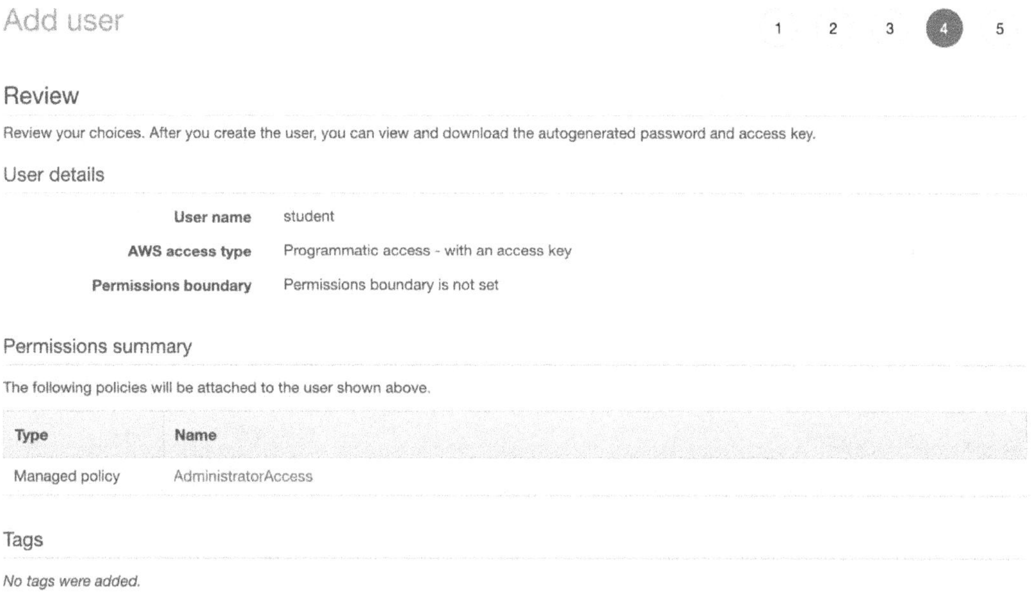

Add user

1   2   3   **4**   5

Review

Review your choices. After you create the user, you can view and download the autogenerated password and access key.

User details

User name	student
AWS access type	Programmatic access - with an access key
Permissions boundary	Permissions boundary is not set

Permissions summary

The following policies will be attached to the user shown above.

Type	Name
Managed policy	AdministratorAccess

Tags

No tags were added.

Figure 0.12: Reviewing information and settings

9.   The final step in this process will show you the access key ID and the secret access key, which are credentials that allow your IAM user to access or control your AWS services. You will need these two keys to configure the AWS CLI, so it's best to copy them or write them down:

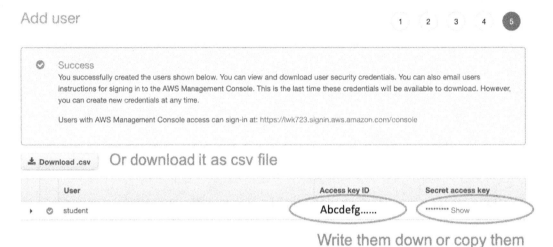

Figure 0.13: User access key ID and secret access key

10. If you have successfully added a new IAM user, you will see the new user in your IAM console as follows:

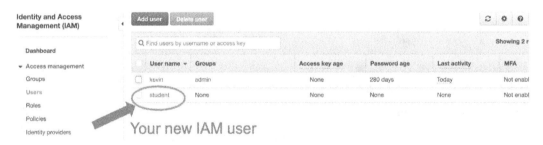

Figure 0.14: New user details

INSTALLING THE AWS CLI

Before installing the AWS CLI, you will need Python and **pip**. If you have installed Anaconda, then you will already have Python and **pip**. If not, then please install Python and **pip**:

1. **Installation**: Please go to https://anaconda.org/conda-forge/awscli and follow the instructions to install the AWS CLI.

2. **Configuration**: We assume that you have created an IAM user and obtained its access key. If not, please go back to the previous step to create an IAM user and its access key. Then, go to https://docs.aws.amazon.com/cli/latest/userguide/cli-chap-configure.html and follow the instructions to configure the AWS CLI.

INSTALLING AN AWS PYTHON SDK — BOTO

The AWS Python SDK we will use is called **Boto**, which also requires Python and **pip**. Follow the instructions at this URL: https://aws.amazon.com/sdk-for-python/.

For installation information, please visit https://pypi.org/project/boto3/.

If you need to install it using **conda**, please visit https://anaconda.org/conda-forge/boto3.

INSTALLING MYSQL CLIENT

We will use the MySQL Client to connect to our AWS RDS instance via a public IP. In order to do that, we need to install the MySQL Client on our local machine. Please refer to https://cloud.google.com/sql/docs/mysql/connect-admin-ip#install-mysql-client to install the MySQL Client.

INSTALLING PYTEST

pytest is a testing framework that makes unit testing easy. To execute it, you just need to change your working directory and run **$ pytest** to run all of the unit tests in that directory.

To install **pytest**, please go to https://docs.pytest.org/en/latest/getting-started.html to install it.

> **NOTE**
>
> Before running **pytest**, please install the other packages required for the given exercise, such as Airflow.

INSTALLING MOTO

Moto is a library that enables your tests to easily mock out AWS services.

To install **moto**, please go to http://docs.getmoto.org/en/latest/docs/getting_started.html.

INSTALLING PYTORCH

PyTorch is a Python-based scientific computing package for deep learning research. It's an open source machine learning library based on the Torch library, and it's usually used for applications such as computer vision and natural language processing.

To install PyTorch, please go to https://pytorch.org/get-started/locally/.

INSTALLING GYM

Gym is a toolkit for developing and comparing reinforcement learning algorithms. It makes no assumptions about the structure of your agent and is compatible with any numerical computation library, such as TensorFlow or Theano.

To install Gym, please go to https://gym.openai.com/docs/#installation.

INSTALLING DOCKER

Docker is a tool for building images, which can be distributed and deployed to various environments.

Follow the installation instructions at https://docs.docker.com/install/ for your operating system.

KUBERNETES – MINIKUBE

Kubernetes is an engine that can host Docker containers. To run a local Kubernetes cluster, we use the Minikube tool.

Follow the installation instructions at https://kubernetes.io/docs/tasks/tools/install-minikube/ for your operating system.

> ### NOTE
>
> Minikube provisions its own VM using a hypervisor and sets up all the Kubernetes components inside it. If you wish to install Minikube inside a local or cloud VM, you will need to instruct Minikube to set up a Kubernetes cluster without using virtualization, because setting up a VM inside a VM is tricky.
>
> To start up Minikube in a virtualized environment, add the `--driver=none` flag when starting up the cluster for the first time. When Minikube is running without a VM, all `minikube` and `kubectl` commands need to be run with root privileges. For this, you can add `sudo` before each of these commands. Alternatively, you can switch to root user by using `sudo -s` and entering the admin password when prompted.

INSTALLING MAVEN

We will use Maven 3 as the build server for our Java applications. Install Maven by following the instructions at https://maven.apache.org/install.html.

INSTALLING JDK

We will use the Java programming language to build a Flink streaming job in *Chapter 12, Productionizing Your AI applications*. There is a good chance that **Java Development Kit (JDK)** 8 or higher is already installed on your machine. If not, there are several ways to install it; follow the instructions for your operating system at https://openjdk.java.net/install/.

INSTALLING NETCAT

The Netcat tool is used to send data over a network with the TCP/IP protocol. Install it by following one of the following installation instructions:

- **Linux (Ubuntu/Debian)**: Type the following command in Terminal:

  ```
  sudo apt install netcat
  ```

- **macOS**: Type the following command in Terminal:

  ```
  brew install netcat
  ```

- **Windows**: Download and install the tool from
  https://joncraton.org/blog/46/netcat-for-windows/.

  Alternatively, type the following in Anaconda Prompt:

  ```
  sudo apt install netcat
  ```

INSTALLING LIBRARIES

`pip` comes pre-installed with Anaconda. Once Anaconda is installed on your machine, all the required libraries can be installed using `pip`, for example, `pip install numpy`. Alternatively, you can install all the required libraries using `pip install -r requirements.txt`. You can find the **requirements.txt** file at https://packt.live/31E4pP0.

The exercises and activities will be executed in Jupyter Notebooks. Jupyter is a Python library and can be installed in the same way as the other Python libraries – that is, with `pip install jupyter`, but fortunately, it comes pre-installed with Anaconda. To open a notebook, simply run the command `jupyter notebook` in the Terminal or Command Prompt.

ACCESSING THE CODE FILES

You can find the complete code files of this book at https://packt.live/3fpBOmh.

If you have any issues or questions about installation, please email us at `workshops@packt.com`.

1

DATA STORAGE FUNDAMENTALS

OVERVIEW

In this chapter, we will explore the broad range of capabilities of AI and look at some of the fields that it is changing. We will cover four areas in which AI is used in detail: medicine, language translation, subtitle generation, and forecasting. Then we will dive into a text classification example where you will build your first AI system – a basic text classifier that can identify when a news headline is regarded as "clickbait." We will look at optimization – an important topic for most machine learning systems that need to operate on a large scale. Finally, we will examine different kinds of hardware, including memory, processes, and storage, and will also see how we can reduce costs when renting this hardware from a cloud vendor.

By the end of this chapter, you will understand what kind of tasks machine learning can be used to perform. You will be able to build your own basic machine learning systems, using a popular Python library, **sklearn**. You will also be able to optimize the hardware of large systems and reduce costs while storing your data in a logical way.

INTRODUCTION

Machine learning, which is a subset of Artificial Intelligence (AI), has had a major influence on nearly every field you can imagine and can solve a wide variety of problems and tasks. Do you want to detect cancer better? You can train an image classifier to inspect mammograms. Do you want to communicate with people in other languages? Machine translation will help you. From ambitious projects such as self-driving cars and astronomical discoveries to fixing minor annoyances such as email spam, machine learning has taken the world by storm, and those who understand what it can do and how to build machine learning systems will be at the forefront of human advancement.

At the heart of any machine learning project is data. Many people, on first coming across the concept of machine learning, assume that it is possible to take mounds of data, shove it into a machine, and have the machine autonomously learn. But it's not so simple. Instead, machines need meticulously structured, organized, and clean data, often in huge quantities. The more data there is, the more difficult it becomes to store, process, and analyze the data, and it is therefore vital to optimize data storage at all stages of storage and usage.

This is a problem. How can we build efficient machine learning systems that do not waste our time or resources?

This course will show you practical real-world examples of how to do exactly that. You will learn how to make your data work for you as efficiently as possible, often by example.

We assume that you are no novice at working with data and that you understand and have used various filesystems, file formats, databases, and storage solutions for digital data. In this book, we will focus specifically on data for machine learning and show how this differs from storing general-purpose data.

Machine learning comes in many different forms, but concepts from linear algebra are core to many of the most important machine learning algorithms. In classical computer science, the focus is often on data structures such as arrays, linked lists, hash tables, and trees. In machine learning, while these structures are still important, you will more often need to work with data in the form of vectors, matrices, and tensors.

Because of this focus on data structures from linear algebra, other components of storage solutions have nuances too. Some processors are optimized at the hardware level for vectorized operations. Some file formats handle this kind of data better too, and there are specialized data structures to store data in this form efficiently as well.

PROBLEMS SOLVED BY MACHINE LEARNING

Before we get our hands too dirty with learning how to store and process machine learning data in efficient ways, let's take a step back. What kinds of real-world problems can we solve using machine learning?

Machine learning is not a new concept, but with new algorithms and better hardware, it has seen a resurgence over the last few years. This means it has received significant attention from many diverse fields. Although the applications of AI are almost uncountable, nearly all of these stem from a far smaller number of subfields within machine learning.

With that in mind, we'll start by examining one problem that machine learning can solve in each of the main subfields of image processing, text and language processing, audio processing, and time series analysis. Some problems, such as navigation systems, need to combine many of these fields, but the fundamental concepts remain very similar. We'll start by looking at image processing: we do not fully understand all the complexities behind how humans see, so helping computers 'see' is a particularly challenging task.

IMAGE PROCESSING – DETECTING CANCER IN MAMMOGRAMS WITH COMPUTER VISION

For classification tasks, the goal is to look at some data and decide which class it belongs to. In simple cases, there are only two classes: positive and negative. When a doctor looks at a mammogram to assess whether a patient has cancer, the doctor is looking for specific patterns and signs. The doctor then makes a diagnosis based on the patterns.

As a slight simplification, a doctor looks at a mammogram X-ray and classifies it into one of two classes: 'cancer' (positive) or 'healthy' (negative). Using image processing and machine learning, a computer can be trained to do the same thing. The computer is fed thousands or millions of X-rays in the form of digital images. Interpreting these as a set of matrices with associated labels, the computer uses a machine-learning algorithm to learn which patterns are indicative of cancer and which are not.

It's an emerging field, but a very promising-looking one. In January 2020, Google published a paper titled *International evaluation of an AI system for breast cancer screening*, which showed results that indicated their AI system was able to identify cancer in mammograms not only faster but also more reliably than human doctors.

Although images and language may seem very different, many techniques from image processing can be used to help machines better understand and learn human languages. Let's take a look at how AI has advanced the field of **Natural Language Processing** (**NLP**).

TEXT AND LANGUAGE PROCESSING – GOOGLE TRANSLATE

While computers are great at repetitive mechanical tasks such as solving well-defined equations, and humans are better at more creative tasks such as drawing, it is likely that if computers and humans could work better together, their complementary skills would be more valuable than they are individually. How can we help machines and humans work better together? A desirable approach is to allow computers to act more like humans, fostering closer collaboration.

To this end, we have tried to make computers take on traditionally 'human' characteristics, such as the following:

- **Look like us**: In 2016, David Hanson created a human-like 'social' robot named Sophia, which, apart from having a transparent skull, looks like a human female, and is partially modeled on Audrey Hepburn:

Figure 1.1: Sophia – The first robot citizen at the AI for
Good Global Summit 2018 (ITU pictures)

- **Walk like us**: Shortly after Sophia was shown to the world, Agility Robotics released 'Cassie' – a robot that looks far less human than Sophia but can walk on two legs in a very similar way to humans:

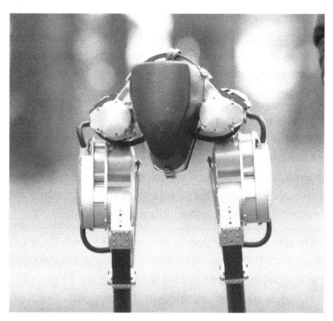

Figure 1.2: Cassie, a walking robot, photo by Oregon State University (CCSA)

- **Play a wide variety of games**: Computers can now beat even the best humans at rock, paper, scissors; chess; Go; and Super Mario Bros:

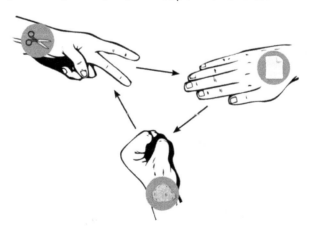

Figure 1.3: Rock, paper, scissors (OpenClipart-Vectors)

But making computers talk like us is hard, and making them understand us is still an unsolved problem and an area of active research.

That said, there is strong progress, especially in the field of machine translation, which is one form of language understanding. Machine translation algorithms can take a written text in one language and output the equivalent text in another language – for example, if you want to read a news article in French but you can only speak English, you can simply paste the article into Google Translate and it will spit out almost perfect English.

As with other machine learning systems, a vital ingredient for machine translation is a huge dataset. And hand-in-hand with a huge dataset, we need optimized data structures and storage methods to successfully create such a system. There are thousands of reasons why you might want to read a text in a language that you do not understand, from ordering at a foreign restaurant to studying old literature to conducting business with people in other countries.

A good example of machine translation in action is eBay, which improved its automatic translation capabilities in 2014. Imagine being a native Spanish speaker based in Latin America buying goods online from a native English speaker based in the US. You'd want to search for products in the language that you are most comfortable with, and you would want to read the details about the product, its condition, and shipping possibilities in Spanish too. Due to the large amount of eBay sales between Latin America and the USA, eBay tried to solve exactly this problem using AI. After improving its machine translation systems, eBay – as shown in the study "Does Machine Translation Affect International Trade? Evidence from a Large Digital Platform" – saw a 10.9 percent increase in purchases where the seller and buyer spoke different languages.

The translation of text is complicated, but at least writing is consistent. Spoken language can be even more complicated due to the complexities of sound waves, different accents, and different voice pitches: let's take a look at audio processing.

AUDIO PROCESSING – AUTOMATICALLY GENERATED SUBTITLES

Subtitles on videos are very useful. They help deaf people access video content and also allow video content to be shared across language barriers. The problem with subtitles is that they are difficult to create. Traditionally, to create subtitles, a person with specialized knowledge had to watch an entire video, potentially multiple times, typing out every audible word. Then each word had to be carefully aligned to the correct timestamp in the video file. How could we create subtitles for every video on YouTube? Once again, AI can come to our aid.

Years ago, Google introduced YouTube videos with automatically generated captions, and these have steadily improved in quality. Being able to read what people are saying as they talk is useful for millions of hard of hearing people and billions of people listening to audio or video content in their second or third language.

Similarly, California State University has used automatic captions to make their content available for deaf people.

We have now seen how AI can help computers act more like humans, but AI can also help computers be more efficient at other tasks, such as mathematics and analysis, including time series analysis, which is used across many fields. Let's study it in the next section.

TIME SERIES ANALYSIS

Seeing how machines can help us with health, communication, and disabilities might already make AI seem almost magical, but another area where AI shines is predicting the future. A common method for forecasting is time series analysis, which involves studying historical data, looking for trends, and assuming that these will hold in the future as well.

In an arguably less noble pursuit than medical advances, one of the most popular applications for time series analysis is in financial trading. If we can predict the rise and fall of stock prices, then we can be rich (as long as we don't share our knowledge too widely).

Despite decades of research and many attempts, it is not completely clear whether machines can reliably turn data directly into money by trading on global stock markets. Nonetheless, billions and potentially trillions of dollars change hands automatically every day, powered by AI predicting which assets will be valuable in the future.

OPTIMIZING THE STORING AND PROCESSING OF DATA FOR MACHINE LEARNING PROBLEMS

All of the preceding uses for artificial intelligence rely heavily on optimized data storage and processing. Optimization is necessary for machine learning because the data size can be huge, as seen in the following examples:

- A single X-ray file can be many gigabytes in size.

- Translation corpora (large collections of texts) can reach billions of sentences.

- YouTube's stored data is measured in exabytes.

- Financial data might seem like just a few numbers; these are generated in such large quantities per second that the New York Stock Exchange generates 1 TB of data daily.

While every machine learning system is unique, in many systems, data touches the same components. In a hypothetical machine learning system, data might be dealt with as follows:

Description	Hardware used
Data captured from a source	Network cables, hard drive (database or filesystem)
Data cleaned and transformed before being stored	Hard drive (database or blob storage)
Data loaded onto a powerful machine for training	Local hard drive
Data vectorized and transformed into batches	Random Access Memory (RAM)
Data iterated through for training	GPU, CPU, TPU
A model stored for later use	Hard drive
Backed up	Hard drive or magnetic tape

Figure 1.4: Hardware used in a hypothetical machine learning system

Each of these is a highly specialized piece of hardware, and although not all of them store data for long periods in the way traditional hard disks or tape backups do, it is important to know how data storage can be optimized at each stage. Let's dive into a text classification AI project to see how optimizations can be applied at some stages.

DIVING INTO TEXT CLASSIFICATION

Let's take a look at a practical use for the machine learning theory described in the preceding section. If you have spent any time reading news articles online, you'll probably have noticed that many sites take advantage of so-called "clickbait" – the practice of publishing headlines that deliberately withhold crucial information and imply that something exceptional happened to make readers click on an otherwise fairly boring article.

For example, "17 Insanely Awesome Starbucks You Need To See" is an example of a real headline that we can call "clickbait." It uses several tricks to try to make readers click through to the full article, even though the article itself is not very interesting: it uses an exact number (17), invoking curiosity to find out what all 17 are; it uses exaggeration ("insanely"), although there is nothing actually "insane" about the Starbucks in question; and it claims that you "need" to see them, although you can probably do just fine without.

On the other hand, publications with stronger commitments to ethical journalism would publish a headline such as "Ralph Nader enters US presidential race as independent" (another real headline). This headline, in direct contrast to the other one, is not clickbait. It is stating a simple fact; it is giving as much relevant information as possible upfront, and it is not trying to mislead the reader in any way.

To a computer, these headlines are difficult to tell apart. They both use standard English words, they are both similar in length, and there are not any specific rules that let you say with certainty "this is how to identify that a headline can be classified as clickbait."

This is a great example to use for machine learning – we, as humans, can tell which headlines are clickbait and which are not, but it is difficult to express this distinction as specific rules. Therefore, it makes sense to show a machine thousands of labeled examples – telling it 'these are clickbait, and these are not' – and see whether the machine can infer the rules on its own.

There are some important fundamentals and terminologies you need to be familiar with to fully follow along. For convenience, we'll summarize these here, but if you are not familiar with vectorization, training data, labels, classification, and evaluation, please note that these are complicated topics and that you may need to spend some time reading more about these in third-party resources.

Let's start by taking a look at TF-IDF vectorization.

LOOKING AT TF-IDF VECTORIZATION

Humans are used to reading and writing text, but computers prefer working with numerical data. For machines to be able to meaningfully process natural language text, we need to first convert this text into a meaningful binary format. There are many different ways of doing this, but a simple one is **TF-IDF** or **Term Frequency, Inverse Document Frequency**.

The fundamental idea of TF-IDF is that the more often a word appears in a text, the more important that word is. So, in an article about "electric cars," it is likely that the words "electric" and "car" will appear often, indicating that these words should be given more attention in any analysis that we do. Unfortunately, there are many common words, such as "the," and even though these words appear *frequently*, they are not *important*. To compensate for this, we don't only look at term frequency, but also *inverse document frequency*. A word that often appears in a single article but does not appear in many different articles is more important than a term that appears often in all articles. The exact weighting equation is not too important, and our Python library, **sklearn**, will take care of it for us, but out of interest, instead of using simple frequency counts as in the previous example, we will use the following equation:

word_freq(w, d) x log (N/doc_freq(w))

Here:

- *word_freq(w,d)* means the count of word *w* in document *d*.

- *N* means the total number of documents in our collection.

- *doc_freq(w)* means the number of documents that the word *w* appears in.

The point of vectorization is to transform the text into vectors, or an array of numbers that can be processed by a machine.

Term frequency, the first part of TF-IDF, relates to how often specific words are used. We'll start by looking at a manual example of vectorization using only term frequency, and then see how we can use a standard Python library for the full version of TF-IDF.

Counter-intuitively, we can ignore the order that the words in a given text are presented in and look only at their frequency. For example, we have two very short sentences in two documents, shown as follows:

1. "a cat and a dog"

2. "a cat and a fish"

We could first create a mapping table, assigning a single number to each word across all of our documents. This would look as follows:

'a' = 0

'cat' = 1

'and' = 2

'dog' = 3 (We skip the second "a" in the first document, as we already assigned it to a number.)

'fish'= 4 (We skip all the words before fish in the second document as they are all already assigned a number.)

The numbers map to what can be used as indices in an array. We will create a single array, containing only numbers to represent our document. The zeroth element of the array will indicate how many times the word "a" appears, as this was assigned the index "0." Once we have this mapping, we can represent both documents entirely numerically with arrays, as shown in the following figure:

Document 1

values	2	1	1	1	0
indices	0	1	2	3	4

Document 2

values	2	1	1	0	1
indices	0	1	2	3	4

Figure 1.5: Vectorized example – values and indices

The **2** at the zeroth index of the first array indicates that the word *a* appears twice in our first document, and the next three ones indicate that the words *cat*, *and*, and *dog* appear once each. *Fish* doesn't appear at all in the first document, so the 4th index of the array is a **0**. The second array looks very similar, but there is a **0** at the 3rd index to indicate that the *dog* doesn't appear and a **1** at the fourth index to indicate that *fish* appears once.

Note that the ordering is lost. The documents *a cat dog and* and *a cat and a dog* look the same now, but surprisingly this is hardly ever a problem in text processing.

We have seen how to convert text into a vectorized form for computers to read, which is an important first step. Before we get to use this in a practical example, we will define some basic terminology in machine learning classification tasks.

LOOKING AT TERMINOLOGY IN TEXT CLASSIFICATION TASKS

In a classification problem, we have **data** and **labels** – in our case, the data is the collection of headlines (clickbait and non-clickbait) and the labels are the indication of whether a specific headline is in fact "clickbait" or is "not clickbait."

We also have the terms **training** and **evaluation**. In the first part of the project, we'll feed both the data and the labels into our machine learning algorithm and it will try to derive a function that maps the data to the labels. We evaluate our model using different metrics, but a common and simple one is accuracy, which is how often the machine can predict the correct label without having access to it.

We'll be using two supervised machine learning algorithms in our project:

- **Support Vector Machine** (**SVM**): SVMs project data into higher dimensional space and look for decision boundaries.

- **Multi-layer perceptron** (**MLP**): MLPs are in some ways similar to SVMs but are loosely inspired by human brains, and contain a network of "neurons" that can send signals to each other.

The latter is a form of neural network, the model that has become the poster-child of machine learning and artificial intelligence.

We'll also be using a specialized data structure called a **sparse matrix**. For matrices that contain many zeros, it is not efficient to store every zero. We can, therefore, use a specialized data structure that stores only the non-zero values, but that nonetheless behaves like a normal matrix in many scenarios. Sparse matrices can be many times smaller than dense or normal matrices.

In the next exercise, you'll load a dataset, vectorize it using TF-IDF, and train both an SVM and an MLP classifier using this dataset.

EXERCISE 1.01: TRAINING A MACHINE LEARNING MODEL TO IDENTIFY CLICKBAIT HEADLINES

In this exercise, we'll build a simple clickbait classifier that will automatically classify headlines as "clickbait" or "normal." We won't have to write any rules to tell the algorithm how to do this, as it will learn from examples.

We'll use the Python `sklearn` library to show how to train a machine learning algorithm that can differentiate between the two classes of "clickbait" and "normal" headlines. Along the way, we'll compare different ways of storing data and show how choosing the correct data structures for storing data can have a large effect on the overall project's feasibility.

We will use a clickbait dataset that contains 10,000 headlines: 5,000 are examples of clickbait while the other 5,000 are normal headlines.

The dataset can be found in our GitHub repository at https://packt.live/2C72sBN

You need to download the **clickbait-headlines.tsv** file from the GitHub repository.

Before proceeding with the exercises, we need to set up a Python 3 environment with **sklearn** and Anaconda (for Jupyter Notebook) installed. Please follow the instructions in the *Preface* to install it.

Perform the following steps to complete the exercise:

1. Create a directory, **Chapter01**, for all the exercises of this chapter. In the **Chapter01** directory, create two directories named **Datasets** and **Exercise01.01**.

    > **NOTE**
    >
    > If you are downloading the code bundle from https://packt.live/3fpBOmh, then the *Dataset* folder is present outside *Chapter01* folder.

2. Move the downloaded **clickbait-headlines.tsv** file to the **Datasets** directory.

3. Open your Terminal (macOS or Linux) or Command Prompt (Windows), navigate to the **Chapter01** directory, and type **jupyter notebook**. The Jupyter Notebook should look like the following screenshot:

Figure 1.6: The Chapter01 directory in Jupyter Notebook

4. Create a new Jupyter Notebook. Read in the dataset file and check its size as shown in the following code:

```
import os
dataset_filename = "../../Datasets/clickbait-headlines.tsv"

print("File: {} \nSize: {} MBs"\
        .format(dataset_filename, \
        round(os.path.getsize(\
        dataset_filename)/1024/1024, 2)))
```

> **NOTE**
>
> Make sure you change the path of the TSV fie (highlighted) based on where you have saved it on your system. The code snippet shown here uses a backslash (\ ) to split the logic across multiple lines. When the code is executed, Python will ignore the backslash, and treat the code on the next line as a direct continuation of the current line.

You should get the following output:

```
File: ../Datasets/clickbait-headlines.tsv
Size: 0.55 MBs
```

We first import the **os** library from Python, which is a standard library for running operating system-level commands. Further, we define the path to the dataset file as the **dataset_filename** variable. Lastly, we print out the size of the file using the **os** library and the **getsize()** function. We can see in the output that the file is less than 1 MB in size.

5. Read the contents of the file from disk and split each line into data and label components, as shown in the following code:

```
import csv

data = []
labels = []

with open(dataset_filename,  encoding="utf8") as f:
    reader = csv.reader(f, delimiter="\t")
```

```
    for line in reader:
        try:
            data.append(line[0])
            labels.append(line[1])
        except Exception as e:
            print(e)

print(data[:3])
print(labels[:3])
```

You should get the following output:

```
["Egypt's top envoy in Iraq confirmed killed",
'Carter: Race relations in Palestine are worse than apartheid',
'After Years Of Dutiful Service, The Shiba Who Ran A Tobacco Shop
Retires']
['0', '0', '1']
```

We import the **csv** Python library, which is useful for processing our file and is in the **tab-separated values** (**TSV**) file format. We then define two empty arrays, **data,** and **labels**. We open the file, create a CSV reader, and indicate what kind of delimiter (**"\t"**, or a tab character) is used. Then, loop through each line of the file and add the first element to the data array and the second element to the labels array. If anything goes wrong, we print out an error message to indicate this. Finally, we print out the first three elements of each of our arrays. They match up, so the first element in our data array is linked to the first element in our labels array. From the output, we see that the first two elements are **0** or "not clickbait," while the last element is identified as **1**, indicating a clickbait headline.

6. Create **vectors** from our text data using the **sklearn** library, while showing how long it takes, as shown in the following code:

```
%%time
from sklearn.feature_extraction.text import TfidfVectorizer

vectorizer = TfidfVectorizer()
vectors = vectorizer.fit_transform(data)
print("The dimensions of our vectors:")
print(vectors.shape)
print("- - -")
```

You should get the following output:

```
The dimensions of our vectors:
(10000, 13169)
- - -
Wall time: 294 ms
```

> **NOTE**
>
> Some outputs for this exercise may vary from the ones you see here.

The first line is a special Jupyter Notebook command saying that the code should output the total time taken. Then we import a **TfidfVectorizer** from the **sklearn** library. We initialize **vectorizer** and call the **fit_transform()** function, which assigns each word to an index and creates the resulting vectors from the text data in a single step. Finally, we print out the shape of the vectors, noticing that it is **10,000** rows (the number of headlines) by **13,169** columns (the number of unique words across all headlines). We can see from the timing output that it took a total of around 200 ms to run this code.

7. Check how much memory our vectors are taking up in their sparse format compared to a dense format vector, as shown in the following code:

```
print("The data type of our vectors")
print(type(vectors))
print("- - -")
print("The size of our vectors (MB):")
print(vectors.data.nbytes/1024/1024)
print("- - -")
print("The size of our vectors in dense format (MB):")
print(vectors.todense().nbytes/1024/1024)
print("- - - ")
print("Number of non zero elements in our vectors")
print(vectors.nnz)
print("- - -")
```

You should get the following output:

```
The data type of our vectors
<class 'scipy.sparse.csr.csr_matrix'>
- - -
The size of our vectors (MB):
0.6759414672851562
- - -
The size of our vectors in dense format (MB):
1004.7149658203125
- - -
Number of non zero elements in our vectors
88597
- - -
```

We printed out the type of the vectors and saw that it was **csr_matrix** or a **compressed sparse row matrix**, which is the default data structure used by **sklearn** for vectors. In memory, it takes up only 0.68 MB of space. Next, we call the **todense()** function, which converts the data structure to a standard dense matrix. We check the size again and find the size is over 1 GB. Finally, we output the **nnz** (number of non-zero elements) and see that there were around 88,000 non-zero elements stored. Because we had **10,000** rows and **13,169** columns, the total number of elements is 131,690,000, which is why the dense matrix uses so much more memory.

8. For machine learning, we need to split our data into a train portion for training and a test portion to evaluate how good our model is, using the following code:

```
from sklearn.model_selection import train_test_split

X_train, X_test, \
y_train, y_test = train_test_split(vectors, \
                    labels, test_size=0.2)

print(X_train.shape)
print(X_test.shape)
```

You should get the following output:

```
(8000, 13169)
(2000, 13169)
```

We imported the **train_test_split** function from **sklearn** and split our two arrays (**vectors** and **labels**) into four arrays (**X_train**, **X_test**, **y_train**, and **y_test**). The **y** prefix indicates labels and the **X** prefix indicates vectorized data. We use the **test_size=0.2** argument to indicate that we want 20% of our data held back for testing. We then print out each shape to show that 80% (**8,000**) of the headlines are in the training set and that 20% (**2,000**) of the headlines are in the test set. Because each dataset was vectorized at the same time, each still has **13,169** dimensions or possible words.

9. Initialize the SVC classifier, train it, and generate predictions with the following code:

```
%%time

from sklearn.svm import LinearSVC

svm_classifier = LinearSVC()
svm_classifier.fit(X_train, y_train)

predictions = svm_classifier.predict(X_test)
```

You should get the following output:

```
Wall time: 55 ms
```

> **NOTE**
>
> The preceding output will vary based on your system configuration.

We import the **LinearSVC** model from **sklearn** and initialize an instance of it. Then we give it the training data and training labels (note that it does not have access to the test data at this stage). Finally, we give it the testing data, but without the testing labels, and ask it to guess which of the headlines in the held-out test set are clickbait. We call these **predictions**. To get some insight into what is happening, let's take a look at some of these predictions and compare them to the real labels.

10. Output the first **10** headlines along with their predicted class and true class by running the following code:

```
print("prediction, label")
for i in range(10):
    print(y_test[i], predictions[i])
```

You should get the following output:

```
prediction, label
1 1
1 1
0 0
0 0
1 1
1 1
0 1
0 1
1 1
0 0
```

We can see that for the first 10 cases, our predictions were spot on. Let's see how we did overall for the test cases.

11. Evaluate how well the model performed using the following code:

```
from sklearn.metrics \
import accuracy_score, classification_report

print("Accuracy: {}\n"\
      .format(accuracy_score(y_test, predictions)))
print(classification_report(y_test, predictions))
```

You should get the following output:

```
Accuracy: 0.965

              precision    recall  f1-score   support

           0       0.95      0.98      0.97      1000
           1       0.98      0.95      0.96      1000

    accuracy                           0.96      2000
   macro avg       0.97      0.96      0.96      2000
weighted avg       0.97      0.96      0.96      2000
```

Figure 1.7: Looking at the evaluation results of our model

> **NOTE**
>
> To access the source code for this specific section, please refer to https://packt.live/2ZlQnSf.

We achieved around 96.5% accuracy, which means around 1,930 of the 2,000 test cases were correctly classified by our model. This is a good summary score, but for a fuller picture, we have printed the full classification report. The model could be wrong in different ways: either by classifying a clickbait headline as normal or by classifying a normal headline as clickbait. Because the precision and recall scores are similar, we can confirm that the model is not biased toward a specific kind of mistake.

By completing the exercise, you have successfully implemented a very basic text classification example, but it highlighted several essential ideas around data storage. A lot of the data that we worked on was a *small* dataset, and we took some shortcuts that we would not be able to do with large data in a real-world setting:

- We read our entire dataset from a single file on disk into memory before saving it to a local disk. If we had more data, we would have had to read from a database, potentially over a network, in smaller chunks.

- We loaded all of the data back into memory and turned it into vectors. We naively did this, again keeping everything in memory simultaneously. With more data, we would have needed to use a larger machine or a cluster of machines, or a smart algorithm to handle processing the data sequentially as a stream.

- We converted our sparse matrix to a dense one for illustrative purposes. At 1,500 times the size, you can imagine that this would not be possible even with a slightly larger dataset.

In the rest of the book, you will examine each of the concepts that we touched on in more detail, through several case studies and real-world use cases. For now, let's take a look at how hardware can help us when dealing with larger amounts of data.

DESIGNING FOR SCALE – CHOOSING THE RIGHT ARCHITECTURE AND HARDWARE

In the examples we looked at, we used relatively small datasets, and we could do all of our analysis on a single commodity machine without any specialized hardware. If we were to use a larger dataset, such as the entire collection of English articles on Wikipedia, which come to many gigabytes of text data, we would need to pay careful attention to exactly what hardware we used, how we used different components of specialized hardware in combination, and how we optimized data flow throughout our system.

By the end of this section, you will be able to make calculated trade-offs in setting up machine learning solutions with specialized hardware. You will be able to do the following:

- Optimize hardware in terms of processing, volatile storage, and persistent storage.

- Reduce cloud costs by using long-running reserved instances and short-running *spot* instances as appropriate.

You will especially gain hands-on experience with running vectorized operations, seeing how much faster code can run on modern processors using these specialized operations compared to a traditional **for** loop.

OPTIMIZING HARDWARE – PROCESSING POWER, VOLATILE MEMORY, AND PERSISTENT STORAGE

We usually think of a computer processor as a central processing unit or CPU. This is circuitry that can perform fundamental calculations such as basic arithmetic and logic. All general-purpose computers such as laptops and desktops come with a standard CPU, and this is what we used in training the model for our text classifier. Our normal CPU was able to execute the billions of operations we needed to analyze the text and train the model in a few seconds.

At scale, when we need to process more data in even shorter time frames, one of the first optimizations we can look to is specialized hardware. Modern CPUs are general-purpose, and we can use them for multiple tasks. If we are willing to sacrifice some of the flexibility that CPUs provide, we can look to alternative hardware components to perform calculations on data. We already saw how CPUs can perform specialized operations on matrices as they are optimized for vectorized processing, but taking this same concept further leads us to hardware components such as **Graphical Processing Units (GPUs)**, **Tensor Processing Units (TPUs)**, and **Field Programmable Gate Arrays (FPGAs)**. GPUs are widely used for gaming and video processing – and now, more commonly, for machine learning too. TPUs are rarer: developed by Google, they are only available by renting infrastructure through Google's cloud. FPGAs are the least generalizable and are therefore not as widely used outside of specialized use cases.

GPUs were designed to carry out calculations on images and graphics. When processing graphical data, it is very common to need to do the same operation in parallel on many blocks of data (for example, to simultaneously move all of the pixels that make up an image or a piece of video into a frame buffer). Although GPUs were originally designed only for rendering graphical data, advances from the early 2000s and onward made it practical to use this hardware for non-rendering tasks too. Because graphical data is also usually represented using matrices and relies on fundamental structures and algorithms from linear algebra, there is an overlap between machine learning and graphical rendering, though at first, they might seem like very different fields. **General Purpose Computing on Graphical Processing Units**, or **GPGPU**, which is the practice of doing non-graphics related calculations on GPUs, is an important advance in being able to train machine learning models efficiently.

Nearly all modern machine learning frameworks provide some level of support for optimizing machine learning algorithms by accelerating some or all of the processing of vectorized data on a GPU.

As an extension of this concept, Google released TPUs in 2016. These chips are specifically designed to train neural networks and can in many cases be more efficient than even GPUs.

In general, we notice a trade-off. We can use specialized hardware to execute specific algorithms and specific data types more efficiently but at the cost of flexibility. While a CPU can be used to solve a wide variety of problems by running a wide variety of algorithms on a wide variety of data structures, GPUs and TPUs are more restricted in exactly what they can do.

A further extension of this is the **Field-Programmable Gate Array** (**FPGA**), which is specialized in specific use cases at the hardware level. These chips again can see big increases in efficiency, but it is not always convenient to build specialized and customized hardware to solve one specific problem.

Optimizing how calculations are carried out is important, but memory and storage can also become a bottleneck in a system. Let's take a look at some hardware options relating to data storage.

OPTIMIZING VOLATILE MEMORY

There are fewer hardware specializations in terms of volatile memory, where RAM is used in nearly all cases. However, it is important to optimize this hardware component nonetheless by ensuring the correct amount of RAM and the correct caching setup.

Especially with the advent of **solid-state drives** (**SSDs**), explored in more detail later, virtual memory is a vital component in optimizing data flow. Because the processing units examined previously can only store very small amounts of data at any given time, it is important that the next chunks of data queued for processing are waiting in RAM, ready to be bussed to the processing unit once the previous chunks have been processed. Since RAM is more expensive than flash memory and other memory types usually associated with persistent storage, it is common to have a page table or virtual memory. This is a piece of the hard disk that is used in the same way as RAM once the physical RAM has been fully allocated.

When training machine learning models, it is common for RAM to be a bottleneck. As we saw in *Exercise 1.01*, *Training a Machine Learning Model to Identify Clickbait Headlines*, matrices can grow in size very quickly as we multiply them together and carry out other operations. Because of this, we often need to rely on virtual RAM, and if you examine your system's metrics while training neural networks or other machine learning models, you will probably notice that your RAM, and possibly your hard disk, are used to full or almost full capacity.

The easiest way to optimize machine learning algorithms is often by simply adding more physical RAM. If this is not possible, adding more virtual RAM can also help.

Volatile storage is useful while data is actively being used, but it's also important to optimize how we store data on longer time frames using persistent storage. Let's look at that next.

OPTIMIZING PERSISTENT STORAGE

We have now discussed optimizing volatile data flow. In the cases of volatile memory and processor optimization, we usually consider storing data for seconds, minutes, or hours. But for machine learning solutions, we need longer-term storage too. First, our training datasets are usually large and need somewhere to live. Similarly, for large models, such as Google Translate or a model that can detect cancer in X-rays, it is inefficient to train a new model every time we want to generate predictions. Therefore, it's important to save these trained models persistently.

As with processing chips, there are many different ways to persistently store data. SSDs have become a standard way to store large and small amounts of data. These drives contain fast flash memory and offer many advantages over older **hard disk drives** (**HDDs**), which have spinning magnetic disks and are generally slower.

No matter what kind of hardware is used to store data persistently, it becomes challenging to store large amounts of data. A single hard drive can usually store no more than a few **terabytes** (**TBs**) of data, and it is important to be able to treat many hard drives as a single storage unit to store data larger than this. There are many databases and filesystems that aim to solve the problem of storing large amounts of data consistently, each with its advantages and disadvantages.

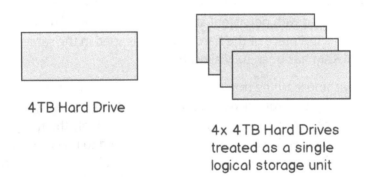

4 TB Hard Drive

4x 4 TB Hard Drives
treated as a single
logical storage unit

Figure 1.8: Linking units of hardware to simulate a larger storage capacity

As you work with larger and larger datasets, you will come across both **horizontal** and **vertical** scaling solutions, and it is important to understand when each is appropriate. Vertical scaling refers to adding more or better hardware to a single machine, and this is often the first way that scaling is attempted. If you find that you do not have enough RAM to run a particular algorithm on a particular dataset, it's often easy to try a machine that has more RAM. Similarly, for constraints in storage or processing capacity, it is often simple enough to add a bigger hard drive or a more powerful processor.

At some point, you will be using the most powerful hardware that money can buy, and it will be important to look at horizontal scaling. This refers to adding more machines of the same type and using them in conjunction with each other by working in parallel or sharing work and load in sophisticated ways.

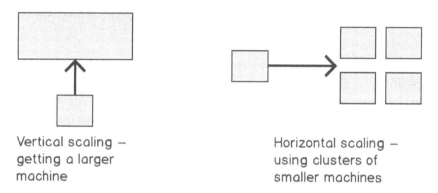

Vertical scaling – getting a larger machine

Horizontal scaling – using clusters of smaller machines

Figure 1.9: Vertical and horizontal scaling

Once again, cloud services can help us abstract away many of these problems, and most cloud services offer both virtual databases and so-called **Binary Large Object** (**BLOB**) storage. You will gain hands-on experience with both in later chapters of this book.

Optimizing hardware to be as powerful as possible is often important, but it also comes at a cost. Cost optimization is another important factor in optimizing systems.

OPTIMIZING CLOUD COSTS – SPOT INSTANCES AND RESERVED INSTANCES

Cloud services have made it much easier to *rent* specialized hardware for short periods, instead of spending large amounts of capital upfront during research and development phases. Companies such as Amazon (with AWS), Google (with GCP), and Microsoft (with Azure) allow you to rent virtual hardware and pay by the hour, so it is feasible to spin up a powerful machine and train your machine learning models in several hours, instead of waiting days or weeks for your laptop to crunch the numbers.

There are two important cost optimizations to be aware of when renting hardware from popular cloud providers: either by renting hardware for a very short time or by committing to rent it for a very long time. Specifically, because most cloud providers have some amount of unused hardware at any given moment, they usually *auction* it for short-term use.

For example, **Amazon Web Services** (**AWS**), the largest cloud provider currently, offers *spot* instances. If they have virtual machines attached to GPUs that no one has bought, you can take part in a live auction to use these machines temporarily at the fraction of the usual cost. This is often very useful for training machine learning models as the training can take place in a few hours or days, and it does not matter if there is a small delay in the beginning while you wait for an optimal price in the auction.

On the other side of the optimization scale, if you know that you are going to be using a specific kind of hardware for several years, you can usually optimize costs by making an upfront commitment about how long you will rent it. For AWS, these are termed *reserved instances*, and if you commit to renting a machine for 1, 2, or 3 years, you will pay less per hour than the standard hourly rate (though in most cases still more than the *spot* rate described previously).

In cases when you know you will run your system for many years, a reserved instance often makes sense. If you are training a model over a few hours or even days, spot instances can be very useful.

That's enough theory for now. Let's take a look at how we can practically use some of these optimizations. Because it is difficult to go out and buy expensive hardware just to learn about optimizations, we will focus on the optimizations offered by modern processors: vectorized operations.

USING VECTORIZED OPERATIONS TO ANALYZE DATA FAST

The core building blocks in all programmers' toolboxes are looping and conditionals – usually materialized as a `for` loop or an `if` statement respectively. Almost any programming problem in its most fundamental form can be broken down into a series of conditional operations (only do something *if* a specific condition is met) and a series of iterative operations (carry on doing the same thing *until* a condition is met).

In machine learning, vectors, matrices, and tensors become the basic building blocks, taking over from arrays and linked lists. When we are manipulating and analyzing matrices, we often want to apply a single operation or function to the entire matrix.

Programmers coming from a traditional computer science background will often use a `for` loop or a `while` loop to do this kind of analysis or manipulation, but they are inefficient.

Instead, it is important to become comfortable with vectorized operations. Nearly all modern processors support efficiently modifying matrices and vectors in parallel by executing the same operation to each element simultaneously.

Similarly, many software packages are optimized for exactly this use case: applying the same operator to many rows of a matrix.

But if you are used to writing **for** loops, it can be difficult to get out of the habit. So, we will compare the **for** loop with the vectorized operation to help you understand the reason to avoid using a **for** loop. In the next exercise, we'll use our headlines dataset again and do some basic analysis. We'll do each piece of analysis twice: first using a **for** loop, and then again using a vectorized operation. You'll see the speed differences even on this relatively small dataset, but these differences will be even more important on the larger datasets that we previously discussed.

While some languages have great support for vectorized operations out of the box, Python relies mainly on third-party libraries to take advantage of these. We'll be using **pandas** in the upcoming exercise.

EXERCISE 1.02: APPLYING VECTORIZED OPERATIONS TO ENTIRE MATRICES

In this exercise, we'll use the **pandas** library to load the same clickbait dataset and we'll carry out some descriptive analysis. We'll do each piece of analysis twice to see the efficiency gains of using vectorized operations compared to **for** loops.

Perform the following steps to complete the exercise:

1. Create a new directory, **Exercise01.02**, in the **Chapter01** directory to store the files for this exercise.

2. Open your Terminal (macOS or Linux) or Command Prompt (Windows), navigate to the **Chapter01** directory, and type **jupyter notebook**.

3. In the Jupyter notebook, click the **Exercise01.02** directory and create a new notebook file with a Python3 kernel.

4. Import the **pandas** library and use it to read the dataset file into a DataFrame, as shown in the following code:

```
import pandas as pd
df = pd.read_csv("../Datasets/clickbait-headlines.tsv", \
                 sep="\t", names=["Headline", "Label"])
df
```

You should get the following output:

	Headline	Label
0	Egypt's top envoy in Iraq confirmed killed	0
1	Carter: Race relations in Palestine are worse ...	0
2	After Years Of Dutiful Service, The Shiba Who ...	1
3	In Books on Two Powerbrokers, Hints of the Future	0
4	These Horrifyingly Satisfying Photos Of "Baby ...	1
...	...	...
9995	What Is Your Weirdest Fear	1
9996	Felipe Massa wins 2008 French Grand Prix	0
9997	Bottled water concerns health experts	0
9998	Death of Nancy Benoit rumour posted on Wikiped...	0
9999	US Dept. of Justice IP address blocked after '...	0

10000 rows × 2 columns

Figure 1.10: Sample of the dataset in a pandas DataFrame

We import the **pandas** library and then use the **read_csv()** function to read the file into a DataFrame called **df**. We pass the **sep** argument to indicate that the file uses tab (**\t**) characters as separators and then pass in the column names as the **names** argument. The output is summarized to show only the first few entries and the last few, followed by a description of how many rows and columns there are in the entire DataFrame.

5. Calculate the length of each headline and print out the first 10 lengths using a **for** loop, along with the total performance timing, as shown in the following code:

```
%%time

lengths = []

for i, row in df.iterrows():
    lengths.append(len(row[0]))
print(lengths[:10])
```

You should get the following output:

```
[42, 60, 72, 49, 66, 51, 51, 58, 57, 76]
CPU times: user 1.82 s, sys: 50.8 ms, total: 1.87 s
Wall time: 1.95 s
```

We declare an empty array to store the lengths, then loop through each row in our DataFrame using the **iterrows()** method. We append the length of the first item of each row (the headline) to our array, and finally, print out the first 10 results.

6. Now re-calculate the length of each row, but this time using vectorized operations, as shown in the following code:

```
%%time
lengths = df['Headline'].apply(len)
print(lengths[:10])
```

You should get the following output:

```
0      42
1      60
2      72
3      49
4      66
5      51
6      51
7      58
8      57
9      76
Name: Headline, dtype: int64
CPU times: user 6.31 ms, sys: 1.7 ms, total: 8.01 ms
Wall time: 7.76 ms
```

We use the **apply()** function to apply **len** to every row in our DataFrame, without a **for** loop. Then we print the results to verify they are the same as when we used the **for** loop. From the output, we can see the results are the same, but this time it took only **16.3** milliseconds instead of over 1 second to carry out all of these calculations. Now, let's try a different calculation.

7. This time, find the average length of all clickbait headlines and compare this average to the length of normal headlines, as shown in the following code:

```
%%time
from statistics import mean

normal_lengths = []
clickbait_lengths = []

for i, row in df.iterrows():
    if row[1] == 1:  # clickbait
        clickbait_lengths.append(len(row[0]))
    else:
        normal_lengths.append(len(row[0]))

print("Mean normal length is {}"\
      .format(mean(normal_lengths)))
print("Mean clickbait length is {}"\
      .format(mean(clickbait_lengths)))
```

> **NOTE**
>
> The # symbol in the code snippet above denotes a code comment. Comments are added into code to help explain specific bits of logic.

You should get the following output:

```
Mean normal length is 52.0322
Mean clickbait length is 55.6876
CPU times: user 1.91 s, sys: 40.7 ms, total: 1.95 s
Wall time: 2.03 s
```

We import the **mean** function from the **statistics** library. This time, we set up two empty arrays, one for the lengths of normal headlines and one for the lengths of clickbait headlines. We use the **iterrows()** function again to check every row and calculate the length, but this time store the result in one of our two arrays, based on whether the headline is clickbait or not. We then take the average of each array and print it out.

8. Now recalculate this output using vectorized operations, as shown in the following code:

```
%%time

print(df[df["Label"] == 0]['Headline'].apply(len).mean())
print(df[df["Label"] == 1]['Headline'].apply(len).mean())
```

You should get the following output:

```
52.0322
55.6876
CPU times: user 10.5 ms, sys: 3.14 ms, total: 13.7 ms
Wall time: 14 ms
```

In each line, we look at only a subset of the DataFrame: first when the label is 0, and second when it is 1. We again apply the **len** function to each row that matches the condition and then take the average of the entire result. We confirm that the output is the same as before, but the overall time is in milliseconds in this case.

9. As a final test, calculate how often the word **"you"** appears in each kind of headline, as shown in the following code:

```
%%time
from statistics import mean

normal_yous = 0
clickbait_yous = 0

for i, row in df.iterrows():
    num_yous = row[0].lower().count("you")
    if row[1] == 1:  # clickbait
        clickbait_yous += num_yous
    else:
        normal_yous += num_yous

print("Total 'you's in normal headlines
  {}".format(normal_yous))
print("Total 'you's in clickbait headlines
  {}".format(clickbait_yous))
```

You should get the following output:

```
Total 'you's in normal headlines 43
Total 'you's in clickbait headlines 2527
CPU times: user 1.48 s, sys: 8.84 ms, total: 1.49 s
Wall time: 1.53 s
```

We define two variables, **normal_yous** and **clickbait_yous**, to count the total occurrences of the word **you** in each class of headline. We loop through the entire dataset again using a **for** loop and the **iterrows()** function. For each row, we use the **count()** function to count how often the word **you** appear and then add this total to the relevant total. Finally, we print out both results, seeing that **you** appear very often in clickbait headlines, but hardly in non-clickbait headlines.

10. Rerun the same analysis without using a **for** loop and compare the time, as shown in the following code:

```
%%time
print(df[df["Label"] == 0]['Headline']\
        .apply(lambda x: x.lower().count("you")).sum())
print(df[df["Label"] == 1]['Headline']\
        .apply(lambda x:
    x.lower().count("you")).sum())
```

You should get the following output:

```
43
2527
CPU times: user 20.8 ms, sys: 1.32 ms, total: 22.1 ms
Wall time: 27.9 ms
```

We break the dataset into two subsets and apply the same operation to each. This time, our function is a bit more complicated than the **len** function we used before, so we define an anonymous function inline using **lambda**. We lowercase each headline and count how often **"you"** appears and then sum the results. We notice that the performance time, in this case, is again in milliseconds.

> **NOTE**
>
> To access the source code for this specific section, please refer to https://packt.live/2OmyEE2.

In this exercise, the main takeaway we can see is that vectorized operations can be many times faster than using **for** loops. We also learned some interesting things about clickbait characteristics though. For example, the word **"you"** appears very often in clickbait headlines (**2,527** times), but hardly ever in normal headlines (**43** times). Clickbait headlines are also, on average, slightly longer than non-clickbait headlines.

Let's implement the concepts learned so far in the next activity.

ACTIVITY 1.01: CREATING A TEXT CLASSIFIER FOR MOVIE REVIEWS

In this activity, we will create another text classifier. Instead of training a machine learning model to discriminate between clickbait headlines and normal headlines, we will train a similar classifier to discriminate between positive and negative movie reviews.

The objectives of our activity are as follows:

- Vectorize the text of IMDb movie reviews and label these as positive or negative.

- Train an SVM classifier to predict whether a movie review is positive or negative.

- Check how accurate our classifier is on a held-out test set.

- Evaluate our classifier on out-of-context data.

> **NOTE**
>
> We will be using some randomizers in this activity. It is helpful to set the global random seeds to ensure that the results you see are the same as in the examples. **Sklearn** uses the **NumPy** random seed, and we will also use the **shuffle** function from the built-in random library. You can ensure you see the same results by adding the following code:

```
import random
import numpy as np
random.seed(1337)
np.random.seed(1337)
```

We'll use the **aclImdb** dataset of 100,000 movie reviews from **Internet Movie Database** (**IMDb**) – 50,000 each for training and testing. Each dataset has 25,000 positive reviews and 25,000 negative ones, so this is a larger dataset than our headlines one. The dataset can be found in our GitHub repository at the following location: https://packt.live/2C72sBN

You need to download the **aclImdb** folder from the GitHub repository.

Dataset Citation: Andrew L. Maas, Raymond E. Daly, Peter T. Pham, Dan Huang, Andrew Y. Ng, and Christopher Potts. (2011). Learning Word Vectors for Sentiment Analysis. The 49th Annual Meeting of the Association for Computational Linguistics (ACL 2011).

In *Exercise 1.01, Training a Machine Learning Model to Identify Clickbait Headlines*, we had one file, with each line representing a different data item. Now we have a file for each data item, so keep in mind that we'll need to restructure some of our training code accordingly.

> **NOTE**
>
> The code and the resulting output for this exercise have been loaded in a Jupyter notebook that can be found at https://packt.live/3iWYZGH.

Perform the following steps to complete the activity:

1. Import the **os** library and the **random** library, and define where our training and test data is stored using four variables: one for **training_positive**, one for **training_negative**, one for **test_positive**, and one for **test_negative**, each pointing at the respective dataset subdirectory.

2. Define a **read_dataset** function that takes a path to a dataset and a label (either **pos** or **neg**) that reads the contents of each file in the given directory and adds these contents into a data structure that is a list of tuples. Each tuple contains both the text of the file and the label, **pos** or **neg**. An example is shown as follows. The actual data should be read from disk instead of being defined in code:

```
contents_labels = [('this is the text from one of the files',
    'pos'), ('this is another text', 'pos')]
```

3. Use the **read_dataset** function to read each dataset into its variable. You should have four variables in total: **train_pos**, **train_neg**, **test_pos**, and **test_neg**, each one of which is a list of tuples, containing the relative text and labels.

4. Combine the **train_pos** and **train_neg** datasets. Do the same for the **test_pos** and **test_neg** datasets.

5. Use the **random.shuffle** function to shuffle the train and test datasets separately. This gives us datasets where the training data is mixed up, instead of feeding all the positive and then all the negative examples to the classifier in order.

6. Split each of the train and test datasets back into **data** and **labels** respectively. You should have four variables again called **train_data**, **y_train**, **test_data**, and **y_test** where the **y** prefix indicates that the respective array contains labels.

7. Import **TfidfVectorizer** from **sklearn**, initialize an instance of it, fit the vectorizer on the training data, and vectorize both the training and testing data into the **X_train** and **X_test** variables respectively. Time how long this takes and print out the shape of the training vectors at the end.

8. Again, find the execution time, import **LinearSVC** from **sklearn** and initialize an instance of it. Fit the SVM on the training data and training labels, and then generate predictions on the test data (**X_test**).

9. Import **accuracy_score** and **classification_report** from **sklearn** and calculate the results of your predictions. You should get the following output:

```
Accuracy:  0.8772

              precision    recall  f1-score   support

         neg       0.87      0.89      0.88     12500
         pos       0.89      0.87      0.88     12500

    accuracy                           0.88     25000
   macro avg       0.88      0.88      0.88     25000
weighted avg       0.88      0.88      0.88     25000
```

Figure 1.11: Results – accuracy and the full report

10. See how your classifier performs on data on different topics. Create two restaurant reviews as follows:

```
good_review = "The restaurant was really great! "\
              "I ate wonderful food and had a very good time"
bad_review = "The restaurant was awful. "\
             "The staff were rude and "\
             "the food was horrible. "\
             "I hated it"
```

11. Now vectorize each using the same vectorizer and generate predictions for whether each one is negative or positive. Did your classifier guess correctly?

Now that we've built two machine learning models and gained some hands-on experience with vectorized operations, it's time to recap.

> **NOTE**
>
> The solution to this activity can be found on page 580.

SUMMARY

In this chapter, we gained a high-level overview of what machine learning can be used for, starting with image processing, text processing, audio processing, and time series analysis examples. This was followed by a deeper dive into text classification where we built a text classifier to identify clickbait headlines. We then looked at how to scale AI systems, looking at different kinds of hardware and cost-optimization techniques. By using vectorized operations, we were able to gain hands-on experience with optimization. Finally, we shored up our text-classification skills by building a second text classifier to differentiate between positive and negative movie reviews.

In the next chapter, we will explore ways to store large amounts of data for AI systems, looking specifically at data warehouses and data lakes.

2

ARTIFICIAL INTELLIGENCE STORAGE REQUIREMENTS

OVERVIEW

In this chapter, you will learn how to differentiate between traditional data warehousing and modern AI-focused systems. You'll be able to describe the typical layers in an architecture that is suited for building AI systems, such as a data lake, and list the requirements for creating the storage layers for an AI system. Later, you will learn how to define the specific requirements per storage layer for a use case and identify the infrastructure as well as the software systems based on the requirements. By the end of this chapter, you'll be able to identify the requirements for data storage solutions for AI systems based on the data layers.

INTRODUCTION

In the previous chapter, we covered the fundamentals of data storage. In this chapter, we'll dive a little deeper into the architecture of **Artificial Intelligence** (**AI**) solutions, starting with the requirements that define them. This chapter will be a mixture of theoretical content and hands-on exercises, with real-life examples where AI is actively used.

Let's say you are a solution architect involved in the design of a new data lake. There are a lot of technology choices to be made that would have an impact on the people involved and on the long-term operations of the organization. It is great to have a set of requirements at the start of the project that each decision could be based on. Storing data essentially means writing data to disk or memory so that it is safe, secure, findable, and retrievable. There are many ways to store data: on-premise, in the cloud, on disk, in a database, in memory, and so on. Each way fulfills a set of requirements to a greater or lesser extent. Therefore, always think about your requirements before choosing a technology or launching an AI project.

When designing a solution for AI systems, it's important to start with the requirements for data storage. The storage solution (infrastructure and software) is determined by the type of data you want to store and the types of analysis you want to perform. AI-powered solutions usually require high scalability, big data stores, and high-performance access.

IT solutions tend to be either data-intensive or compute-intensive. Data-intensive solutions are "big data" systems that store large amounts of data in a distributed form but require relatively little processing power. An example of a data-intensive system is an online video website that just shows videos, but where no intelligent algorithms are being run to classify them or offer any suggestions about what to watch next to its users. Compute-intensive solutions can have smaller datasets but demand many computing resources from the hardware; for example, language translation software that is continuously being trained with neural networks.

AI projects are not your typical IT projects; they are both data-intensive and compute-intensive. Data scientists need to have access to huge amounts of data to build and train their models. Once trained, the models need to be served in production and fed through a data pipeline. It's possible that these models get their features from a data store that holds the customer data in a type of cache for quick access. Another possibility is that data is continuously loaded from source systems so that it can be stored in a historical overview and queried by real-time dashboards that contain predictive models or other forms of intensive data usage. For example, a retail organization might want to predict trends in their product sales based on previous years. This kind of data cannot be retrieved from the source systems directly since they only keep track of the current state. For each of these scenarios, a combination of data stores must be deployed, filled, and maintained in order to fulfill the business requirements.

Let's have a look at the requirements that need to be evaluated for an AI project. We'll start with a brief list and do a deep dive later in the chapter.

STORAGE REQUIREMENTS

It's crucial to keep track of the requirements of your solution in all phases of the project. Since most projects follow the agile methodology, it's not an option to just define the requirements at the start of the project and then "get to work."

The agile methodology requires team members to continuously reflect on the initial plan and requirements provided in the Deming cycle, as shown in the following figure:

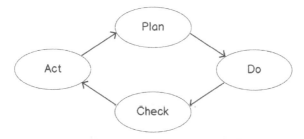

Figure 2.1: The Deming cycle

A list of requirements can be divided into functional and non-functional requirements. The **functional** requirements contain the user stories that explain how to interact with the system; these are not in the scope of this book since they are less technical and more concerned with UX design and customer journeys. The **non-functional** (or technical) requirements contain descriptions of the required workings of the system. The non-functional architecture requirements for an AI storage solution describe the technical aspects and have an impact on technology choices and their way of working. The major requirements of an AI system are as follows:

Requirement	Definition
Scalability	The ability to scale the infrastructure (servers, CPUs, disks) and thereby the costs up and down on demand.
Security	The ability to keep data safe and to administer access.
Performance	The speed with which data can be ingested and retrieved.
Retention	The amount of time that data can be stored before it's deleted.
Cloud integration	The requirements for infrastructure to run cloud-native, on-premise, or hybrid.
Locality of data	The geographical considerations for data storage.
Time travel	The ability to query data in the past.
Metadata and lineage	The ability to store data about the data, and to trace the origin of any data point.
Cost efficiency	The price and resources needed to run a solution.
Flexibility	The modularity and amount of vendor lock-in (the required long-term commitments to a company that provides the technology, such as after-sales support and mandatory updates) a solution provides.
Availability	The uptime, reliability, and other aspects related to "always-on" systems.
Quality	The requirements related to quality aspects of architecture, such as reliability, maintainability, and usability

Figure 2.2: Requirements for AI systems

Since this a very extensive list and some requirements are more important for a certain architectural layer than others, we will list the most important requirements per architecture layer. Before we start with that deep dive, we'll give a brief overview of the architecture of an AI system or data lake.

Throughout this chapter, we'll provide you with an example of a use case that helps translate the abstract concepts in the requirements for data storage to real-world, hands-on content. Although the sample is fictional, it's built on some common projects that we came across in real life. Therefore, the situation, target architecture, and requirements are quite realistic for an AI project.

A bank in the UK (let's say it's called PacktBank) wanted to upgrade its data storage systems to create a better environment for data scientists for AI-related projects. Currently, the data is spread out in various source systems, ranging from an old ERP system to on-premise Oracle databases, to a SaaS solution in the cloud. The new data environment (data lake) must be secure, accessible, high-performing, scalable, and easy to use. The target infrastructure is **Amazon Web Services** (**AWS**), but in the future, the company might switch to other cloud vendors or take a multi-cloud strategy; therefore, the software components should be vendor-agnostic if possible.

THE THREE STAGES OF DIGITAL DATA

It's important to realize that data storage comes in three stages:

- **At rest**: Data that is stored on a disk or in memory for long-term storage; for example, data on a hard disk or data in a database.

- **In motion**: Data that is transferred across a network from one system to another. Sometimes, this is also called *in transit*; for example, HTTP traffic on the internet, or data that comes from a database and is "on its way" to an application.

- **In use**: Data that is loaded in the RAM of an application for short-term usage. This data is only available in the context of the software that is loaded. It can be seen as a cache that is temporarily needed by the software that performs tasks on the data. The data is usually a copy of data at rest; for example, a piece of customer information (let's say, a changed home address) that has been pushed from a website to the server where an API processes the update.

These stages are important to keep in mind when reasoning about technology, security, scalability, and so on. We'll bring them up in this book in several places, so make sure that you understand the differences.

DATA LAYERS

An AI system consists of multiple data storage layers that are connected with **Extract, Transform**, and **Load (ETL)** or **Extract, Load**, and **Transform (ELT)** pipelines. Each separate storage solution has its own requirements, depending on the type of data that is stored and the usage pattern. The following figure shows this concept:

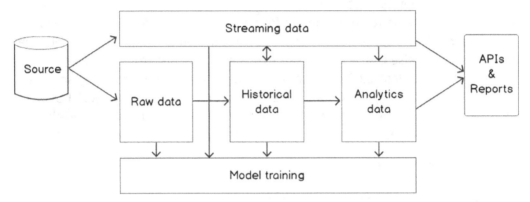

Figure 2.3: Conceptual overview of the data layers in a typical AI solution

From a high-level viewpoint, the backend (and thus, the storage systems) of an AI solution is split up into three parts or layers:

- **Raw data layer**: Contains copies of files from source systems. Also known as the staging area.

- **Historical data layer**: The core of a data-driven system, containing an overview of data from multiple source systems that have been gathered over time. By stacking the data rather than replacing or updating old values, history is preserved and time travel (being able to make queries over a data state in the past) is made possible in the data tables.

- **Analytics data layer**: A set of tools that are used to get access to the data in the historical data layer. This includes cache tables, views (virtual or materialized), queries, and so on.

These three layers contain the data in production. For model development and training, data can be offloaded into a special model development environment such as a DataBricks cluster or SageMaker instance. In that case, an extra layer can be added:

- **Model training layer**: A set of tools (databases, file stores, machine learning frameworks, and so on) that allows data scientists to build models and train them with massive amounts of data.

For scenarios where data is not being ingested by the system per batch but rather streamed in continuously, such as a system that processes sensory machine data from a factory, we must set up specific infrastructure and software. In those cases, we will use a new layer that takes the role of the raw data layer:

- **Streaming data layer**: An event bus that can store large amounts of continuously inflowing data streams, combined with a streaming data engine that is able to get data from the event bus in real time and analyze it. The streaming data engine can also read and write data to data stores in other layers, for example, to combine the real-time data from the event bus with historical data about customers from a historical data view.

Depending on the requirements for data storage and analysis, for each layer, a different technology set can be picked. The data stores don't have to be physical file stores or databases. An in-memory database, graph database, or even a virtual view (just queries) can be considered as a proper data storage mechanism. Working with large datasets in complex machine learning algorithms requires special attention since the models to be trained require the storage and usage of big datasets, but for a relatively short period of time.

To summarize, the data layers of AI solutions are like many other data-driven architecture layers, with the addition of the model training layer and possibly the streaming data layer. But there is a bigger shift happening, namely the one from data warehouses to data lakes. We'll explore that paradigm shift in the next part of this chapter.

FROM DATA WAREHOUSE TO DATA LAKE

The common architecture of a data processing system is shifting from traditional data warehouses that run on-premise toward modern data lakes in the cloud. This shift is being made by a new technology wave that started in the "big data" era with Hadoop in around 2006. Since then, more and more technology has arrived that makes it easier to store and process large datasets and to build models efficiently, making use of advanced concepts such as distributed data, in-memory storage, and **graphical processing units** (**GPU**s). Many organizations saw the opportunities of these new technologies and started migrating their report-driven data warehouses toward data lakes with the aim of becoming more predictive and to get more value out of their data.

Next to a technology shift, what we see is a move toward more virtual data layers. Since computing power has become cheap and parallel processing is now a common practice, it's possible to virtualize the analytics layer instead of storing the data on disk. The following figure illustrates this; while the patterns and layers are almost the same, the technology and approach differ:

Traditional data warehouse:
Source → Staging (raw) → Historical (relational, temporal) → Analytical (schema for reading) → report/API → app

Modern cloud-based data lake:
Source → Staging (raw) → Historical + Analytical → Views → report/API → app

Figure 2.4: Data pipelines in a data warehouse and a data lake

The following table highlights the similarities and differences between the data platforms:

	Data Warehouse	Data Lake
Infrastructure	On-premise	Cloud-native
Data pipeline	ETL	ELT
Raw data layer	Staging area	Staging area
Historical data layer	Relational data vault	Relational data vault
Analytics data layer	Analytical database	Virtual views on top of historical data
Model training layer	-	Machine learning cluster
Streaming data layer	-	Streaming data cluster and stream processing engine

Figure 2.5 Comparison between data warehouses and data lakes

NOTE

The preceding table has been written to highlight the differences between data warehouses and date lakes, but there is a "gray zone." There are many data solutions that don't fall into either of the extremes; for example, ETL pipelines that run in the cloud.

To understand data layers better, let's revisit our bank example. The architects of PacktBank started by defining a new data lake architecture that should be the data source for all AI projects. Since the bank did not foresee any streaming project on short notice, the focus was on batch; every source system had to upload a daily export to the data lake. This data was then transformed into a relational data vault with time-traveling possibilities. Finally, a set of views allowed quick access to the historical data for a set of use cases. The data scientists were building models on top of the data from these views, which was temporarily stored in a model development environment (Amazon SageMaker). Where needed, the data scientists could also get access to the raw data. All these access points were regulated with role-based access and were monitored extensively. The following figure shows the final architecture of the solution:

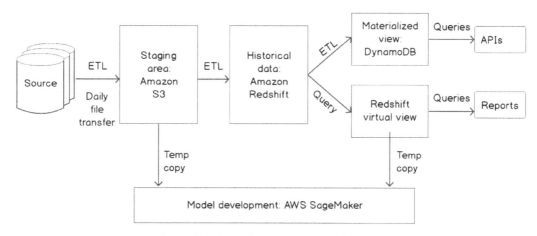

Figure 2.6: Sample data lake architecture

EXERCISE 2.01: DESIGNING A LAYERED ARCHITECTURE FOR AN AI SYSTEM

The purpose of this exercise is to get acquainted with the layered architecture that is common for large AI systems.

For this exercise, imagine that you must design a new system for a telecom organization; let's call it PacktTelecom. The company wants to analyze internet traffic, call data, and text messages on a daily basis and perform predictive analysis on the dataset to forecast the load on the network. The data itself is produced by the clients of the company, who are using their smartphones on the network. The company is not interested in the content of the traffic itself, but most of the information at the meta-level is interesting. The AI system is considered part of a new data lake, which will be created with the aim of supporting many similar use cases in the future. Data from multiple sources should be combined and analyzed in reports and made available to websites and mobile apps through a set of APIs.

Now, answer the following questions for this use case:

1. What is the data source of the use case? Which systems are producing data, and in which way are they sending data to the data lake?

   The prime data source is the smartphones of the clients. The smartphones are continuously connected to the network of the company and send their metadata to the core systems (for example, internet traffic and text messages). These core systems will send a daily batch of data to the data lake; if required, this can be made real-time streaming at a later stage.

2. Which data layers should you use for the use case? Is it a streaming infrastructure or a more traditional data warehousing scenario?

   A raw data layer stores the daily batches that are sent from the core systems. A historical data layer is used to build a historical overview per customer. An analytics layer is used to query the data in an efficient way. At a later stage, a streaming infrastructure can be realized to replace the daily batches.

3. What data preparation steps need to be done in the ETL process to get the raw data in shape so that it's useful to work with?

   The raw data needs to be cleaned; the content must be removed. Some metadata might have to be added from other data sources, for example, client information or sales data.

4. Are there any models that need to be trained? If so, which layer are they getting their data from?

   To forecast the network load, a machine learning model needs to be created that is trained on the daily data. This data is gathered from the historical data layer.

By completing this exercise, you have reasoned about the layers in an AI system and (partly) designed an architecture.

REQUIREMENTS PER INFRASTRUCTURE LAYER

Let's dive into the requirements for each part of a data solution. Depending on the actual requirements per layer, an architect can choose the technology options for the layer. We'll list the most important requirements per category here.

Some requirements apply to all data layers and can, therefore, be considered generic. For example, scalability and security are always important for a data-driven system. However, we've chosen to list them for each layer separately because each layer has many specific attention points for these generic requirements as well.

The following table highlights the most important requirements per data layer:

Data layer	Important requirements
Raw data	Security
	Scalability
	Time travel
	Retention
	Metadata and lineage
Historical data	Security
	Scalability
	Availability
	Time travel
	Locality
	Metadata and lineage
Streaming data	Security
	Scalability
	Performance
	Availability
	Retention
Analytics data	Security
	Performance
	Cost-efficiency
	Quality
Model development and training	Security
	Scalability
	Retention

Figure 2.7: Important requirements per data layer

RAW DATA

The raw data layer contains the one-to-one copies of files from the source systems. The copies are stored to make sure that any data that arrives is preserved in its original form. After storing the raw data, some checks can be done to make sure that the data can be processed by the rest of the ETL pipeline, such as a checksum.

SECURITY

We'll look at data security first. All modern software and data systems must be secure. By security requirements, we mean all aspects related to ensuring that the data in a system cannot be viewed or deleted by unauthorized people or systems. It entails identity and access management, role-based access, and data encryption.

BASIC PROTECTION

In any data project, security is a key requirement. The basic level of data protection is to require a username-password combination for anyone who can access the data: customers, developers, analysts, and so on. In all cases, the passwords should be evaluated against a strong password policy and must be changed on a regular basis. Passwords should never be stored in plain text; instead, they should always be in a salted and hashed form so that even system administrators cannot retrieve the actual passwords. The security levels themselves depend on the **Availability**, **Integrity**, and **Confidentiality** (**AIC**) rating, which we'll explain in the following paragraph. Suffice to say that highly secure data should not only be protected with a username and password. Multi-factor authentication is a way of adding a security layer by making use of a second or even third authentication method alongside password protection, such as an SMS message on your phone, a fingerprint ID, or a dedicated security token generator.

THE AIC RATING

Data security in organizations can be classified with the AIC rating (sometimes the CIA rating is also used, but this causes confusion with the abbreviation for the Central Intelligence Agency). Each data source and application should be categorized into three dimensions:

- **A = availability**: The level of protection against data loss in cases of system failure or upgrades. If data loss must be prevented at all times, for example, for payment transactions, the level is high. If data loss is annoying but not terrible, for example, spam emails, the level is low.

- **I = integrity**: The level of consistency and accuracy that the dataset must uphold during its life cycle. Some data might be removed or updated without any consequences; those datasets will receive a low integrity rating. However, if it's crucial that data records are preserved for a long time, for example, tax records, the rating will go up.

- **C = confidentiality**: The level of personal details of a person or company that is included in the dataset. If personal details such as names or addresses are stored, it's likely that the rating is high. The highest level of confidentiality is reserved for datasets that contain very private data, such as credit card numbers or passwords.

Each of these dimensions gets a rating from 1 to 3. Thus, a dataset with an AIC rating of 111 is considered to be less risky and vulnerable compared to a dataset with a rating of 333. You could argue that when data from a 111 system falls into the wrong hands (a hacker or competing organization), it's no big deal. However, when data from a 333 system ends up "on the street," you're in serious legal trouble and might even be out of business. When it comes to securing data, each category should imply certain measurements within your company. For example, any dataset with a C rating of 3 can only be accessed with multi-factor authentication.

ROLE-BASED ACCESS

In the raw data layer, all data files must be governed with **role-based access control** (**RBAC**) to ensure that no data falls into the wrong hands. Every principal (human or machine account) has one or more roles. Each role has one or more permissions to a database, file share, table, or other pieces of data. With the right permissions, files can be read by humans (for example, data scientists that require access to the raw data); write access is only available for software that imports the data from the source systems. Every file must be secured. In many cases, the actual security is inherited from a higher directory. The file structure proposed in *Figure 2.6: Sample data lake architecture*, allows security to be configured per source system or per period. It should also be possible to give data scientists access to a part of the data store temporarily, for example, to copy data to a machine learning environment. Modern security frameworks and identity and access management systems such as Microsoft Active Directory and Amazon AWS **Identity and Access Management** (**IAM**) have these options available. The following figure shows these:

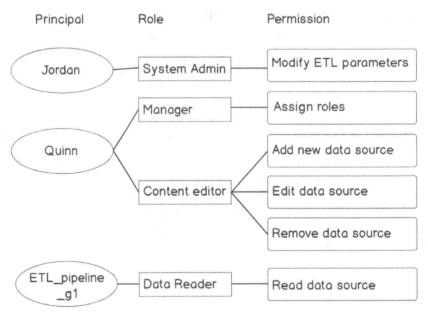

Figure 2.8: Simplified sample of roles and permissions of a data importing system

The preceding figure is a highly simplified role-based access diagram that has been sketched to illustrate the relations between principals, roles, and permissions. In this example, **Jordan** has the role of **System Admin** and is therefore allowed to modify the parameters of an ETL pipeline. However, he cannot read the actual contents of the data sources since he has no permissions. **Quinn** has two roles: **Manager** and **Content Editor**. She can assign roles to other principals and add, edit, and remove data sources. **ETL_pipeline_g1** is a non-human principal, namely an account that a piece of software uses to access data and execute tasks. It has the role of **Data Reader** and therefore has permission to read the data from the data sources.

ENCRYPTION

Data at rest (stored on disk) and in motion (transferred across a network) should be encrypted to prevent third parties and hackers from accessing it. By encrypting data, a hacker who intercepts the data or reads it straight from disk still cannot get to its content; it will just be a scrambled array of characters. To read the data, you must be in possession of a private key that has a relation to the public key that the data was encrypted with (with asymmetric encryption, which is by far the most popular mechanism). These private keys must, therefore, be kept secure at all times, for example, in a special purpose key store. All modern cloud infrastructure providers have such a key store as a service in their offerings.

There are many types of encryption possible. **Advanced Encryption Standard (AES)** and **Rivest-Shamir-Adleman (RSA)** are two examples of popular asymmetric encryption algorithms. These can both be used to encrypt data at rest and in motion, although some performance considerations might apply; RSA is a bit slower than AES. What's more important is to choose the size of the keys; the more bits, the harder it is to break the encryption with a brute-force attack. Currently, 256 bits is considered to be a safe key size.

Keys should never be stored in places with more or less open access, for example, sticky notes or Git repositories. It's good practice to rotate your keys, which means that they alter after a certain period (say, a month). This makes sure that even if keys are accidentally stored in a public place, they can only be used for a limited period of time.

At the core of data security are four basic principles:

- Security starts with basic protection, such as strong and rotating passwords. All users are registered in a central identity and access management system.

- All access to data is regulated with permissions. Permissions can be attached to roles, and roles can be assigned to users. This is called role-based access.

- The data security measurements are related to the AIC rating of a dataset. A higher rating indicates that more security controls should be put into place.

- Data at rest and in motion can be encrypted to protect it from intruders.

Let's understand this better by going through the next exercise.

EXERCISE 2.02: DEFINING THE SECURITY REQUIREMENTS FOR STORING RAW DATA

For this exercise, imagine that you are creating a new data environment for an ambulance control room. The goal of the system is to gather as much useful information as possible from government and open data sources in order to direct ambulances on their way to a 911 call once it arrives. The core data sources that must be stored are the 911 calls; this is combined with maps data, traffic information, local news from the internet, and other sources. It's apparent that such systems are prone to hacking and a wrong/fake call could lead to medical mistakes or the late arrival of emergency personnel.

In this exercise, you will create a security plan for the ambulance control room. The aim of this exercise is to become familiar with the security requirements of a system where data protection plays an important role.

Now, answer the following questions for this use case:

1. Consider the data source of your application. Who is the owner of the data? Where is the data coming from?

   The prime data source is the 911 calls that come from the people who need help. The call data is owned by the person who makes the call.

2. What is a potential security threat? A hacker on the internet? A malicious employee of your company? A physical attack on your data center?

    Potential security threats are hackers on the internet, fake phone callers, employees who might turn against the company, physical attackers on the data center, terrorists, and many more.

3. Try to define the AIC rating of the dataset. What are the levels (from 1 to 3) for the availability, integrity, and consistency of the data?

    The phone calls have an AIC rating of 233. The availability is reasonably high but retrieving the data, in retrospect, is not as important as being able to respond to the calls once they arrive; thus, the overall availability is 2. The infrastructure for making the calls has an availability rating of 3. The integrity rating is 3 since the ambulance control room must be able to rely on the data; the location, time, and call quality are all very important aspects of the data. The confidentiality rating is also 3 since the calls themselves will contain many privacy-related details and confidential information.

4. Regarding the AIC rating, which measurements should you take to secure the data? What kind of identity and access management should you put in place? What kind of data encryption will you use? Consider the roles and permissions for accessing the data, as well as password regulations and multi-factor authentication.

    Considering the high integrity and confidentiality ratings, the security around the call data must be very good. The data should only be accessed by registered and authorized personnel of the control room, who have been given access by a senior manager. The access controls must be audited on a regular basis. Two-factor authentication, a strong password policy, and encryption of all data at rest and in motion must be put in place in order to minimize the risk of security breaches and hacks from outside.

By completing this exercise, you have created a security plan for a demanding system. This helps when setting the requirements for systems in your own organization.

SCALABILITY

Scalability is the ability of a data store to increase in size. Elasticity is the ability of a system to grow and shrink on demand, depending on the need at hand. For the sake of simplicity, we address both scalability and elasticity under one requirement: scalability.

In traditional data warehousing projects, a retention policy was very important in the raw data layer since it prevented the disks from getting full. In modern AI projects, what we see is that it's best to keep the raw data for as long as possible since (file) storage is cheap, scalable, and available in the cloud. Moreover, the raw data often provides the best source for model training in a machine learning environment, so it's valuable to give data scientists access to many historical data files.

To cater for storing this much data on such a large scale, a modern file store such as Amazon S3 or Microsoft Azure ADLS Gen2 is ideally suited. These cloud-based services can be seen as the next generation of Hadoop file stores, where massive parallel file storage is made easily available to its consumers. For an on-premise solution, Hadoop HDFS is still a good solution.

Using the same example of PacktBank, the new data lake for AI must start small but soon scale to incorporate many data sources of the bank.

The architects of PacktBank defined the following set of requirements for the new system:

- The data store should start very small since we will first do a proof-of-concept with only test data. The initial dataset is about 100 MB in size.

- The data store should be able to expand rapidly toward a size where all the data from hundreds of core systems will be stored. The expected target size is 20 TB.

- There will be a retention policy forced on some parts of the data since privacy regulations enforce that certain sensitive data be removed after 7 years. The data store should be able to shrink back to a smaller size (~15 TB) if needed, and the costs associated with the data store should follow proportionally.

TIME TRAVEL

For many organizations, it is important to be able to query data in the past. It's very valuable and is often required by laws or regulations to be able to answer questions such as "how many customers were in possession of product 3019 one month ago?" or "which employees had access to document 9201 on 14 March 2018?". This ability is called **time travel**, and it can be embedded in data storage systems.

Raw data must be stored in a way so that its origins and time of storage are apparent. Many companies choose to create a directory structure that reflects the daily or hourly import schedule, like so:

- Raw
    - Source_system_1
    - Source_system_2
        - 2019
            - January
            - February
                - 1
                - 2
                    - Daily_export_20190202_ERROR.csv
                    - Daily_export_20190202.csv
                    - Export_import_log_20190202.txt

Figure 2.9: Example of a directory structure for raw data files

In the preceding figure, a file that arrives on a certain date and time gets placed in a directory that reflects its arrival date. We can see that on February 2, 2019, a daily export was stored. There is also a file containing **ERROR** which possibly indicates a failed or incomplete import. The full import log is stored in a text file in the same folder. By storing the raw data in this structure, it's very easy for an administrator to ask questions about the source data in the past; all they must do is browse to the right directory on the filesystem.

RETENTION

Data retention requirements define in what way data is stored to meet laws and regulations or to preserve disk space by deleting old files and data records. Since it's often convenient and useful in a modern AI system to keep storing all the data (after all, data scientists are data-hungry), a retention policy is not always necessary from a scalability perspective. As we saw when exploring the scalability requirement, many data stores can store massive amounts of data cheaply. However, a retention requirement (and therefore, a policy) might be needed because of laws and regulations. For example, in the EU's GDPR regulations, it's stated that data must be stored "for the shortest time possible." Some laws and regulations are specific for industries, for example, call data and metadata in a telecom system may only be stored for 7 years by a telecom provider.

To cater to retention requirements, the infrastructure and software of your data lake should have a means of removing and/or scrambling/anonymizing data periodically. In modern file stores and databases, policies can be defined in the tool itself and the tool automatically implements the retention mechanisms. For example, Amazon S3 supports lifecycle policies in which data owners can specify what should happen with the data over time, as shown in the following figure:

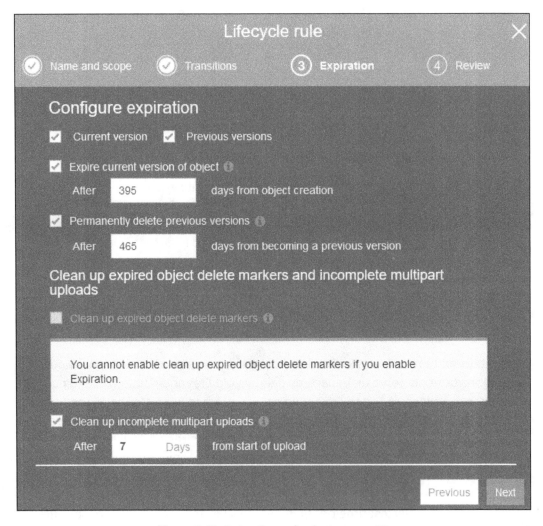

Figure 2.10: Retention rules in Amazon S3

When storing data in an on-premise data store that does not support retention (for example, MySQL or a regular file share), you'll have to write code yourself that periodically runs through all your data and deletes it, depending on the parameters that have been defined for your policy.

METADATA AND LINEAGE

Metadata is data about data. If we store a file on a disk, the actual data is the contents of the file. The metadata is its name, location, size, owner, history, origin, and many other properties. If metadata is correctly and consistently registered for each dataset (a file or database record), this can be used to track and trace data across an organization. This tracking and tracing is called **lineage**, and it can be mostly automated with modern tooling.

If the requirements for metadata management and lineage requirements are in place, every data point in the pipeline must be traced back to its source if required. This can be requested in audits or for internal data checks. For the raw data layer, it implies that for every data source file, we store a set of metadata, including the following:

- **Filename**: The name of the file
- **Origin**: The location or source system where the data comes from
- **Date**: The timestamp when the data entered the data layer
- **Owner**: The data owner who is responsible for the file's contents
- **Size**: The file size

For data streams, we store the same attributes as metadata but they are translated to the streaming world; for example, the stream name instead of a filename. In *Chapter 3, Data Preparation*, we'll discuss lineage in detail and provide an exercise for building lineage into ETL jobs.

A further extension to the security model and metadata-driven lineage, once in place, is **consent management**. Based on the metadata of the raw data files, a data owner can select which roles or individuals have access to its files. In that way, it's possible to create a data-sharing environment where each data owner can take responsibility for the availability of their own data.

Now, we have discussed the main requirements for the raw data layer: security, scalability, time travel, retention, and metadata. Keep in mind that these topics are important for any data storage layer and technology. In the next section, we will explore these requirements for the historical data layer and add availability as an important requirement that is most apparent for the technology and data in that layer.

HISTORICAL DATA

The historical data layer contains data stores that hold all data from a certain point in the past (for example, the start of the company) up until now. In most cases, this data is considered to be important to run a business, and in some cases, even vital for its existence. For example, the historical data layer of a newspaper agency contains sources, reference material, interviews, media footage, and so on, all of which were used to publish news articles. Data is stored in blobs, file shares, and relational tables, often with primary and foreign keys (enforced by the infrastructure or in software). The data can be modeled to a standard such as a data vault to preserve historical information. This data layer is responsible for keeping the truth, which means it is highly regulated and governed. Any data that is inserted into one of the tables in this layer has gone through several checks, and metadata is stored next to the actual data to keep track of the history and manage security.

SECURITY

In general, the same requirements that apply to the raw data store also apply to the historical data layer. Whereas the raw data layer dealt with files, the historical data layer has tables to protect. But that is often not enough granularity since a lot of data can be combined in tables. For example, a company that provides consultancy services to multiple customers could have a table with address information that contains records from these companies. But for the sake of privacy and secrecy, not all account managers in the consultancy organization may have access to each client; they should only see the information of the clients that they are working for directly. For these kinds of cases, it must be possible to apply row-level or column-level security.

Row-Level Security

When setting up tables in a multi-tenant way, containing data from multiple owners, it's necessary to administer the owner per record. Modern databases and data warehouse systems such as Azure Synapse can then assign role-based access security controls to the tables per row; for example, "people that have role A have read-only access to all records where the data owner is O."

Column-Level Security

In a similar way, security can be arranged per column in modern columnar NoSQL databases. It might be beneficial to add columns to a table for a specific client or data owner. In those cases, access to the columns can be arranged with similar role-based access; for example, "people that have role B have read and write access to all data in columns Y and Z."

SCALABILITY

The amount of data in the historical data layer will keep on growing since fresh data will arrive every day and not all data will be part of a retention plan. Business users will also regard the historical data as highly valuable since the information there can be used to train models and generally compare situations of the organization. Therefore, it's crucial to pick technology that can scale. Modern data warehouses in the cloud all cater to scalability and elasticity, for example, Amazon Redshift and Snowflake. The scalability of data stores on-premise is more limited, constrained by your own data center. However, for many organizations, the requirements to scale might be perfectly met by an on-premise infrastructure. For example, local government organizations often deal with complex data from many sources, but the total size of a data lake usually does not surpass the 1 TB mark. In these kinds of cases, setting up a solid infrastructure that can hold this data is perhaps a better choice than to put all the data in the public cloud.

AVAILABILITY

A system or datastore is considered to be reliable and highly available if there is a guarantee that a system keeps on running and no data is lost in the lifespan of the system, even during outages or failures. Usually, this problem is solved by distributing data across an infrastructure cluster (separated in servers/nodes or geographical regions) as soon as it enters the system. In the case of a crash or other malfunction, the data is backed up in multiple locations that seamlessly take over from the main database.

The historical data layer is the heart of the modern data lake and as such, it is primarily concerned with storing data for an (in principle) indefinite duration. Therefore, reliability and robustness are key to selecting the technology components for this layer. Technology components are selected based on the maximum downtime of a system. So, first, let's look into calculating the availability percentage and from there, decide on the technology.

The availability of a system is usually expressed in a percentage, which denotes time (in hours) that the system is up and running as a function of its lifespan. For example, a system with an availability of 90% is expected to be online 168 * 0.9 = 151 hours of a full week of 168 hours, which means that there are 168 - 151 = 17 hours in which the system can be taken offline for maintenance. Unfortunately, an availability of 90% is very poor nowadays.

The following table gives an overview of availability as a percentage, and the amount of downtime related to the percentage:

Availability	Downtime per week	Downtime per year
90%	17 hours	37 days
95%	8 hours	18 days
99%	2 hours	4 days
99.9%	10 minutes	9 hours
99.99%	1 minute	53 minutes
99.999%	26 seconds	5 minutes
99.9999%	0.5 seconds	32 seconds

Figure 2.11: Availability percentages explained in downtime

It's possible to calculate the availability of a system using the following formula:

$$\text{Availability} = \frac{\text{Available hours} - \text{Downtime}}{\text{Available hours}} \times 100\%$$

Figure 2.12: Availability formula

For example, if a data store was down for maintenance for 3 hours with available hours being 5040, the availability of that system was as follows:

$$\frac{5040 - 3}{5040} \times 100\% = 99.94\%$$

Figure 2.13: Availability calculation

Let's understand this better with an exercise.

EXERCISE 2.03: ANALYZING THE AVAILABILITY OF A DATA STORE

For this exercise, imagine that you work as an operations engineer for TV broadcasting company PacktNet. The core system of the company must be available so that clients can binge-watch their favorite series at all times. The system, therefore, has received an availability rating of 99.99%.

The aim of this exercise is to become familiar with the availability formula and to practice it in a real-world scenario.

Now, answer the following questions for this use case:

1. Is it allowed to bring the system down for 5 minutes per month for maintenance?

    An availability of 99.99% means that the system can be down for 1 minute per week or about 4 minutes per month. So, a downtime of 5 minutes per month is *not* allowed.

2. In the previous year, there were only a few minor incidents and almost no scheduled maintenance. The system was offline for 1 hour in total. What was the availability during that year?

    There are 365 x 24 = 8760 hours in a year. The availability in the previous year was $\frac{8760 - 1}{8760} \times 100\% = 99.99\%$ (rounded to two decimals).

By completing this exercise, you have successfully calculated the availability of a data store. This is an important requirement for any system.

AVAILABILITY CONSEQUENCES

Once the availability requirements of a system have been determined, a matching infrastructure should be chosen. There are many different data stores (file shares and databases) that all have their own availability percentage. The following table contains the percentages for a selection of popular cloud-based data stores. Note that some services offer a stepped approach, where more uptime costs more. Also, note that the availability can be drastically increased if the data is parallelized across different data centers and different regions:

Service	Availability
Amazon S3	99.99%
Amazon EC2	99.99%
Azure Cosmos DB	99.99%
Azure Databricks	99.95%

Figure 2.14: Availability of a few popular data stores in the cloud

Some cloud services offer high availability but don't express this in a percentage. For example, the documentation of Amazon SageMaker states that it is designed for high availability and runs on Amazon's scalable and robust infrastructure (with availability zones, data replication, and so on), but does not give a guaranteed maximum downtime percentage.

When working with data on-premise, the calculation differs a bit. The correct way to calculate the entire availability of a system is to multiply the availability of the infrastructure (servers) by the availability of the software.

Some considerations when writing down the availability requirements for data storage are as follows:

- There is planned downtime versus unplanned downtime; planned downtime is for scheduled upgrades and other maintenance on the system that must be done on a regular basis. Planned downtime counts as non-available, but obviously, it can be controlled, managed, and communicated better (for example, an email to all users of the system) than unplanned downtime, which is when there are unexpected crashes of the system.

- Once availability becomes very high, it's often described in the "number of nines." This indicates the number of nines in the percentage; 99.99% is four nines availability.

- A system that is 100% available might still be unusable, for example, if the performance of the interface is very slow. So, keep in mind that the availability percentage does not express the entire scope of the reliability and availability of the system; it's just a helpful measurement.

- Next to measuring availability, it can be even more important to measure data loss. If a system is down for 1 hour, but all the data that was entered into the system during that period is lost, there could be major implications. Therefore, it's good practice to focus on the distribution (and thereby redundancy) of data; as soon as data is stored, it should be replicated across multiple nodes. Furthermore, there should be backups in place for emergency scenarios where all the nodes fail.

- High availability always comes at a cost. There is a trade-off between the "number of nines" and the cost of development, infrastructure, and maintenance. This is ultimately a business decision; how mission-critical is the data and the system, and at what price?

TIME TRAVEL

One of the key requirements of the historical data layer is the ability to "time travel." That means it should be possible to retrieve the status of a record or table from any moment in the past, providing the data was there. Tables that have this ability are called **temporal tables**. This can be achieved by applying a data model that allows time travel, for example, a data vault. Essentially, these data models are append-only; no updates or deletes are allowed. With every new record that enters the system, a "valid from" and optionally "valid to" timestamp is set. Let's understand this better with an example.

The following table contains an example of a table with company addresses. The data format is according to the "facts and dimensions" model, where a relatively static dimension (for example, company, person, or object) is surrounded by changing facts (for example, address, purchase, call). The company with ID 51 recently changed address from Dallas to Seattle, so a new record was added (record ID: 4). The old address of the same company is still preserved in the table (record ID: 3). So, now, we have two address rows for the same company, which is fine since only one can be valid at any given moment:

ID	CompanyId	Address	PostalCode	City	ValidFrom
2	45	Main St. 34	10982	New York	2013-10-15 12:18
3	51	Old St. 41	92028	Dallas	2011-02-18
4	51	6th Avenue 1002	87108	Seattle	2019-11-01

Figure 2.15: Example of a table that preserves historical data and allows time travel

Suppose the government needs to have a report with a list of offices that have shut down in Dallas and their new locations. In such cases, time travel is a very important requirement. A query that retrieves all these addresses (current and historical) of the company is as follows:

```
SELECT * FROM Addresses WHERE CompanyId = 51;
```

A query that retrieves the current address of a company is as follows:

```
SELECT TOP 1 * FROM Addresses WHERE CompanyId = 51 ORDER BY ValidFrom DESC;
```

The same requirement can also be fulfilled by using a **ValidTo** column; if that is empty (**NULL**), the record is the most actual one. The downside of this approach is that it requires updates to a table, not just inserts, so the ETL code can become more complex. Using both **ValidFrom** and **ValidTo** is also possible and provides better querying options but adds some complexity and requires the **insert** and **update** statements to be in sync.

If time travel is a key feature for your use case, for example, a healthcare system that needs to keep track of all the medicine that was provided to patients, you might consider a database where these kinds of timestamps are a native element for all data entry; for example, Snowflake.

Another way to achieve the possibility to time-travel your data is with a mechanism called **event sourcing**. This is a relatively new method of storing data, where each change that's made to the data is stored separately rather than as a result of the change. For example, an **UPDATE** statement in a traditional database results in the overwriting of a record. With event sourcing, an **UPDATE** statement would not alter the record itself but rather add a new record to the table, along with information about the change. In this way, there is automatically a trail of events that leads from the original record to the latest one. This trail can be quite long and is therefore mostly used in the historical data layer, not in the analytics layer of an AI application. To get the latest record, all events must be replayed and calculated over the original event. This can include events that cancel each other out; for example, the combination of an **INSERT** and **DELETE** statement.

The data of PacktBank has great value if it can be queried from a historical perspective. Since many source systems only store the "present" situation, it's important that the new data lake preserves the data's history. To that extent, the bank chooses to create a historical data warehouse that AI systems can benefit from. On a daily basis, the current state of the source systems is appended to the database tables, which are arranged in a data vault model. Data scientists and analysts can now request access to perform time-series analysis, for example, of the earnings and spending of a customer to forecast their ability to afford a loan.

LOCALITY OF DATA

When data is stored in an international organization, it's important to think about the physical location of data storage. We are used to systems that respond instantaneously and smoothly; a lag of 1 second when visiting a web page is already considered to be annoying. When data is stored in one continent, it can take some time (up to a few seconds) to reach another continent, due to the time it takes on the network. This kind of delay (latency) is not acceptable to clients who are working with the data, for example, the visitors of websites. Therefore, data must be stored close to the end users to minimize the amount of network distance. Furthermore, there might be laws and regulations that constrain the possible physical locations of data storage. For example, a government organization might require that all data is stored in its own country.

Most cloud-based data storage services offer the option to store data in specific regions or locations. Amazon and Microsoft provide geographical regions across the globe for their cloud offerings (AWS and Azure), in which customers can choose to put their data. If needed, there is a guarantee that the data will not leave the chosen region. For the sake of availability and robustness, it's best to distribute data across regions.

METADATA AND LINEAGE

Metadata management in the historical data layer is important but quite difficult to realize. For every table and record (row), there should be a metadata entry that lists the origin, timestamp, owner, transformations, and so on. Usually, this metadata is stored in a separate table or database, or in a dedicated metadata management system. Since data is usually entered into the database via an ETL process, there is a big responsibility for the ETL tools to keep updating the metadata. Once in place, the metadata repository will be a valuable asset in the data lake, since it allows questions such as the following to be answered:

- What are the sources of the aggregated calculation in my report?

- At what date and time were the records from source system X last updated?

- How often on average is the data in the employee table refreshed?

In this section, we have looked at the most important requirements for the historical data layer of an AI system. We have looked at security, scalability, availability, time travel, and the locality of data. In the next section, we'll look at the typical requirements that should be considered when working with data streams.

STREAMING DATA

The requirements for a streaming data layer are different from a batch-oriented data lake. Firstly, the time dimension plays a crucial role. Any event data that enters the message bus as a stream must be timestamped. Secondly, performance and latency are more important since it must be certain that data can be processed in due time. Thirdly, the way that analytics and machine learning are applied differs; while the data is being streamed in, the system must analyze it in near-real-time. In general, streaming data software relies more on computing power than storage space; processing speed, low latency, and high throughput are key. Nevertheless, the storage requirements that are in place for a streaming data system are worth considering and are a bit different from "static" batch-driven applications.

SECURITY

A typical streaming datastore is separated into **topics**. A topic is named as such in the popular streaming data store Kafka. These can be considered as being like tables in a traditional database. In other frameworks, they might have different names; for example, in the cloud-based Amazon Kinesis data store, the topics are called shards. For the sake of clarity, we will continue to call them topics in the remainder of this chapter. Topics can be secured by administering role-based access. For example, in a bank, there are many kinds of streaming data: financial transactions, page visits of clients, stock market traders, and others. Each of these kinds of data gets its own topic in a streaming data store, with its own data format, security, access role, and availability rating.

Data at rest (stored in an event bus) and in motion (incoming and outgoing traffic) should be encrypted with the mechanisms we explained in the *Security* subsection within the *Raw Data* section.

PERFORMANCE

When analyzing data streams, it's crucial to select technology and write code that can handle thousands or even millions of records per second. For example, systems that work with **Internet of Things (IoT)** or sensory data must handle massive amounts of events. There are two important performance requirements for these systems:

- The amount of data (the number of events per second and bytes per second) that the event bus and stream processing engine is able to handle; this is the base figure that tells us whether there is a risk of overloading the system. In frameworks such as Kafka, Spark, and Flink, this is scalable; roughly speaking, just add more hardware to process more events.

- The amount of data that the software runs as jobs on the stream processing engine that it is able to handle. The software should run fast enough to be able to process all events per time window before a new calculation is required. Therefore, the software that performs the aggregations and eventually more complex event processing, such as machine learning, must be optimized and carefully tested.

AVAILABILITY

When a streaming engine crashes or must be taken offline for maintenance, it should only temporarily stop processing the never-ending stream of data and reprocess any events it missed. Also, there must be a guarantee that no data is lost; even when the system is down temporarily, data should be replayed into the streaming engine to make sure that all the events go through the system. To that extent, modern streaming engines such as Spark and Flink offer **savepoints** and **checkpoints**. These are backups of the in-memory state of the streaming engine to disk. If there is a crash or scheduled maintenance, the latest checkpoint or `savepoint` is reloaded into memory from disk, and the data isn't lost. In combination with Kafka offsets (the latest point that is read from a data source topic by a consumer), it's clear that all data is replayed if necessary.

There are three main semantics when configuring availability and preventing the data loss of a streaming system:

- **At-least-once**: The guarantee that any event is processed at least once, but it's possible that one event goes through the system multiple times in the event of failures.

- **At-most-once**: The guarantee that any event is never processed more than once.

- **Exactly-once**: The guarantee that any event is processed exactly once by the streaming engine; this can only be accomplished with tight integration between the event bus (using offsets) and the streaming engine (using checkpoints and `savepoints`).

For example, one of the requirements of PacktBank is to analyze the financial transactions and online user activity of its customers in real time. Use cases that should be supported include real-time fraud detection and customer support (based on clickstreams and actions in the mobile app). A streaming engine was designed and developed with state-of-the-art technology, including Apache Kafka and Apache Flink. The requirement for the availability of the system was clear: in the case of maintenance or bugs, the system should not lose any data, since all transactions have to be processed. It's better to have a little delay and to keep customers waiting for a few minutes more than to miss fraudulent transactions altogether. Therefore, the architecture of the system was designed with an at-least-once guarantee of data availability. For every streaming job that handles the customer event data, an offset in Kafka keeps track of the latest data that has been read. Once data has entered a job in Flink, it's backed up in `savepoints` and checkpoints to make sure that no data has been lost.

RETENTION

In a streaming data system, the data is used when it's "fresh." Old data is only used for training models and generating historical (aggregated) reports. Therefore, the retention of a streaming data topic in an event bus can usually be set to a few days or a few weeks at the most. This saves storage space and other resources. When setting this requirement, think carefully about the aggregation step; perhaps it's useful to store the averages per hour or the results of the window calculations in your stream. As a typical example, the clickstream data of an online news website is only valuable when it's less than 1 day old. After all, news that's 1 day old is not very relevant anymore and the customers who have read the articles have already moved on!.

Retention is also related to the amount of data that is expected. Sometimes, the number of events is just too many to be stored for a long period of time. When reasoning about data retention, it's advised to estimate the average and peak load of a system first. This can be done by multiplying the number of concurrent data sources that produce event data with the number of events that are being produced. For example, a payment processing engine at PacktBank has 1 million users in total, which all make 3 payments per day on average with a peak of 20 per day. The average load of the system is 1 million x 3 = 3 million payments per day, which is about 2,000 payments per minute or 35 per second. At peak times, this can rise to 250 or so per second. A streaming data store that handles these events should be able to store these amounts of data and set a retention period in such a way that the disks will not become full.

EXERCISE 2.04: SETTING THE REQUIREMENTS FOR DATA RETENTION

For this exercise, imagine that you are building a real-time marketing engine for an online clothing distribution company. Based on the online behavior of (potential) customers, you want to create advertisements and personalized offerings to increase your sales. You will get real-time clickstreams (page visits) as your prime event data source. On average, 200,000 individuals visit your website per day. They spend about 20 minutes on your site and usually visit the home page, their favorites, about 75 clothing items, and their shopping basket.

The aim of this exercise is to become familiar with the concept of data retention.

Now, answer the following questions for this use case:

1. What is a reasonable number of events that the system should be able to handle per minute? Are there peak times, and how would you handle them?

    A quick estimation: 200,000 visits per day is 8,333 per hour on average. But we expect the evenings to be much busier than the mornings and nights, so we aim for a load of 25,000 concurrent users. They visit at least 75 items plus some other pages, so a reasonable clickstream size is 100-page visits per 20 minutes, which is 5 visits per minute. So, the total load is 5 x 25,000 = 125, 000 events per minute.

2. What retention policy would you attach to the event data? How long would the events be useful for in their raw form? Would you still require a report or other form of historical insight into the old data?

The page visits will probably be valuable data for a week or so. After a week, other items will have sparked the interest of the clients, so the real-time information about the old events won't be as valuable anymore. Of course, this depends on the frequency of visits; if someone only logs in once per month, the historical data might be valuable over a longer period of time.

In this section, we discussed the typical requirements for streaming data storage: security, performance, availability, and retention. In the next section, we'll explore the requirements for the analytics layer, where data is stored for quick access in APIs and reports.

ANALYTICS DATA

The responsibility of the analytics layer of an AI system is to make data fast and available for machine learning models, queries, and so on. This can be achieved by caching data efficiently or by virtualizing views and queries where needed to materialize these views.

PERFORMANCE

The data needs to be quickly available for ad hoc queries, reports, machine learning models, and so on. Therefore, the data schema that is chosen should reflect a "schema-on-read" pattern rather than a "schema-on-write" one. When caching data, it can be very efficient to store the data in a columnar **NoSQL** database for fast access. This would mean the duplication of data in many cases, but that's all right since the analytics layer is not responsible for maintaining "one version of the truth." We call these caches **data marts**. They are usually specific for one goal, for example, retrieving the sales data of the last month.

In modern data lakes, the entire analytics layer can be virtualized so that it just consists of queries and source code. When doing so, regular performance testing should be done to make sure that these queries are still delivering data as quickly as expected. Also, monitoring is crucial since queries may be being used in inappropriate ways, for example, setting parameters to such values (dates in **WHERE** clauses that span too many days) that the entire system becomes slow to respond. It's possible to set maximum durations for queries.

COST-EFFICIENCY

The queries that run in the analytics layer can become very resource-intensive and keep running for hours. Any compute action in a cloud-based environment costs money, so it's crucial to keep the queries under control and to limit the amount of resources that are spent. A few ways to make the environment more cost-effective are as follows:

- Limit the maximum duration of queries to (for example) 10 minutes.

- Develop more specific queries for reports, APIs, and so on, rather than having a few parameterized queries that are "one size fits all." Large, complicated queries and views are more difficult to maintain, debug, and tune.

- Apply good database practices to the tables where possible: indexes, partitioning, and so on.

- Analyze the usage of the data and create caches and/or materialized views for the most commonly used queries.

QUALITY

Maintaining the data and software in an analytics cluster is difficult but necessary. The quality of the environment becomes higher when traditional software practices are being applied to the assets in the environment, which are as follows:

- DTAP environments (development → test → acceptance → production)

- Software development principles (SOLID, KISS, YAGNI, clean code, and so on)

- Testing (unit tests, regression tests, integration tests, security tests)

- Continuous integration (version control, code reviews)

- Continuous delivery (controlled releases)

- Monitoring and alerting

- Proper and up-to-date documentation

For example, PacktBank stores its data about products, customers, sales, and employees in the new data lake. The analytics layer of the data lake provides business users and data analysts access to the historical data in a secure and controlled way. Since the results of queries and views must be trusted by management, any updates to the software must go through an extensive review and testing pipeline before they're deployed to the production environment. The models and code are part of a continuous integration and delivery cycle, where the release pipeline and an enforced "4-eyes principle" (the mechanism that ensures that development and operations are separated) makes sure that no software goes to production before a set of automatic and manual checks. When writing code, the engineers often engage in pair programming to keep the code quality high and to learn from each other. Models are documented and explained as carefully as possible and reviewed by a central risk management team.

In this section, we have discussed some important requirements for the analytics layer: performance, cost-efficiency, and quality. Keep in mind that other requirements for data storage that were described in the other layers, such as scalability, metadata, and retention, also play an important role. In the next and final section, we will dive into the specific requirements for the model development and training layer.

MODEL DEVELOPMENT AND TRAINING

Data that is used for developing and training machine learning models is temporarily stored in a model development environment. The data store itself can be physical (a file share or database) or in memory. The data is a copy of one or more sources in the other data layers. Once the data has been used, it should be removed to free up space and to prevent security breaches. When developing reinforcement learning systems, it's necessary to merge this environment with the production environment; for example, by training the models directly on the data in the historical data layer.

In our example of PacktBank, the model development environment of the new data lake is used by data scientists to build and train new risk models. Whereas the old way of forecasting whether clients could afford a loan was purely based on rules, the new management wants to become more data-driven and rely on algorithms that have been trained on historical data. The historical data in this example is the combination of clients' history with the number of missed payments (defaults) for each client. This information can be used in the model development environment to create machine learning models.

SECURITY

The model development environment must be secured with a very strict access policy since it can contain a lot of production data. A common practice is to make the environment only available to a select set of data scientists, who only have access to a few datasets that are temporarily available.

Datasets that are copied from a production environment into the model development environment fall under the responsibility of the data scientist who is developing and training the model. They should govern access to the datasets that are temporarily acquired in this way. The level of security is set is on the dataset level and not a row or table level. Data scientists usually require a full dataset and it would make no sense to have fine-grained access once the data has been copied into the secure model environment where only data scientists have access.

AVAILABILITY

The required availability of data in a model development environment is usually not very high. Since the data is only a copy of the actual production data, the data doesn't have to be available at all times. If the environment becomes unavailable due to a crash or regular maintenance, the data scientists working with the data will just have a productivity loss (for example, they will not be able to work on new models for a few days) but not a production incident (for example, a mission-critical system not working for a few days).

RETENTION

The data in a model development environment is essentially copied here temporarily, for data scientists to work on. This data should be treated as a developer cache. This means the data should be automatically or manually deleted as soon as the models have been trained. For any subsequent training that's required, the data can be loaded as a copy of the production environment again. It's considered to be good practice to only keep the data in the model development environment for a short period in order to prevent data leaks and to keep resources available. For example, once the data scientists of PacktBank have created a new model for the acceptance of a credit card, the data that was used to train the model can be removed from the model development environment. It's important to store the metadata of the model so that the data can be loaded from the historical archive again if needed.

ACTIVITY 2.01: REQUIREMENTS ENGINEERING FOR A DATA-DRIVEN APPLICATION

Now, we have covered the major requirements that should be considered when building AI systems. In this activity, you will combine the concepts you've learned and defined the requirements of so far for a new set of data-driven applications for a national taxi organization. The customers, taxi rides, drivers, financial information, and other data must all be captured in one data lake. Historical data such as rides in the past should be combined with real-time data, such as the current location of the taxis. The aim of the data lake is to serve as the new data source for financial reports, client investigations, marketing and sales, real-time updates, and more. Algorithms and machine learning models should be trained on the data to provide advice to taxi drivers and planners about staffing and routes. Imagine that you're the architect who is responsible for setting the requirements for the data lake and choosing the technology.

The aim of this activity is to combine all the requirements that you've learned into one set that makes it possible to select the technology for a data-driven solution.

Answer the following questions to complete this activity:

1. Write down a list of data sources that you will need to import data from, either via batch or streaming.

2. With the business aim and the data sources in mind, select the layers of the solution that you will require.

3. What would your ETL data pipelines look like? How often would they be required to run? Are there any streaming data pipelines?

4. What metadata will be captured when importing and processing the data? To what extent can the metadata be used in the extension to the raw and transformed data?

5. What are the security, scalability, and availability requirements of the solution (per layer)?

6. How important are time travel, retention, and the locality of the data?

7. When selecting technology, how will you judge the cost-efficiency and quality (maintainability, operability, and so on) of the solution?

8. What are the requirements for an environment that will serve data scientists who are building forecasting models on top of the historical and real-time data?

Now that you've completed this activity, you have mastered the skill of designing a high-level architecture and can reason about the requirements for AI solutions.

> **NOTE**
>
> The solution to this activity can be found on page 585.

SUMMARY

In this chapter, we discussed the non-functional requirements for data storage solutions. It has become clear that a data lake, which is an evolution of a data warehouse, consists of multiple layers that have their own requirements and thus technology. We have discussed the key requirements for a raw data store where primarily flat files need to be stored in a robust way, for a historical database where temporal information is saved, and for analytics data stores where fast querying is necessary. Furthermore, we have explained the requirements for a streaming data engine and for a model development environment. In all cases, requirements management is an ongoing process in an AI project. Rather than setting all the requirements in stone at the start of the project, architects and developers should be agile, revisiting and revising the requirements after every iteration.

In the next chapter, we will connect the layers of the architecture we have explored in this chapter by creating a data processing pipeline that transforms data from the raw data layer to the historical data layer and to the analytics layer. We will do this to ensure that all the data has been prepared for use in machine learning models. We will also cover data preparation for streaming data scenarios.

3

DATA PREPARATION

OVERVIEW

In this chapter, we will focus on the data preparation that has to be done before an AI project can start with model training and evaluation. You will practice **ETL** (**Extract**, **Transform**, and **Load**) or **ELT** (**Extract**, **Load**, and **Transform**), data cleaning, and any other data prep work that is commonly required by data engineers. We will cover batch jobs, streaming data ingestion, and feature engineering. By the end of this chapter, you will have knowledge and some hands-on experience of data preparation techniques.

INTRODUCTION

In the previous chapter, we discussed the layers of a data-driven system and explained the important storage requirements for each layer. The storage containers in the data layers of AI solutions serve one main purpose: to build and train models that can run in a production environment. In this chapter, we will discuss how to transfer data between the layers in a pipeline so that the data is prepared to be used to train a model to create an actual forecast (called the *execution* or *scoring* of the model).

In an **Artificial Intelligence** (**AI**) system, data is continuously updated. Once data enters the system via an upload, **application program interface** (**API**), or data stream, it has to be stored securely and typically goes through a few ETL steps. In systems that handle streaming data, the incoming data has to be directed into a stable and usable data pipeline. Data transformations have to be managed, scheduled, and orchestrated. Further, the lineage of the data has to be stored to trace back the origins of a data point in a report or application. This chapter explains all *data preparation* (sometimes called *pre-processing*) mechanisms that ensure raw data can be used for machine learning by data scientists. This is important since raw data is hardly in a form that can be used by models. We will elaborate on the architecture and technology as explained by the layered model in *Chapter 1, Data Storage Fundamentals*. To start with, let's dive into the details of ETL.

ETL

ETL is the standard term that is used for **Extracting**, **Transforming**, and **Loading** data. In traditional data warehousing systems, the entire data pipeline consists of multiple ETL steps that follow after each other to bring the data from the source to the target (usually a report on a dashboard). Let's explore this in more detail:

E: Data is extracted from a source. This can be a file, a database, or a direct call to an API or web service. Once loaded with a query, the data is kept in memory, ready to be transformed. For example, a daily export file from a source system that produces client orders is read every day at 01:00.

T: The data that was captured in memory during the extraction phase (or in the loading phase with ELT) is transformed using calculations, aggregations, and/or filters into a target dataset. For example, the customer order data is cleaned, enriched, and narrowed down per region.

L: The data that was transformed is loaded (stored) into a data store.

This completes an ETL step. Similarly, in ELT, all the extracted data gets stored in the data store and then later transformed.

The following figure is an example of a full data pipeline, from a source system to a trained model:

Figure 3.1: An example of a typical ETL data pipeline

In modern systems such as data lakes, the ETL chain is often replaced by ELT. Rather than having, say, five ETL steps, where data is slowly refined and made ready for analysis, from a raw format to a queryable form, all the data is loaded into one large data store. Then, a series of transformations that are mostly virtual (not stored on disk) runs directly on top of the stored data to produce a similar outcome for analytics. In this way, a gain in storage space and performance can be achieved since modern (cloud-based) storage systems are capable of handling massive amounts of data. The data pipeline becomes somewhat simpler, although the various T(Transform) steps still have to be managed as separate software pieces:

Figure 3.2: An example of an ELT data pipeline

In the remainder of this chapter, we will look in detail at the ETL and ELT steps. Use the text and exercises to form a good understanding of the possibilities for preparing your data. Remember that there is no silver bullet; every use case will have specific needs when it comes to data processing and storage. There are many tools and techniques that can be used to get data from A to B; pick the ones that suit your company best, and whatever you pick, never forget the best practices of software development, such as version control, test-driven development, clean code, documentation, and common sense.

DATA PROCESSING TECHNIQUES

In *Chapter 2, Artificial Intelligence Storage Requirements,* we discussed the layers of a modern data lake and the requirements and possible data storage options for each layer. It became clear that data has to be sent to different data stores to maximize the abilities of AI: building a historical overview and a high-performing queryable source. This means that some work needs to be done with the data before it's suitable for a machine learning model. These data transfers usually happen as ETL steps in a data pipeline. We'll dive into the specifics and possibilities of batch processing in the following paragraphs.

Transactions

In databases, a transaction is a fixed set of instructions that either fail or succeed. Transactions are very useful for data processing since they are reliable and produce no undesirable outcomes. Use them when certain steps are related, or have to be done in a certain order. If a transaction is composed of a hundred steps and the last one fails, all the previous steps are either not executed or rolled back. Transactions that guarantee this are called atomic and consistent. If two transactions are executed at the same time on the same database, for example, updating a large table, they should not influence each other; this property is called isolation. Transactions also should be durable: a completed transaction cannot be undone. If the transactions of a database are guaranteed to be atomic, consistent, isolated, and durable, the database is ACID-compliant. The opposite is a database that is BASE: Basically Available, Soft state, Eventually consistent.

Simple data processing

In many cases, building a simple script or another piece of software is enough to get data from A to B. Each infrastructure, tool, and environment has its way of copying and filtering data. Some examples of simple data processing are:

- Bash or PowerShell scripts

- Python scripts

- SQL scripts

There are also ways of processing data that are discouraged, for various reasons. Building an Excel macro in **Visual Basic** (**VB**) is one of those methods. It might work, but it's difficult to test, distribute, and scale. Whenever you want to keep it simple, remember that all the follow-up actions should also be simple. In the following exercise, you'll build a simple data transformation step on CSV files with a Bash script.

> **NOTE**
>
> If you are facing challenges installing PySpark on your system, refer the following link: https://medium.com/tinghaochen/how-to-install-pyspark-locally-94501eefe421.

EXERCISE 3.01: CREATING A SIMPLE ETL BASH SCRIPT

> **NOTE**
>
> This exercise requires the BASH shell which is available on Linux and Unix-based systems (including MacOS). If you have Windows 10, you can install the shell through **Windows Subsystem for Linux** (**WSL**). To install WSL, you can follow the instructions detailed in the following article: https://docs.microsoft.com/en-us/windows/wsl/install-win10.

In this exercise, we're going to write a simple Bash script that reads data from a file (extract), does some data filtering (transform), and writes it to a new file (load). From the dataset, we're only interested in movies that are currently available in a certain country, we'll filter out the other records and remove a set of columns. This filter is the data transformation step. We'll use the standard CSV parsing library GNU awk (**gawk**) to process the file. After filtering, we'll simply write the result to a new file on disk.

We will be using a sample dataset of Netflix movies and TV series that was collected from Flixable, which is a third-party Netflix search engine. The dataset can be found in our GitHub repository at the following location: https://packt.live/2C72sBN.

You can download the `netflix_titles_nov_2019.csv` file from the GitHub repository.

Before proceeding to the exercise, we need to install the **gawk** utility in the local dev environment. Please follow the instructions in the *Preface*.

Perform the following steps to complete the exercise:

1. Create a directory called **Chapter03** for all the exercises in this chapter. In the **Chapter03** directory, create the **Exercise03.01** directory to store the files for this exercise.

2. Move the **netflix_titles_nov_2019.csv** file to the **Exercise03.01** directory.

3. Open your Terminal (macOS or Linux) or Command Prompt (Windows), navigate to the **Chapter03** directory, and type **jupyter notebook**.

4. Select the **Exercise03.01** directory, then click on **New** -> **Python3** to create a new Python 3 notebook.

> **NOTE**
>
> If you have not installed **gawk** from the *Preface*, then you can type the installation commands in the first cell of the Jupyter Notebook. They are not included in the steps.

5. Read the file using the Bash **head** command with an exclamation mark (**!**), as shown in the following code:

```
!head netflix_titles_nov_2019.csv
```

> **NOTE**
>
> The **!** sign at the start of the preceding step notifies that we can run the BASH **head** command in Jupyter Notebook. To run the same command in the command line shell directly, exclude the **!** sign.

You should get the following output:

```
show_id,title,director,cast,country,date_added,release_year,rating,duration,listed_in,description,type
81193313,Chocolate,,"Ha Ji-won, Yoon Kye-sang, Jang Seung-jo, Kang Bu-ja, Lee Jae-ryong, Min Jin-woong, Kim Won-ha
e, Yoo Teo",South Korea,"November 30, 2019",2019,TV-14,1 Season,"International TV Shows, Korean TV Shows, Romantic
TV Shows","Brought together by meaningful meals in the past and present, a doctor and a chef are reacquainted when
they begin working at a hospice ward.",TV Show
81197050,Guatemala: Heart of the Mayan World,"Luis Ara, Ignacio Jaunsolo",Christian Morales,,"November 30, 2019",2
019,TV-G,67 min,"Documentaries, International Movies","From Sierra de las Minas to Esquipulas, explore Guatemala's
cultural and geological wealth, including ancient Mayan cities and other natural wonders.",Movie
81213894,The Zoya Factor,Abhishek Sharma,"Sonam Kapoor, Dulquer Salmaan, Sanjay Kapoor, Sikander Kher, Angad Bedi,
Koel Purie, Pooja Bhamrah, Manu Rishi Chadha",India,"November 30, 2019",2019,TV-14,135 min,"Comedies, Dramas, Inte
rnational Movies","A goofy copywriter unwittingly convinces the Indian cricket team that she's their lucky mascot,
to the dismay of their superstition-shunning captain.",Movie
81082007,Atlantics,Mati Diop,"Mama Sane, Amadou Mbow, Ibrahima Traore, Nicole Sougou, Amina Kane, Mariama Gassama,
Coumba Dieng, Ibrahima Mbaye, Diankou Sembene","France, Senegal, Belgium","November 29, 2019",2019,TV-14,106 min,"
Dramas, Independent Movies, International Movies","Arranged to marry a rich man, young Ada is crushed when her tru
e love goes missing at sea during a migration attempt — until a miracle reunites them.",Movie
80213643,Chip and Potato,,"Abigail Oliver, Andrea Libman, Briana Buckmaster, Brian Dobson, Chance Hurstfield, Domi
nic Good, Emma Jayne Maas, Evan Byarushengo, Scotia Anderson, Alessandro Juliani","Canada, United Kingdom",,2019,T
V-Y,2 Seasons,Kids' TV,"Lovable pug Chip starts kindergarten, makes new friends and tries new things — with a litt
le help from Potato, her secret mouse pal.",TV Show
81172754,Crazy people,Moses Inwang,"Ramsey Nouah, Chigul, Sola Sobowale, Ireti Doyle, Ben Touitou, Francis Onwoche
i, Ememobong Nkana, Emem Inwang, Patrick Onyeke",Nigeria,"November 29, 2019",2018,TV-14,107 min,"Comedies, Interna
tional Movies, Thrillers","Nollywood star Ramsey Nouah learns that someone is impersonating him and breaks out of a
mental institution to expose the imposter.",Movie
81120982,I Lost My Body,Jérémy Clapin,"Hakim Faris, Victoire Du Bois, Patrick d'Assumçao, Dev Patel, Alia Shawkat,
George Wendt",France,"November 29, 2019",2019,TV-MA,81 min,"Dramas, Independent Movies, International Movies","Rom
ance, mystery and adventure intertwine as a young man falls in love and a severed hand scours Paris for its owner
in this mesmerizing animated film.",Movie
81227195,Kalushi: The Story of Solomon Mahlangu,Mandla Dube,"Thabo Rametsi, Thabo Malema, Welile Nzuza, Jafta Mama
bolo, Louw Venter, Pearl Thusi",South Africa,"November 29, 2019",2016,TV-MA,107 min,"Dramas, International Movie
s","The life and times of iconic South African liberation fighter Solomon Mahlangu, who battled the forces of apar
theid, come into focus.",Movie
70205672,La Reina del Sur,,"Kate del Castillo, Cristina Urgel, Alberto Jiménez, Juan José Arjona, Humberto Zurita,
Dagoberto Gama, Christian Tappán, Miguel de Miguel, Salvador Zerboni, Carmen Navarro, Santiago Meléndez, Juan Carl
os Solarte","United States, Spain, Colombia, Mexico",,2019,TV-14,2 Seasons,"Crime TV Shows, International TV Show
s, Spanish-Language TV Shows","This compelling show tells the story of the legendary Teresa Mendoza, a courageous
woman who is perceived as conquering the world of drug trafficking.",TV Show
```

Figure 3.3: Contents of the file with Netflix titles

> **NOTE**
>
> The output of the Jupyter Notebook is not visible in GitHub. Please download the Jupyter Notebook file locally and execute it.

You might notice that some fields are delimited by double quotes ("), and in those fields are commas as well. We should not treat those commas as field separators.

6. Open a text editor and create a file called **netflix.awk** in the **Exercise3.01** directory.

7. Write the following code inside the **netflix.awk** file:

```
BEGIN {
    FS=","
}


{
    print "NF = " NF
    for (i = 1; i <= NF; i++) {
        printf("#%d = %s\n", i, $i)
    }
}
```

This file will be used to parse the CSV file. In the **BEGIN** statement, we have specified the parameters: for now, we just state that the separator is a comma (**FS=","**). In the remaining section, we have specified the output of the script. It now shows how many fields there are per line (**NF**) and then lists all the values in a **print** statement. The **gawk** utility can parse files or receive input from the prompt.

8. Send the first 10 lines of a file as input to the **gawk** command using the pipe command (**|**), as shown in the following code:

```
!head netflix_titles_nov_2019.csv | gawk -f netflix.awk
```

> **NOTE**
>
> This is only a temporary step, and so the output is not in the notebook on GitHub since we will make modifications to the **awk** file.

You should get the following output:

```
NF = 12
#1 = show_id
#2 = title
#3 = director
#4 = cast
#5 = country
#6 = date_added
#7 = release_year
#8 = rating
#9 = duration
#10 = listed_in
#11 = description
#12 = type
NF = 23
#1 = 81193313
#2 = Chocolate
#3 =
#4 = "Ha Ji-won
#5 =  Yoon Kye-sang
```

Figure 3.4: Output of the parsing of a comma-separated file with Netflix titles

As you can see, there is some comma-separation being done. We can see that there are 12 fields (**NF=12**) in the first line and the line contains the names of the fields as headers. However, the next line that is being parsed already contains 23 fields (**NF=23**). The reason is that the list of actors in the **cast** column contains commas as well. So, we can use the **FPAT** parameter instead of the **FS** parameter to gain more control over the separation of fields.

9. Replace **FS=","** with **FPAT = "([^,]+)|(\"[^\"]+\")"** in the **netflix.awk** file, as shown in the following code:

```
BEGIN {
    FPAT = "([^,]+)|(\"[^\"]+\")"
}


{
    print "NF = " NF
    for (i = 1; i <= NF; i++) {
        printf("#%d = %s\n", i, $i)
    }
}
```

The **FPAT** parameter contains a regular expression that indicates what the contents of a field are. In our case, the fields are all text that is not a comma (indicated by ^,) or are delimited by double quotes (^\"). For example, the string `this is,"a test",and this,"also, right?"` would split into the fields `"this is"`, `"a test"`, `"and this"`, and `"also, right?"`.

10. Now check the output again by running the following command:

```
!head netflix_titles_nov_2019.csv | gawk -f netflix.awk
```

> **NOTE**
>
> This is only a temporary step and so the output is not in the notebook on GitHub since we will make modifications to the **awk** file.

You should get the following output:

```
NF = 12
#1 = show_id
#2 = title
#3 = director
#4 = cast
#5 = country
#6 = date_added
#7 = release_year
#8 = rating
#9 = duration
#10 = listed_in
#11 = description
#12 = type
NF = 11
#1 = 81193313
#2 = Chocolate
#3 = "Ha Ji-won, Yoon Kye-sang, Jang Seung-jo, Kang Bu-ja, Lee Jae-ryong, Min Jin-woong, Kim Won-hae, Yoo Teo"
#4 = South Korea
#5 = "November 30, 2019"
```

Figure 3.5: Output of the parsing of a comma-separated file with field delimiters

This output shows that the value in the **cast** column now is a comma-separated list of actors, just as expected. We also see double quotes in some values (for example, in the second line for the **cast** column), but we will leave them as-is for now.

There are two more things to fix. When looking at the data file and the output, you might notice that some lines contain less than 12 fields while there are 12 headers. The reason is that some fields in the source file are empty, resulting in commas next to each other (, ,). This is something that we can easily change in our script by replacing all values of , , with , , with a space in between the commas. Another peculiarity of the file is that quotes within quoted strings are represented with two double quotes (" "). So, let's replace them with a single quote ('). There is a **gsub ()** function within **gawk** that replaces all the instances of text with something else.

> **NOTE**
>
> The replacement will result in spaces for empty values; if at a later stage we decide on a different value to represent these **null** values, spaces can of course be changed to something else.

11. Add **gsub (",,", ", ,")**, **gsub (",\"\"\"", ",\"'")**, and **gsub ("\"\"", "'")** in the **netflix.awk** file, as shown in the following code:

```
BEGIN {
    FPAT = "([^,]+)|(\"[^\"]+\")"
}


{
    gsub (",,", ", ,")

    gsub (",\"\"\"", ",\"'")
    gsub ("\"\"", "'")

    print "NF = " NF
    for (i = 1; i <= NF; i++) {
        printf("#%d = %s\n", i, $i)
    }
}
```

12. Now check the output again by running the following command:

```
!head netflix_titles_nov_2019.csv | gawk -f netflix.awk
```

> **NOTE**
>
> This is only a temporary step and so the output is not in the notebook on GitHub since we will make modifications to the **awk** file.

You should get the following output:

```
NF = 12
#1 = show_id
#2 = title
#3 = director
#4 = cast
#5 = country
#6 = date_added
#7 = release_year
#8 = rating
#9 = duration
#10 = listed_in
#11 = description
#12 = type
NF = 12
#1 = 81193313
#2 = Chocolate
#3 =
#4 = Ha Ji-won, Yoon Kye-sang, Jang Seung-jo, Kang Bu-ja, Lee Jae-ryong, Min Jin-woong, Kim Won-hae, Yoo Teo
#5 = South Korea
#6 = November 30, 2019
#7 = 2019
#8 = TV-14
#9 = 1 Season
#10 = International TV Shows, Korean TV Shows, Romantic TV Shows
#11 = Brought together by meaningful meals in the past and present, a doctor and a chef are reacquainted when they be
gin working at a hospice ward.
#12 = TV Show
NF = 12
#1 = 81197050
#2 = Guatemala: Heart of the Mayan World
#3 = Luis Ara, Ignacio Jaunsolo
#4 = Christian Morales
#5 =
#6 = November 30, 2019
```

Figure 3.6: Output of the parsing of a comma-separated file with empty fields replaced and double quotes removed

You'll see that the output now has 12 columns for every line. With this step, we have completed our transformation step of ETL. This is great, but in the end, we don't want to get console output but rather filter the columns and write the output to a file. So, let's replace the **print** statement and the **for** loop in the script with the columns you want to have.

13. To create a file with the show's title, year, rating, and type, add the following commands in the **netflix.awk** file:

```
BEGIN {
    FPAT = "([^,]+)|(\"[^\"]+\")"
}


{
    gsub(",,,,,", ",,,,,")
    gsub(",,,,", ",,,,")
    gsub(",,,", ",,,")
    gsub(",,", ",,")

    gsub(",\"\"\"\"", ",\"'")
    gsub("\"\"\"", "'")


    print $2","$7","$8","$12
}
```

> **NOTE**
>
> You can download the AWK file by visiting this link: https://packt.live/3ep6KBF.

The digits after the dollar signs refer to the indexes of the fields. In this example, we print the fields with numbers **2**, **7**, **8**, and **12**.

14. Create a Bash script called **parse.sh** in the **Exercise03.01** directory and add the following code:

```
head netflix_titles_nov_2019.csv | gawk -f netflix.awk
cat netflix_titles_nov_2019.csv| gawk -f netflix.awk >
netflix_filtered.csv
```

The Bash script sends its output to the disk. We have replaced the **head** (just the first 10 lines) of the file with the entire thing using the **cat** command. The pipe commands (**|**) send the input to the **gawk** command, and the **>** writes the output to a file.

15. Open your Terminal (macOS or Linux) or Command Prompt window (Windows), navigate to the **Chapter03/Exercise03.01** directory, and run the following command:

```
sh parse.sh
```

You should get the following output:

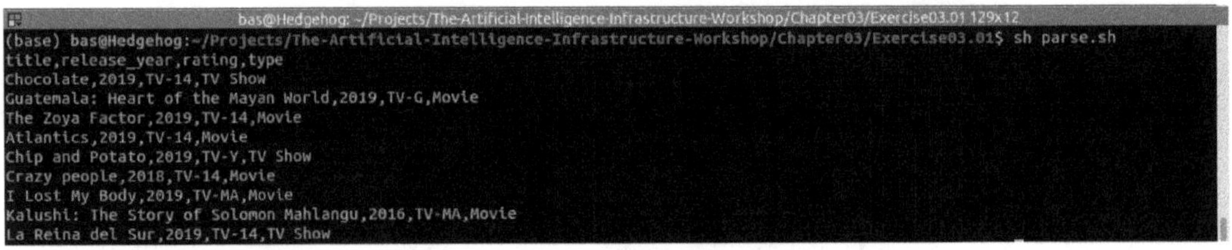

Figure 3.7: Output of the parse command that displays the top 10 Netflix titles and creates an output file

The data is stored in a CSV file named **netflix_filtered.csv** in the **Exercise03.01** directory.

16. Open the **netflix_filtered.csv** file in Jupyter Notebook using the following command:

```
!cat netflix_filtered.csv
```

```
title,release_year,rating,type
Chocolate,2019,TV-14,TV Show
Guatemala: Heart of the Mayan World,2019,TV-G,Movie
The Zoya Factor,2019,TV-14,Movie
Atlantics,2019,TV-14,Movie
Chip and Potato,2019,TV-Y,TV Show
Crazy people,2018,TV-14,Movie
I Lost My Body,2019,TV-MA,Movie
Kalushi: The Story of Solomon Mahlangu,2016,TV-MA,Movie
La Reina del Sur,2019,TV-14,TV Show
Lagos Real Fake Life,2018,TV-14,Movie
Payday,2018,TV-MA,Movie
Sugar Rush Christmas,2019,TV-PG,TV Show
The Accidental Spy,2017,TV-14,Movie
The Charming Stepmom,2019,TV-14,TV Show
The Island,2018,TV-14,Movie
The Movies That Made Us,2019,TV-MA,TV Show
Holiday Rush,2019,TV-PG,Movie
Levius,2019,TV-14,TV Show
```

Figure 3.8: Displaying the contents of the output file

> **NOTE**
>
> The output of the Jupyter Notebook is not visible in GitHub. Please download the Jupyter Notebook file locally and execute it.

We have successfully created a simple ETL Bash script.

> **NOTE**
>
> To access the source code for this specific section, please refer to https://packt.live/38TUKqA.

By completing this exercise, you processed a CSV file with some simple tools. We have shown that using a few standard Bash commands and an open CSV-parsing library is sometimes enough to process a file. At the same time, you have experienced that working with a simple CSV file is more complex than you would consider at first sight; trivial things such as empty values and quotes sometimes can be challenging. In the next paragraphs, we'll explore other methods of ETL that are suitable when simple scripting is not enough.

TRADITIONAL ETL WITH DEDICATED TOOLING

Many companies choose to acquire a software tool with the specific purpose of providing an ETL environment where batch jobs can be built, tested, and run in production. Examples of these tools include IBM DataStage, Informatica, Microsoft Azure Data Factory, and AWS Glue. Although these systems have their origin in the business intelligence and data warehousing domain, there is still a huge market for them. One of the benefits of these tools is that they provide a large collection of data transformations and connectors out of the box. After installation, the tools just work, and with a few configuration settings, developers can start creating ETL pipelines. They usually provide a rich graphical user interface with drag-n-drop functionality. There are also downsides to these tools. To truly understand them and work efficiently, engineers have to be trained and gain some experience with them. In the world of software engineering, this is often considered to be a too-specific career path. Rather than investing in skills for only one purpose (for example, ETL with a dedicated tool), many IT professionals consider it better to become senior in slightly more generic skills such as functional programming or database interfaces. Further, dedicated ETL tools such as DataStage and Informatica can become expensive since they are usually heavily licensed.

When choosing technology for data preparation and processing, it's worth a look inside your organization at existing ETL tools. Although it might feel more modern to write code in Python, Scala, or Go with a cool and hip framework, having a well-established team write an ETL job in a properly managed environment could be a more reasonable solution. Do bear in mind that these tools were often not created for scalability, portability, and performance in the big data and AI era.

As an alternative to traditional ETL with dedicated tooling, it's possible to write ETL in code. Many programming languages and frameworks are suited to this. In the next section, we'll focus on a modern big data engine that is popular for ETL: Apache Spark.

DISTRIBUTED, PARALLEL PROCESSING WITH APACHE SPARK

Over the past few years, data processing has become more and more important in enterprises as the size of data has grown tremendously. The term *big data* was coined around the year 2000 to indicate data where the volume, variety, or velocity was too large to handle with normal infrastructure and software. An important moment was the introduction and open-source publication of the Apache Hadoop framework, with which Yahoo! could store and process their massive amounts of data. Since then, the importance of big data and the amount of attention it attracts have grown tremendously. Companies such as Google, Facebook, and Alibaba continue to increase their data needs and have published papers and software about it. In the academic world, data analytics and machine learning have increased tremendously in popularity. We live in the big data era, which is impacting many organizations. For our evaluation of data processing methods, we can now consider a large set of tools and methods that help to make ETL processes possible even when dealing with big datasets.

Currently, Apache Spark is one of the leading big data frameworks, designed to allow massive data processing. It works by parallelizing data across the nodes of a cluster and executing tasks on the distributed data in memory. Spark quickly became one of the most popular data tools after Hadoop MapReduce, which is a slower and more low-level programming framework.

The main concept of Spark is the **resilient distributed dataset** (**RDD**). Since Spark 1.3, RDDs have evolved into DataFrames, and Spark 1.6 introduced the concept of Datasets. These are data structures that represent an RDD in a table-like format with rows and columns, making it even easier to work with big data.

The beauty of Spark is that it has an API that abstracts away the complexity of distributing data and bringing it back together again. Programming in Spark feels like ordinary (functional) programming, where functions such as **map** and **filter** can be applied to data collections. Only, with Spark, it doesn't matter if that data collection is 10 MB or 10 TB in size. You'll learn more about the concepts and workings of Spark in *Chapter 7, Introduction to Analytics Engine (Spark) for Big Data*. For now, we'll go through a simple exercise to get familiar with Spark for ETL.

EXERCISE 3.02: BUILDING AN ETL JOB USING SPARK

In this exercise, we'll use Spark to process the same dataset as used in *Exercise 3.01, Creating a Simple ETL Bash Script*. We'll download the data, apply a filter, transform a column into a machine-readable format, and store the resulting dataset on disk. This illustrates the process of writing ETL code in Spark. Since most of our examples in this book are programmed in Python, we will use the **pyspark** library, which uses Python to connect to Spark.

> ### NOTE
>
> In many cases, using Python might not be obvious since Spark is built in Scala. However, once you have learned the basics of Spark with PySpark, it's easy to switch to another programming language, such as Java or Scala.

Before proceeding to the exercise, we need to set up Java and PySpark in the local environment. Please follow the instructions in the *Preface* to install them.

Java is needed to run Spark since the Spark framework makes use of some standard Java libraries. It expects the **java** command to be available on your system, and the **JAVA_HOME** environment variable to be set. So, make sure that Oracle Java 8, 9, or 10 is installed on your system. Java 11 is not supported by Spark yet. Also, make sure that your **JAVA_HOME** environment variable is set; if not, do so by pointing it to your local Java folder, for example, **export JAVA_HOME=/usr/lib/jvm/java-8-oracle**.

Perform the following steps to complete the exercise:

1. Create a directory called **Exercise03.02** in the **Chapter03** directory to store the files for this exercise.

2. Open your Terminal (macOS or Linux) or Command Prompt (Windows), navigate to the **Chapter03** directory, and type **jupyter notebook**.

3. Select the **Exercise03.02** directory, then click on **New** -> **Python3** to create a new Python 3 notebook.

4. First, you need to have Spark installed (link is provided in the *Preface*). You can then install PySpark by referring the *Preface*, or by running the following commands in a Jupyter Notebook cell.

```
import sys
!conda install --yes --prefix {sys.prefix} \
-c conda-forge pyspark
```

You should get the following output:

```
Collecting package metadata (current_repodata.json): done
Solving environment: done

## Package Plan ##

  environment location: /home/bas/anaconda3

  added / updated specs:
    - pyspark


The following packages will be downloaded:

    package                    |            build
    ---------------------------|-----------------
    conda-4.8.2                |           py37_0         3.0 MB  conda-forge
    ---------------------------------------------------------------
                                           Total:         3.0 MB

The following NEW packages will be INSTALLED:

  py4j               conda-forge/noarch::py4j-0.10.7-py_1
  pyspark            conda-forge/noarch::pyspark-2.4.5-py_0

The following packages will be SUPERSEDED by a higher-priority channel:

  conda                                      pkgs/main --> conda-forge



Downloading and Extracting Packages
conda-4.8.2          | 3.0 MB    | ################################## | 100%
Preparing transaction: done
Verifying transaction: done
Executing transaction: done
```

Figure 3.9: Installing PySpark with Anaconda

> **NOTE**
>
> Alternatively, you can also install PySpark using the following command in Terminal (macOS or Linux) or Command Prompt (Windows):
>
> **pip install pyspark**

5. Connect to a Spark cluster or a local instance using the following code:

```
from pyspark.sql import SparkSession
from pyspark.sql.functions import col, split, size
spark = SparkSession.builder.appName("Packt").getOrCreate()
```

These commands produce no output if they are successful; you'll only see that the cell in the Jupyter Notebook has completed running.

The **getOrCreate** command of the **SparkSession.builder** object creates a new Spark session, which is the main context of a Spark job. When this code is run in a production environment, it might contain hundreds of servers. For now, it's just your laptop.

6. Load and show the contents of the dataset in a Spark DataFrame using the following code:

```
data = spark.read.csv(\
        '../../Datasets/netflix_titles_nov_2019.csv', \
        header='true')
data.show()
```

> **NOTE**
>
> The CSV file used in the preceding step can be downloaded from https://packt.live/2C72sBN. Make sure you change the path (highlighted) based on where you have saved the file locally.

You should get the following output:

```
+---------+--------+--------+--------------------+--------------------+----------------+------------------+-----------+--
---------+----------+--------+--------------------+--------------------+-------+
| show_id|         title|        director|            cast|         country|    date_added|re
lease_year|rating| duration|          listed_in|      description|  type|
+---------+--------+--------+--------------------+--------------------+----------------+------------------+-----------+--
---------+----------+--------+--------------------+--------------------+-------+
|81193313|        Chocolate|            null|Ha Ji-won, Yoon K...|      South Korea|November 30, 2019|
2019| TV-14| 1 Season|International TV ...|Brought together ...|TV Show|
|81197050|Guatemala: Heart ...|Luis Ara, Ignacio...|   Christian Morales|            null|November 30, 2019|
2019| TV-G|  67 min|Documentaries, In...|From Sierra de la...|  Movie|
|81213894|   The Zoya Factor|  Abhishek Sharma|Sonam Kapoor, Dul...|           India|November 30, 2019|
2019| TV-14| 135 min|Comedies, Dramas,...|A goofy copywrite...|  Movie|
|81082007|        Atlantics|            Mati Diop|Mama Sane, Amadou...|France, Senegal, ...|November 29, 2019|
2019| TV-14| 106 min|Dramas, Independe...|Arranged to marry...|  Movie|
|80213643|   Chip and Potato|            null|Abigail Oliver, A...|Canada, United Ki...|            null|
2019|  TV-Y|2 Seasons|          Kids' TV|Lovable pug Chip ...|TV Show|
|81172754|      Crazy people|     Moses Inwang|Ramsey Nouah, Chi...|         Nigeria|November 29, 2019|
2018| TV-14| 107 min|Comedies, Interna...|Nollywood star Ra...|  Movie|
|81120982|     I Lost My Body|    Jérémy Clapin|Hakim Faris, Vict...|          France|November 29, 2019|
2019| TV-MA|  81 min|Dramas, Independe...|Romance, mystery ...|  Movie|
|81227195|Kalushi: The Stor...|      Mandla Dube|Thabo Rametsi, Th...|    South Africa|November 29, 2019|
2016| TV-MA| 107 min|Dramas, Internati...|The life and time...|  Movie|
|70205672|   La Reina del ...|            null|Kate del Castillo...|United States, Sp...|            null|
2019| TV-14|2 Seasons|Crime TV Shows, I...|This compelling s...|TV Show|
|81172841|Lagos Real Fake Life|   Mike Ezuruonye|Nonso Diobi, Mike...|            null|November 29, 2019|
2018| TV-14| 118 min|Comedies, Interna...|Two mooching frie...|  Movie|
|81172899|          Payday|     Cheta Chukwu|Baaj Adebule, Ebi...|         Nigeria|November 29, 2019|
2018| TV-MA| 110 min|Comedies, Indepen...|After an expensiv...|  Movie|
|81094391|Sugar Rush Christmas|            null|Hunter March, Can...|   United States|November 29, 2019|
2019| TV-PG| 1 Season|       Reality TV|"It's everything ...|TV Show|
|81172908|  The Accidental Spy|    Roger Russell|Ramsey Nouah, Chr...|         Nigeria|November 29, 2019|
2017| TV-14| 104 min|Action & Adventur...|Nursing a broken ...|  Movie|
|81152641|The Charming Stepmom|            null|Shahkrit Yamnarm,...|            null|November 29, 2019|
2019| TV-14| 1 Season|International TV ...|A quirky fashion ...|TV Show|
|81172901|        The Island|    Toka McBaror|Sambasa Nzeribe, ...|         Nigeria|November 29, 2019|
2018| TV-14|  93 min|Dramas, Internati...|When a colonel un...|  Movie|
|80990849|The Movies That M...|            null|            null|   United States|November 29, 2019|
2019| TV-MA| 1 Season|       Docuseries|These blockbuster...|TV Show|
|81033086|      Holiday Rush|    Leslie Small|Romany Malco, Son...|   United States|November 28, 2019|
2019| TV-PG|  94 min|Children & Family...|A widowed radio D...|  Movie|
|80156701|          Levius|            null|Nobunaga Shimazak...|           Japan|November 28, 2019|
2019| TV-14| 1 Season|Anime Series, Int...|Young Levius rise...|TV Show|
|81161538|    Lugar de Mulher|            null|            null|          Brazil|November 28, 2019|
2019| TV-MA| 1 Season|International TV ...|Four comedians fr...|TV Show|
|80997965|Merry Happy Whatever|            null|Dennis Quaid, Bri...|   United States|November 28, 2019|
2019| TV-PG| 1 Season|      TV Comedies|A struggling LA m...|TV Show|
+---------+--------+--------+--------------------+--------------------+----------------+------------------+-----------+--
---------+----------+--------+--------------------+--------------------+-------+
only showing top 20 rows
```

Figure 3.10: Reading data from a CSV file into a Spark DataFrame

The **spark.read.csv** command reads the contents of a CSV file into memory. The resulting object (**data** in our case) is a DataFrame, which is a distributed dataset that is available for querying and other analysis. Since we specified **header='true'** in the **read.csv** command, PySpark interprets the first line as the header of the dataset. Notice that the difficulties we had in *Exercise 3.01, Creating a Simple ETL Bash Script*, with separating the commas and quotes are handled by PySpark for us automatically. There are 12 columns as expected, and empty strings are denoted as **null**.

The **data** object now contains an in-memory representation of the dataset in a Spark DataFrame object (made visible in the **data.show()** statement). If you run this on a Spark cluster, the DataFrame will be distributed across the nodes

7. Apply the **data.filter()** function to filter the movies of **2019**, as shown in the following code:

```
movies = data.filter((col('type') == 'Movie') \
         & (col('release_year') == 2019))
movies.show()
```

You should get the following output:

```
+--------+--------------------+--------------------+--------------------+--------------------+----------------+--
---------+------+--------+--------------------+--------------------+-----+
| show_id|               title|            director|                cast|             country|      date_added|re
lease_year|rating|duration|           listed_in|         description| type|
+--------+--------------------+--------------------+--------------------+--------------------+----------------+--
---------+------+--------+--------------------+--------------------+-----+
|81197050|Guatemala: Heart ...|Luis Ara, Ignacio...|   Christian Morales|                null|November 30, 2019|
2019|  TV-G|  67 min|Documentaries, In...|From Sierra de la...|Movie|
|81213894|     The Zoya Factor|     Abhishek Sharma|Sonam Kapoor, Dul...|               India|November 30, 2019|
2019| TV-14| 135 min|Comedies, Dramas,...|A goofy copywrite...|Movie|
|81082007|           Atlantics|           Mati Diop|Mama Sane, Amadou...|France, Senegal, ...|November 29, 2019|
2019| TV-14| 106 min|Dramas, Independe...|Arranged to marry...|Movie|
|81120982|        I Lost My Body|      Jérémy Clapin|Hakim Faris, Vict...|              France|November 29, 2019|
2019| TV-MA|  81 min|Dramas, Independe...|Romance, mystery ...|Movie|
|81033086|         Holiday Rush|        Leslie Small|Romany Malco, Son...|       United States|November 28, 2019|
2019| TV-PG|  94 min|Children & Family...|A widowed radio D...|Movie|
|81194544|Evvarikee Cheppoddu|Basava Shankar Eeday|Rakesh Varre, Gar...|               India|November 27, 2019|
2019| TV-14| 134 min|Comedies, Interna...|When caste differ...|Movie|
|80995081|Little Singham: M...|       Prakash Satam|Saumya Daan, Sona...|                null|November 27, 2019|
2019| TV-Y7|  69 min|Children & Family...|In a journey back...|Movie|
|81177504|The Body Remember...|Elle-Máijá Tailfe...|Violet Nelson, El...|     Canada, Norway|November 27, 2019|
2019| TV-MA| 106 min|Dramas, Independe...|After a traumatic...|Movie|
|80175798|        The Irishman|     Martin Scorsese|Robert De Niro, A...|       United States|November 27, 2019|
2019|    R| 209 min|              Dramas|Hit man Frank She...|Movie|
|81062293|Mike Birbiglia: T...|        Seth Barrish|     Mike Birbiglia|       United States|November 26, 2019|
2019| TV-MA|  86 min|     Stand-Up Comedy|Comedian Mike Bir...|Movie|
|80235524|Super Monsters Sa...|          Steve Ball|Elyse Maloway, Vi...|       United States|November 26, 2019|
2019|  TV-Y|  24 min|Children & Family...|It's Christmas Ev...|Movie|
|81035121| True: Winter Wishes|                null|Michela Luci, Jam...|                null|November 26, 2019|
2019|  TV-Y|  46 min|              Movies|An ice crystal fr...|Movie|
|81215481|             Pranaam|      Sanjiv Jaiswal|Rajeev Khandelwal...|               India|November 25, 2019|
2019| TV-MA| 117 min|Action & Adventur...|Aspiring to fulfi...|Movie|
|81005044|What the F* Is Go...|Marta Jaenes, Ros...|                null|                null|November 25, 2019|
2019| TV-MA|  87 min|Documentaries, In...|Featuring extensi...|Movie|
|81218079|               Awake|Aleksandr Chernya...|Jonathan Rhys Mey...|       United States|November 24, 2019|
2019| TV-MA|  92 min|            Thrillers|After an accident...|Movie|
|80198859|             Brother|      Julien Abraham|MHD, Darren Musel...|              France|November 22, 2019|
2019| TV-MA|  97 min|Dramas, Independe...|Thrust from a vio...|Movie|
|81218074|     Shelby American|Nate Adams, Adam ...|     Carroll Shelby|       United States|November 22, 2019|
2019| TV-14| 119 min|Documentaries, Sp...|Featuring intervi...|Movie|
|81026188|The Knight Before...|     Monika Mitchell|Vanessa Hudgens, ...|       United States|November 21, 2019|
2019| TV-14|  93 min|Children & Family...|Medieval magic se...|Movie|
|80221584|Bikram: Yogi, Gur...|          Eva Orner|    Bikram Choudhury|       United States|November 20, 2019|
2019| TV-MA|  86 min|        Documentaries|This documentary ...|Movie|
|81217738|           Dorasaani|       KVR Mahendra|Anand Deverakonda...|               India|November 20, 2019|
2019| TV-14| 134 min|Dramas, Internati...|A village landlor...|Movie|
+--------+--------------------+--------------------+--------------------+--------------------+----------------+--
---------+------+--------+--------------------+--------------------+-----+
only showing top 20 rows
```

Figure 3.11: Filtering a Spark DataFrame

8. Transform the list of actors into a number to show how many main actors were listed in the dataset using the following code:

```
transformed = movies.withColumn(\
              'count_cast', size(split\
              (movies['cast'], ',')))
```

We have created a new column, **'count_cast'**. To calculate it, we first split the string in the **'cast'** column and then take the size of the resulting array; this counts the number of actors and saves it in the **transformed** variable. As the dataset contains a lot of columns, we will only select a few columns.

9. Select a subset of columns for the 2019 movies using the following code:

```
selected = transformed.select('title', 'director', \
        'count_cast', 'cast', 'rating', \
        'release_year', 'type')
```

10. Read the filtered data using the following code:

```
selected.show()
```

You should get the following output:

```
+--------------------+--------------------+----------+--------------------+------+------------+-----+
|               title|            director|count_cast|                cast|rating|release_year| type|
+--------------------+--------------------+----------+--------------------+------+------------+-----+
|Guatemala: Heart ...|Luis Ara, Ignacio...|         1|    Christian Morales|  TV-G|        2019|Movie|
|     The Zoya Factor|      Abhishek Sharma|         8|Sonam Kapoor, Dul...| TV-14|        2019|Movie|
|           Atlantics|           Mati Diop|         9|Mama Sane, Amadou...| TV-14|        2019|Movie|
|       I Lost My Body|       Jérémy Clapin|         6|Hakim Faris, Vict...| TV-MA|        2019|Movie|
|         Holiday Rush|        Leslie Small|         9|Romany Malco, Son...| TV-PG|        2019|Movie|
| Evvarikee Cheppoddu|Basava Shankar Eeday|         5|Rakesh Varre, Gar...| TV-14|        2019|Movie|
| Little Singham: M...|       Prakash Satam|         5|Saumya Daan, Sona...| TV-Y7|        2019|Movie|
|The Body Remember...|Elle-Máijá Tailfe...|        10|Violet Nelson, El...| TV-MA|        2019|Movie|
|        The Irishman|     Martin Scorsese|        20|Robert De Niro, A...|     R|        2019|Movie|
|   Mike Birbiglia: T...|       Seth Barrish|         1|    Mike Birbiglia| TV-MA|        2019|Movie|
|  Super Monsters Sa...|         Steve Ball|        12|Elyse Maloway, Vi...|  TV-Y|        2019|Movie|
|  True: Winter Wishes|                null|         8|Michela Luci, Jam...|  TV-Y|        2019|Movie|
|             Pranaam|     Sanjiv Jaiswal|         8|Rajeev Khandelwal...| TV-MA|        2019|Movie|
| What the F* Is Go...|Marta Jaenes, Ros...|        -1|                null| TV-MA|        2019|Movie|
|               Awake|Aleksandr Chernya...|         6|Jonathan Rhys Mey...| TV-MA|        2019|Movie|
|             Brother|     Julien Abraham|         7|MHD, Darren Musel...| TV-MA|        2019|Movie|
|     Shelby American|Nate Adams, Adam ...|         1|     Carroll Shelby| TV-14|        2019|Movie|
| The Knight Before...|     Monika Mitchell|         8|Vanessa Hudgens, ...| TV-14|        2019|Movie|
|   Bikram: Yogi, Gur...|          Eva Orner|         1|    Bikram Choudhury| TV-MA|        2019|Movie|
|           Dorasaani|        KVR Mahendra|         3|Anand Deverakonda...| TV-14|        2019|Movie|
+--------------------+--------------------+----------+--------------------+------+------------+-----+
only showing top 20 rows
```

Figure 3.12: Transforming a Spark DataFrame

11. Write the contents of our still in-memory DataFrame to a comma-separated file using the following code:

```
selected.write.csv('transformed' , header='true')
```

This produces a directory called **transformed** that contains some metadata and the actual output in CSV format. The output file has a name like part-00000-...csv. The first line of the output file contains the header since we specified **header='true'** in a similar way as when loading the source file.

NOTE

Alternatively, we can add the complete code of the ETL process so far in a Python script and run it through the Terminal (macOS or Linux) or Command Prompt (Windows). We have created the same **spark_etl. py** Python script at the following location: https://packt.live/2ATRPSt. Run this by opening a Terminal (macOS or Linux) or Command Prompt (Windows) or Anaconda Prompt in the same folder as the script and executing the following command: **python spark_etl.py**

12. Open the comma-separated file in the **transformed** directory in Jupyter Notebook using the following command:

```
# note: the actual name of the csv file
#('part-....') differs on each run

!head transformed/part-00000-\
96ee95ae-9a80-4e88-b876-7f73893c2f21-c000.csv
```

NOTE

The aforementioned command will only work on Linux and UNIX-based systems. On Windows, you can browse the same file via Windows Explorer.

You should get the following output:

```
title,director,count_cast,cast,rating,release_year,type
Guatemala: Heart of the Mayan World,"Luis Ara, Ignacio Jaunsolo",1,Christian Morales,TV-G,2019,Movie
The Zoya Factor,Abhishek Sharma,8,"Sonam Kapoor, Dulquer Salmaan, Sanjay Kapoor, Sikander Kher, Angad Bedi, Koel P
urie, Pooja Bhamrah, Manu Rishi Chadha",TV-14,2019,Movie
Atlantics,Mati Diop,9,"Mama Sane, Amadou Mbow, Ibrahima Traore, Nicole Sougou, Amina Kane, Mariama Gassama, Coumba
Dieng, Ibrahima Mhaye, Diankou Sembene",TV-14,2019,Movie
I Lost My Body,Jérémy Clapin,6,"Hakim Faris, Victoire Du Bois, Patrick d'Assumçao, Dev Patel, Alia Shawkat, George
Wendt",TV-MA,2019,Movie
Holiday Rush,Leslie Small,9,"Romany Malco, Sonequa Martin-Green, Darlene Love, Deon Cole, La La Anthony, Deysha Ne
lson, Amarr M. Wooten, Selena-Marie Alphonse, Andrea-Marie Alphonse",TV-PG,2019,Movie
Evvarikee Cheppoddu,Basava Shankar Eeday,5,"Rakesh Varre, Gargeyi, Vamsi raj Nekkanti, D P Ghani, K Prasanna",TV-1
4,2019,Movie
Little Singham: Mahabali,Prakash Satam,5,"Saumya Daan, Sonal Kaushal, Anamaya Verma, Ganesh Divekar, Neshma Chembu
rkar",TV-Y7,2019,Movie
The Body Remembers When the World Broke Open,"Elle-Máijá Tailfeathers, Kathleen Hepburn",10,"Violet Nelson, Elle-M
áijá Tailfeathers, Charlie Hannah, Barbara Eve Harris, Sonny Surowiec, Jay Cardinal Villeneuve, Tony Massil, Aidan
Dee, James Angus Cowan, Anthony Bolognese",TV-MA,2019,Movie
The Irishman,Martin Scorsese,20,"Robert De Niro, Al Pacino, Joe Pesci, Harvey Keitel, Ray Romano, Bobby Cannavale,
Anna Paquin, Stephen Graham, Stephanie Kurtzuba, Kathrine Narducci, Welker White, Jesse Plemons, Jack Huston, Dome
nick Lombardozzi, Louis Cancelmi, Paul Herman, Gary Basaraba, Marin Ireland, Sebastian Maniscalco, Steven Van Zand
t",R,2019,Movie
```

Figure 3.13: CSV file in the transformed directory

> **NOTE**
>
> To access the source code for this specific section, please refer to https://packt.live/308Xahr.

You have now completed a simple exercise that demonstrates the power of Spark when it comes to working with data. If you choose to write your ETL as source code, Spark can be considered a good option for large datasets. When working with a Spark DataFrame (or DataSet) object, many more transformations are possible: grouping data, aggregation, and so on. It's even possible to write SQL queries on the data. We will learn about these in more detail in *Chapter 7, Introduction to Analytics Engine (Spark) for Big Data*. In the next activity, you'll write an ETL job with Spark in a business-like scenario.

ACTIVITY 3.01: USING PYSPARK FOR A SIMPLE ETL JOB TO FIND NETFLIX SHOWS FOR ALL AGES

You work for a kids' TV channel and plan to launch a new show. You want a list of Netflix shows for 2019 that are for all ages, along with their ratings and other details. Based on the details, you will work on your new show.

In this activity, you'll load the Netflix dataset, filter the shows based on parental guidelines, and store the result in a CSV file.

> **NOTE**
>
> The code and the resulting output for this activity have been loaded in a Jupyter Notebook that can be found here: https://packt.live/3iXSinP.

Perform the following steps to complete the activity:

1. Create a Python file or Jupyter Notebook.

2. Connect to a local Spark cluster by importing the PySpark libraries and building a **SparkSession** object.

3. Read the CSV file, **netflix_titles_nov_2019.csv**, from disk, and load it into a Spark DataFrame.

4. Filter the data: select only the TV shows where the rating is either **TV-G** (for all ages) or **TV-Y** (for children). You should get the following output:

```
+--------+--------------------+--------+--------------------+--------+--------+--------------------+--------------------+------------+
------+---------+--------------------+--------------------+-------+
| show_id|               title|director|                cast|        country|         date_added|release_year|
rating| duration|            listed_in|         description|   type|
+--------+--------------------+--------+--------------------+--------+--------+--------------------+--------------------+------------+
------+---------+--------------------+--------------------+-------+
|80213643|     Chip and Potato|    null|Abigail Oliver, A...|Canada, United Ki...|                null|        2019|
TV-Y|2 Seasons|          Kids' TV|Lovable pug Chip ...|TV Show|
|80117560|Trolls: The Beat ...|    null|Amanda Leighton, ...|       United States|                null|        2019|
TV-G|8 Seasons|Kids' TV, TV Come...|As Queen Poppy we...|TV Show|
|80115338|         Llama Llama|    null|Jennifer Garner, ...|       United States|                null|        2019|
TV-Y|2 Seasons|          Kids' TV|Beloved children'...|TV Show|
|80045925|Bella and the Bul...|    null|Brec Bassinger, C...|       United States|    November 2, 2019|        2015|
TV-G| 1 Season|Kids' TV, TV Come...|The life of cheer...|TV Show|
|70172485|          Victorious|    null|   Victoria Justice,...|       United States|                null|        2013|
TV-G|3 Seasons|Kids' TV, TV Come...|When aspiring sin...|TV Show|
|81184735|Barbie Dreamhouse...|    null|America Young, Ki...|                null|    November 1, 2019|        2019|
TV-Y| 1 Season|          Kids' TV|As the Roberts fa...|TV Show|
|80227818|        Hello Ninja|    null|Lukas Engel, Zoey...|                null|    November 1, 2019|        2019|
TV-Y| 1 Season|          Kids' TV|BFFs Wesley and G...|TV Show|
|80991060|   Flavorful Origins|    null|          Yang Chen|               China|                null|        2019|
TV-G|2 Seasons|Docuseries, Inter...|Delve into the de...|TV Show|
|81034099|          Jeopardy!|    null|        Alex Trebek|       United States|                null|        2019|
TV-G|5 Seasons|        Reality TV|Alex Trebek hosts...|TV Show|
|81185502|A Little Thing Ca...|    null|Lai Kuan-lin, Zha...|               China|    October 26, 2019|        2019|
TV-G| 1 Season|International TV ...|A shy college stu...|TV Show|
|81192130|ChuChu TV Kids So...|    null|                null|                null|                null|        2019|
TV-Y|2 Seasons|          Kids' TV|This educational ...|TV Show|
|81011059|              Booba|    null|        Roman Karev|              Russia|                null|        2019|
TV-Y|3 Seasons|Kids' TV, TV Come...|The world is a my...|TV Show|
|80176872|Little Baby Bum: ...|    null|Chloe Marsden, Aa...|                null|                null|        2019|
TV-Y|2 Seasons|British TV Shows,...|Twinkle, Mia, Jac...|TV Show|
|81020066|Mighty Little Bhe...|    null|Samriddhi Shukla,...|               India|    October 18, 2019|        2019|
TV-Y| 1 Season|Kids' TV, TV Come...|From decorating h...|TV Show|
|81094271|Spirit Riding Fre...|    null|Amber Frank, Bail...|       United States|                null|        2019|
TV-Y|2 Seasons|          Kids' TV|"Find the fun and...|TV Show|
|81154549|      Magical Andes|    null|                null|                null|    October 15, 2019|        2019|
TV-G| 1 Season|Docuseries, Inter...|From Argentina to...|TV Show|
|80212481|YooHoo to the Rescue|    null|Kira Buckland, Ry...|         South Korea|                null|        2019|
TV-Y|2 Seasons|Kids' TV, Korean ...|In a series of ma...|TV Show|
|80124711|     Super Monsters|    null|Elyse Maloway, Vi...|                null|                null|        2019|
TV-Y|3 Seasons|          Kids' TV|Preschool kids wh...|TV Show|
|80218107|Dragons: Rescue R...|    null|Nicolas Cantu, Br...|                null|September 27, 2019|        2019|
TV-Y| 1 Season|Kids' TV, TV Come...|Twins Dak and Ley...|TV Show|
|81073764|Loo Loo Kids: Joh...|    null|                null|                null| September 1, 2019|        2016|
TV-Y| 1 Season|          Kids' TV|Music-loving baby...|TV Show|
+--------+--------------------+--------+--------------------+--------+--------+--------------------+--------------------+------------+
------+---------+--------------------+--------------------+-------+
only showing top 20 rows
```

Figure 3.14: The contents of the file with TV shows filtered by rating

5. Add a new column called **count_lists**, which contains the number of lists that are in the **listed_in** column.

6. Select the **title**, **count_directors**, **director**, **cast**, **rating**, **release_year**, **duration**, **listed_in**, and **description** columns.

7. Write the output of the filtered data to a comma-separated file (**'transformed2.csv'**).

8. View the CSV file to get the following output:

```
title,cast,rating,release_year,duration,count_lists,listed_in,description
Chip and Potato,"Abigail Oliver, Andrea Libman, Briana Buckmaster, Brian Dobson, Chance Hurstfield, Dominic Good,
Emma Jayne Maas, Evan Byarushengo, Scotia Anderson, Alessandro Juliani",TV-Y,2019,2 Seasons,1,Kids' TV,"Lovable pu
g Chip starts kindergarten, makes new friends and tries new things — with a little help from Potato, her secret mo
use pal."
Trolls: The Beat Goes On!,"Amanda Leighton, Skylar Astin, Ron Funches, David Fynn, David Koechner, David Kaye, Sea
n T. Krishnan, Sam Lerner, Patrick Pinney, Kevin Michael Richardson, Kari Wahlgren, Fryda Wolff",TV-G,2019,8 Seaso
ns,2,"Kids' TV, TV Comedies","As Queen Poppy welcomes a new time of peace in Troll Village with parties, sports an
d holiday celebrations, Branch tries to be more fun-loving."
Llama Llama,"Jennifer Garner, Shayle Simons, Vania Gill, Islie Hirvonen, Brendon Sunderland, Austin Abell, Evans J
ohnson, Kathleen Barr, David Hoole",TV-Y,2019,2 Seasons,1,Kids' TV,"Beloved children's book character Llama Llama
springs to life in this heartwarming series about family, friendship and learning new things."
Bella and the Bulldogs,"Brec Bassinger, Coy Stewart, Jackie Radinsky, Buddy Handleson, Lilimar, Haley Tju, Dorien
Wilson, Rio Mangini, Annie Tedesco",TV-G,2015,1 Season,2,"Kids' TV, TV Comedies","The life of cheerleader Bella Da
wson turns upside down when, in a twist of fate, she becomes quarterback for her middle school's football team."
Victorious,"Victoria Justice, Leon Thomas III, Matt Bennett, Elizabeth Gillies, Ariana Grande, Avan Jogia, Daniell
a Monet, Michael Eric Reid, Jake Farrow, Eric Lange",TV-G,2013,3 Seasons,2,"Kids' TV, TV Comedies","When aspiring
singer Tori Vega joins the eccentric students at Hollywood Arts High, she struggles to fit in with the amazingly t
alented teens."
Barbie Dreamhouse Adventures: Go Team Roberts,"America Young, Kirsten Day, Cassandra Morris, Cassidy Naber, Emma G
alvin, Stephanie Sheh, Desirae Whitfield, Cristina Milizia, Lisa Fuson, Greg Chun, Ritesh Rajan, Eamon Brennan",TV
-Y,2019,1 Season,1,Kids' TV,"As the Roberts family heads to Costa Rica to investigate a mermaid legend, Barbie tak
es on a summer job at a water park run by a devious boss."
Hello Ninja,"Lukas Engel, Zoey Siewert, Sam Vincent, Mayumi Yoshida",TV-Y,2019,1 Season,1,Kids' TV,"BFFs Wesley an
d Georgie and their silly cat sidekick Pretzel transform into ninjas and enter a magic world, where they solve pro
blems and save the day."
Flavorful Origins,Yang Chen,TV-G,2019,2 Seasons,2,"Docuseries, International TV Shows","Delve into the delectable
world of Chaoshan cuisine, explore its unique ingredients and hear the stories of the people behind its creation."
Jeopardy!,Alex Trebek,TV-G,2019,5 Seasons,1,Reality TV,"Alex Trebek hosts one of TV's longest-running game shows,
where a trio of players buzz in with their knowledge of history, arts, pop culture and more."
```

Figure 3.15: CSV file in the transformed2 directory

You now have a general understanding of data processing techniques that can be used for any ETL step. You have learned about simple scripts, ETL tools, and Spark. In the remainder of this chapter, we will do a deep dive into each ETL step. Each section refers to one of the steps in the pipeline shown in *Figure 3.1*. We'll start with the first step: importing raw data from a source system.

> **NOTE**
>
> The solution to this activity can be found on page 587.

SOURCE TO RAW: IMPORTING DATA FROM SOURCE SYSTEMS

Typically, data comes to an AI system in the form of files. This may sound old-fashioned, but the truth is that this kind of data transfer is still very effective and universal. Almost all core systems of an organization can export their data in some form, whether CSV, Excel, XML, JSON, or something else. More modern ways to produce batch data are via on-demand APIs that query a system with parameters. However, this way of interfacing is sometimes not the highest priority for software builders.

They understand that systems have to interact and thus provide some form of interface, but it's not in their interests to make it very easy to get data from the system. Moreover, it's expensive to document and maintain an API once it's developed; once consumers rely on the connection, any change has to be managed carefully. Since many core systems are built on standard technology such as a relational database (Oracle, SQL Server, PostgreSQL, MySQL, and so on) and these databases have built-in features to export data already, it's tempting to simply utilize these features rather than to build custom APIs that require software engineering, testing, and management. So, although APIs provide a more stable and robust way of getting data, file transfers are a major way of interacting with source systems and it's not likely that a quick shift will happen in the near future. We often deal with legacy systems or vendor products and cannot influence the way that this software works.

After importing the raw data from a source file, the next step in an ETL pipeline is to clean it. Let's explore that topic in the next section.

RAW TO HISTORICAL: CLEANING DATA

One of the most important tasks for a data engineer who works on AI solutions is cleaning the data. Raw data is notoriously dirty, by which we mean that it's not suitable for use in a model or any other form of consumption, such as displaying on a website. Dirty data can have many forms:

- Missing values, such as **null** values or empty strings

- Inconsistent values, such as the string "007" where integers are expected

- Inaccurate values, such as a date with a value of 31 February 2018

- Incomplete values, such as the string "Customer 9182 has arri"; this can be due to limited field lengths

- Unreadable values, such as "K/dsk2#ksd%9Zs|aw23k4lj0@#$" where a name is expected; this can be due to encryption where it's not expected or wrong data formats (Unicode, UTF-8, and so on)

The ideal point to clean data is when data is processed from the raw data layer into the historical data layer. After all, the historical archive functions as the one version of the truth and should contain ready-to-use records. But we have seen many times that data needs a bit more cleaning, even after querying it from the historical layer or the analytics layer. This can be due to sloppy ETL developers or other reasons. It might also be on purpose; a conscious decision might have been made by management as part of a trade-off discussion to keep the data in a somewhat raw/dirty format since cleaning it simply takes a lot of time and effort.

The way to clean data is to write code that transforms the dirty data into the proper format and values. There is no one good data format or data model; it depends on the organization and use case. Some people might prefer a date-time format of YYYY-MM-DD, while others prefer MM-DD-YYYY. Some might allow storing `null` values, while others require a value, for example, an empty string or default integer. What's important is to make choices, to document them, to communicate them clearly, and to check them. Code that cleans data can be shared and reused among developers to make the work easier and to standardize the outcomes.

The next step in an ETL pipeline is modeling the data. We'll cover that in the next section.

RAW TO HISTORICAL: MODELING DATA

To transform raw data into a form that can be stored in a historical archive, some steps need to be taken depending on the shape of the raw data and the chosen target model. Raw data should be modeled into a shape so that is can be stored for historical analysis. The raw data usually resides in files and has to be mapped to relational database tables. A typical process is to read a CSV file, transform the data in memory, and write the transformed data to a database where the tables are created in the fashion of a data vault. A data vault is a relational model intended for data warehousing, developed by Dan Linstedt. Other popular target models that offer normalized data models are the **Dimensional Data Mart** (**DDM**) approach by Ralph Kimball and the **Corporate Information Factory** (**CIF**) by Bill Inmon.

After modeling the data and creating a historical archive, it's time to start preparing the data for analysis by building up the analytics layer with ETL jobs. The following paragraphs contain two important steps in that phase: filtering and aggregation, and flattening the data.

HISTORICAL TO ANALYTICS: FILTERING AND AGGREGATING DATA

Not all the data in a historical archive will be needed in the analytics layer. Therefore, a filter has to be applied, for example, to only take the data of the past year. This is needed to further reduce the amount of data that is stored in the analytics layer and improve query performance. Some fields can be aggregated (for example, sums, summaries, and averages).

HISTORICAL TO ANALYTICS: FLATTENING DATA

A technique to make data available for efficient querying in the analytics layer of a data-driven solution is to flatten it. Flattening data means to let go of the normalized form of the Linstedt/Kimball/Inmon tables and create one big table with many columns, where some data is repeated instead of being stored in foreign keys.

When an analytics layer is created, the data is prepared for consumption by a machine learning model. The following section explores an important step that is still considered to be part of the ETL pipeline: feature engineering.

ANALYTICS TO MODEL: FEATURE ENGINEERING

The final step in preparing the data for machine learning model development and training is to transform the data records so that the models can consume them. This transformation into model-readable data is called *feature engineering*. Features are the characteristics of a dataset by which a predictive model can be evaluated. A good feature has a lot of predictive power; knowing the value of such a feature gives us immediate results in terms of the predictability of the outcome of a model. For example, if we have a dataset with 100,000 people and have to predict which language each person speaks, a very indicative feature will be the country of residence. In the same dataset, the height or age of the person will be weak features for the language detection algorithm.

Features can be stored in a dedicated feature store, which is part of the analytics layer in a modern data lake. The feature store is just a database that is easy to query when working with machine learning models. It can be considered the final data store in an ETL or ELT pipeline, where the final T step is feature engineering. Features can also be exported to disk as part of the model's code.

Features play an important role in two stages of a machine learning project:

- While developing and training the model, all training data has to be transformed into a feature set. This is called feature engineering.

- While running in production and executing/scoring the model, new data that comes in and has to be analyzed by the model has to go through the same feature transformation steps.

The code that transforms the source data into features, therefore, has to be deployed to two locations: the machine learning environment and the production environment. The source code in these two places has to be the same, otherwise, the trained models will produce different results than the models in production. Therefore, it's important to carefully maintain and manage this code. Version control, continuous delivery, documentation, and monitoring are all best practices that ensure that no mistakes are made. A good way to ensure consistency is to store the models together with the feature engineering code; they belong together and thus should have the same version, release plan, and documentation.

Features can be derived in four ways. A column with features in a dataset for model training can have one of the following origins:

- Data column: Data that feeds a model may already be in the right shape.

- The translation of one source column, for example, the usage of the number in the age column in a dataset of people.

- The combination of other columns, for example, the sum of all values in three source columns with product prices.

- From an external source: Sometimes, the data in a source column only contains a reference or pointer to an external data source. In those cases, a query has to be made to an external system, for example, a cache database with customer records or an API call to a core system. This will introduce a dependency on an external resource, which has to be managed and monitored. Since this step is resource-intensive, especially for large datasets, it's also crucial to carefully test the performance.

In many cases, depending on the choice of the machine learning algorithm, features have to be normalized. This is the process of getting numerical features in the same order of magnitude to be able to optimize the algorithm. The usual practice is to bring all numbers into the range 0 to 1 by looking at the minimum and maximum values that are in a dataset. For example, the values in the height column of a dataset of people might initially range from 45 to 203 centimeters. By executing the function $h \rightarrow (h - 45) / 158$ on each value h, all values will range from 0 (the initial 45 cm) to 1 (the initial 203 cm).

Alongside feature engineering, it's important to split a dataset for training and testing models. We'll explore that topic in the next section.

ANALYTICS TO MODEL: SPLITTING DATA

The final data preparation step in a data pipeline is to split the data in a train and test dataset. If you have a big dataset, a model needs to be trained on only a part of it (for example, 70%). The remainder of the same dataset (30%) is kept aside for validating or testing the model. Common ratios are in the range of 80-20 to 70-30. It's essential that the datasets for training and testing are both from the same original dataset, and that the division is done randomly. Otherwise, it's possible that the model will be overfitted to the training dataset and the results on the test dataset (and all other forthcoming predictions) will not be accurate or precise.

The work of splitting the original dataset is usually done by a data scientist. All modern data frameworks and tools contain methods to do this. For example, the following code is for splitting a dataset in the popular Python framework scikit-learn:

```
from sklearn.model_selection import train_test_split
X_train, X_test, \
Y_train, Y_test = train_test_split(X, y, \
                  test_size=0.25, random_state=42)
```

In the preceding statement, the **train_test_split** function is called. This function is a standard method in the Python scikit-learn (**sklearn**) library to split a dataset (**X**) with output variables (**y**) into a random training part and a random testing part. The training set is called **X_train**, and the output variables are in **Y_train**, and they contain 100% – 25% = 75% of the data.

STREAMING DATA

This chapter so far has explored data preparation methods for batch-driven ETL. You have learned the steps and techniques to get raw data from a source system, transform it into a historical archive, create an analytics layer, and finally do feature engineering and data splitting. We'll now make a switch to streaming data. Many of the concepts you have learned for batch processing are also relevant for stream processing; however, things (data) move a bit more quickly and timing becomes important.

When preparing streaming event data for analytics, for example, to be used in a model, some specific mechanisms come into play. Essentially, a data stream goes through the same steps as raw batch data: it has to be loaded, modeled, cleaned, and filtered. However, a data stream has no beginning and ending, and time is always important; therefore, the following patterns and practices need to be applied:

- Windows

- Event time

- Watermarks

We'll explain these topics in the next sections.

WINDOWS

In many use cases, it's not possible to process all the data of a never-ending stream. In those cases, it is required to split up the data stream into chunks of time. We call these chunks windows of time. To window a data stream means to split it up into manageable pieces, for example, one chunk per hour.

There are several types of windows:

- A *tumbling* window has a fixed length and does not overlap with other windows. Events (the individual records) in a stream with tumbling windows only belong to one window.

- A *sliding window* has a fixed length but overlaps other sliding windows for a certain duration. Events in a stream with sliding windows belong to multiple windows since one window will not be completed yet when other windows begin. The window size is the length of the window; the window slide determines the frequency of windows.

- A *session* window is a window that is defined by the events themselves. The start and end of a session window are determined by data fields, for example, the login and log out actions of a website. It's also possible to define session windows by a series of following events with a gap in between. Session windows can overlap, but events in a stream with session windows can only belong to one session.

The following figure displays these kinds of windows in an overview. Each rectangle depicts a window that is evaluated in the progression of time. Each circle represents an event that occurred at a certain moment in time. For example, the first tumbling window contains two events, labeled **1** and **2**. It's clear to see that sliding windows and session windows are more complex to manage than tumbling windows: events can belong to multiple windows and it's not always clear to which window an event belongs.

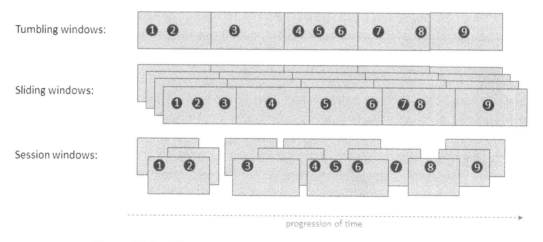

Figure 3.16: Different types of time windows for streaming data

Modern stream processing frameworks such as Apache Spark Structured Streaming and Apache Flink provide built-in mechanisms for windowing, and customizable code to allow you to define your time windows.

While windowing the data in your stream is not necessary, it provides a great way to work with streaming data.

EVENT TIME

It's important to realize that events contain multiple timestamps. We can identify three important ones:

- The *event time*: The timestamp that the event occurs. For example, the moment a person clicks on a button on a web page (say, at 291 milliseconds after 16:45).

- The *ingestion time*: The timestamp at which the event enters the stream processing software engine. This is slightly later than the event time since the event has to travel across the network and perhaps a few data stores (for example, a message bus such as Kafka). An example of the ingestion time is the moment that the mouse-click of the customer enters the stream processing engine (say, at 302 milliseconds after 16:45).

- The *processing time*: The timestamp at which the event is evaluated by the software that runs as a job within the stream processing engine. For example, the moment that the mouse-click is compared to the previous mouse-click in a window, to determine whether a customer is moving to another section of a website (say, at 312 milliseconds after 16:45).

In most event data, the event time is included as a data field. For example, money transactions contain the actual date and time when the transfer was made. This should be fairly accurate and precise, for example, in the order of milliseconds. The processing time is also available in the data stream processor; it's just the server time, defined by the clock of the infrastructure where the software runs. The ingestion time is somewhat rare and hardly used; it might be included as an extra field in events if you know that the processing of events takes a long time and you want to perform performance tests.

When analyzing events, the event time is the most useful timestamp to work with. It's the most accurate indication of the event. When the order of events is important, the event time is the only timestamp that guarantees the right order, since latency can cause out-of-order effects, as shown in the following figure:

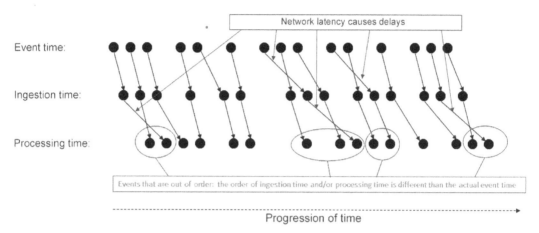

Figure 3.17: The timestamps and out-of-order effects in a data stream

In *Figure 3.17* the dark blue circles represent timestamps. The blue arrows indicate the differences in time between the timestamps. For each event, event time always occurs first, followed by the ingestion time, and finally the processing time. Since these latencies can differ per event, as indicated by the highlighted circles, the out-of-order effect can occur.

LATE EVENTS AND WATERMARKS

At the end of each window, the events in that window are evaluated. They can be processed as a batch in a similar way as normal ETL processes: filtering, enriching, modeling, aggregating, and so on. But the end of a window is an arbitrary thing. There should be a timed trigger that tells the data stream processor to start evaluating the window. For that trigger, the software could look at the processing time (its server clock). Or, it could look at the event data that comes in and trigger the window evaluation once an event comes in with a timestamp that belongs to the next window. All these methods are slightly flawed; after all, we want to look at the event time, which might be very different from the processing time due to network latency. So, we can never be sure that a trigger is timed well; there might always be events arriving in the data stream that belong to the previous window. They just happened to arrive a bit later. To cater to this, the concept of *watermarking* was introduced.

A *watermark* is an event that triggers the calculation of a window. It sets the evaluation time for the window a bit later than the actual end-time of the window, to allow late events to arrive. This bit of slack is all that is needed to make sure that most of the events are evaluated in the right window where they belong, and not in the next window or even ignored (see *Figure 3.18*).

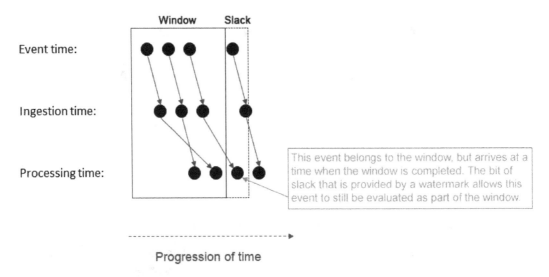

Figure 3.18: Late events that are still evaluated in the correct window based on their event time

In the next exercise, you'll build a stream processing job with Spark where the concepts of windows, event time, and watermarks are used.

EXERCISE 3.03: STREAMING DATA PROCESSING WITH SPARK

In this exercise, we are going to connect to a data stream and process the events. You'll connect to a real Twitter feed and aggregate the incoming data by specifying a window and counting the number of tweets. We'll use the Spark Structured Streaming library, which is rich in functionality and easy to use.

Since we are going to connect to Twitter, you'll need a Twitter developer account:

1. Go to https://developer.twitter.com/ and create an account (or log in with your regular Twitter account if you already have one).

2. Apply for an API by clicking **Apply** and selecting the purpose of **Exploring the API**.

3. Fill in the form where you have to state the usage of Twitter data.

4. After approval, you'll see an app in the **Apps** menu. Select the new app and navigate to the **Keys and tokens** tab. Click **Generate** to generate your access token and secret. Copy these and store them in a safe location, for example, a local file.

5. Click **Close** and also copy the **API key** and **API secret key** in the local file. Never store these on the internet or in source code.

6. Make sure that Oracle Java 8, 9, or 10 is installed on your system. Java 11 is not supported by Spark yet. Also, make sure that your **JAVA_HOME** environment variable is set; if not, do so by pointing it to your local Java folder, for example, **export JAVA_HOME=/usr/lib/jvm/java-8-oracle**.

7. We'll use the **tweepy** library to connect to Twitter. Get it by following the installation instructions in the *Preface*.

There are two parts to this exercise. First, we will connect to Twitter and get a stream of tweets. We'll write them to a local socket. This is not supported for production, but it will do fine in a local development scenario. In production, it's common to write the raw event data to a message bus such as Kafka.

Perform the following steps to complete the exercise:

1. Create a new Python file in your favorite editor (for example, PyCharm or VS Code) or a Jupyter Notebook and name it **get_tweets.py** or **get_tweets.ipynb**.

2. If you're using Jupyter Notebook, install the **tweepy** library by entering the following lines in a cell and running it:

```
import sys
!conda install --yes --\
prefix {sys.prefix} -c conda-forge tweepy
```

3. If you're using an IDE, make sure that **tweepy** is installed by typing the following command in your Anaconda console or shell:

```
pip install tweepy
```

4. Connect to the Twitter API by entering the following code:

```
import socket
import tweepy
from tweepy import OAuthHandler

# TODO: replace the tokens and secrets with your own Twitter API
  values

ACCESS_TOKEN = ''
ACCESS_SECRET = ''
CONSUMER_KEY = ''
CONSUMER_SECRET = ''
```

Replace the values in the strings with the keys and secrets that you wrote down in the preparation part of this exercise.

5. To connect to Twitter, the **tweepy** library first requires us to log in with the **Oauth** credentials that are in the keys and secrets. Let's create a dedicated method for this:

```
def connect_to_twitter():
    auth = OAuthHandler(CONSUMER_KEY, CONSUMER_SECRET)
    auth.set_access_token(ACCESS_TOKEN, ACCESS_SECRET)
    api = tweepy.API(auth)
```

6. Now let's handle the events by adding some lines to the **connect_to_twitter** function. Put the following lines in the cell with the **connect_to_twitter** function:

```
my_stream_listener = MyStreamListener()
my_stream = tweepy.Stream(auth=api.auth,
    listener=my_stream_listener)

# select a (limited) tweet stream
my_stream.filter(track=['#AI'])
```

The **connect_to_twitter** function now authenticates us at the Twitter API, hooks up the **MyStreamListener** event handler, which sends data to the socket and starts receiving data from Twitter in a stream with the **#AI** search keyword. This will produce a limited stream of events since the Twitter API will only pass through a small subset of the actual tweets.

7. Next, the **tweepy** library needs us to inherit the **StreamListener** class. We have to specify what to do when a tweet comes in an override of the **on_data()** function. In our case, we send all data (the entire tweet in JSON format) to the local socket:

```
class MyStreamListener(tweepy.StreamListener):

    def on_error(self, status_code):
        if status_code == 420:
            return False

    def on_data(self, data):
        print(data)

        # send the entire tweet to the socket on localhost where
          pySpark is listening
        client_socket.sendall(bytes(data, encoding='utf-8'))
        return True
```

We will now set up a local socket that has to be connected to Twitter. Python has a library called **socket** for this, which we imported already at the top of our file.

8. Write the following lines to set up a **socket** at port **1234**:

```
s = socket.socket()
s.bind(("localhost", 1234))
print("Waiting for connection...")

s.listen(1)
client_socket, address = s.accept()
print("Received request from: " + str(address))
```

9. The final step is to call the **connect_to_twitter** method at the end of the file. We are still editing the same file in the IDE; add the following lines at the end:

```
# now that we have a connection to pySpark, connect to Twitter
connect_to_twitter()
```

10. Run the **get_tweets.py** file from a Terminal (macOS or Linux) or Command Prompt (Windows)or Anaconda Prompt using the following command:

```
python get_tweets.py
```

You should get the following output:

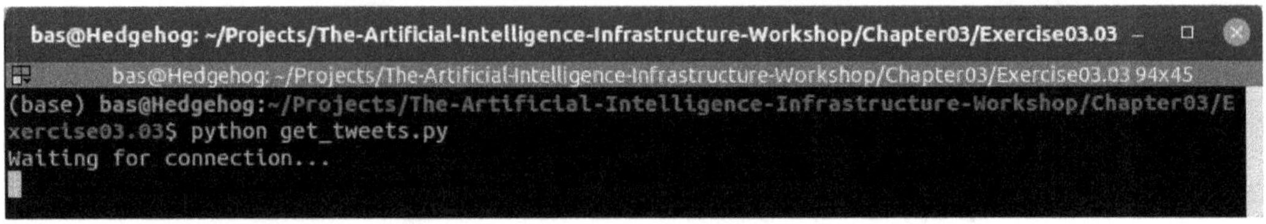

Figure 3.19: Setting up a connection to the Twitter API from a terminal

The first part is now done. Run this file to set up the socket and the connection to Twitter. If all is well, there will be a Python job running with no output but **Waiting for connection**.... Since the socket has to respond to a caller, we have to create a socket client that listens to the same port.

The next part is to connect a Spark job to the socket on localhost (or a Kafka topic in production), get the tweets as a data stream, and perform a window operation on them.

11. Create a new Python file or Jupyter Notebook and name it **spark_twitter.py** or **spark_twitter.ipynb**.

> **NOTE**
>
> The **spark_twitter** Jupyter Notebook can be found here:
> https://packt.live/302eWms.

12. If you have done *Exercise 3.02, Building an ETL Job Using Spark*, PySpark is already installed on your local machine. If not, install PySpark with the following lines:

```
import sys
!conda install --yes --prefix {sys.prefix} \
-c conda-forge pyspark
```

13. We first have to connect to a Spark cluster or a local instance. Enter the following lines in the file, notebook, or Python shell:

```
from pyspark.sql import SparkSession
from pyspark.sql.functions import from_json, \
window, to_timestamp
from pyspark.sql.types import StructType, \
StructField, StringType
```

14. Enter the following line to create a Spark session:

```
spark = SparkSession.builder.appName('Packt').getOrCreate()
```

15. To connect to the socket on localhost, enter the following line:

```
raw_stream = spark.readStream.format('socket')\
            .option('host', 'localhost')\
            .option('port', 1234).load()
```

Now, we have a stream of raw strings. These should be converted from JSON to a useful format, namely, a **text** field with the tweet text and a **timestamp** field that contains the event time. Let's do that in the next step.

16. We'll define the JSON schema and add the string format that Twitter uses for its timestamps:

```
tweet_datetime_format = 'EEE MMM dd HH:mm:ss ZZZZ yyyy'
schema = StructType([StructField('created_at', \
        StringType(), True),\
        StructField('text', StringType(), True)])
```

17. We can now convert the JSON strings with the **from_json** PySpark function:

```
tweet_stream = raw_stream.select(from_json('value', schema).
alias('tweet'))
```

18. The **created_at** field is still a string, so we have to convert it to a timestamp with the **to_timestamp** function:

```
timed_stream = tweet_stream\
                .select(to_timestamp('tweet.created_at', \
                                    tweet_datetime_format)\
                        .alias('timestamp'),\
                        'tweet.text')
```

19. At this moment, you might want to check whether you're receiving the tweets and doing the parsing right. Add the following code:

```
query = timed_stream.writeStream.outputMode('append')\
        .format('console').start()
query.awaitTermination()
```

> **NOTE**
>
> This function is not complete yet. We'll add some code to it in the next step.

20. Run the file on a Terminal (macOS or Linux) or Command Prompt (Windows) or Anaconda Prompt using the following command:

```
python get_tweets.py
```

You should get a similar output to this:

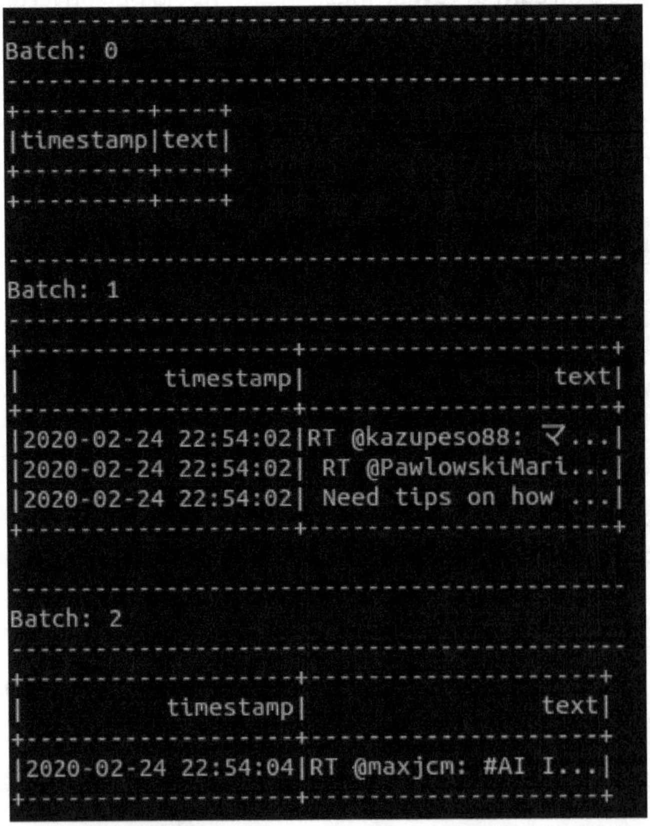

Figure 3.20: Output for the get_tweets.py file

21. If all is fine, let's remove the last two lines of Python code and continue with the windowing function. We can create a sliding window of **1 minute** and a slide of **10 seconds** with the following statement:

```
windowed = timed_stream \
    .withWatermark('timestamp', '2 seconds') \
    .groupBy(window('timestamp', '1 minute', '10 seconds'))
```

As you can see, the code also contains a watermark that ensures that we have a slack of **2 seconds** before the window evaluates. Now that we have a windowed stream, we have to specify the evaluation function of the window. In our case, this is a simple count of all the tweets in the window.

22. Enter the following code to count all the tweets in the window:

```
counts_per_window = windowed.count().orderBy('window')
```

There are two more lines to get our stream running. First, we have to specify the output mode (or sink) for the stream. In many cases, this will be a Kafka topic again, or a database table. In this exercise, we'll just output the stream to the console. The **awaitTermination()** call has to be done to signal to Spark that it should start executing the stream:

```
query = counts_per_window.writeStream.outputMode('complete')\
        .format('console').option("truncate", False).start()
query.awaitTermination()
```

Now we have created a windowed stream of tweets, where for each window of 1 minute we count the total number of tweets that come in with hashtag **#AI**. The final output will be similar to this:

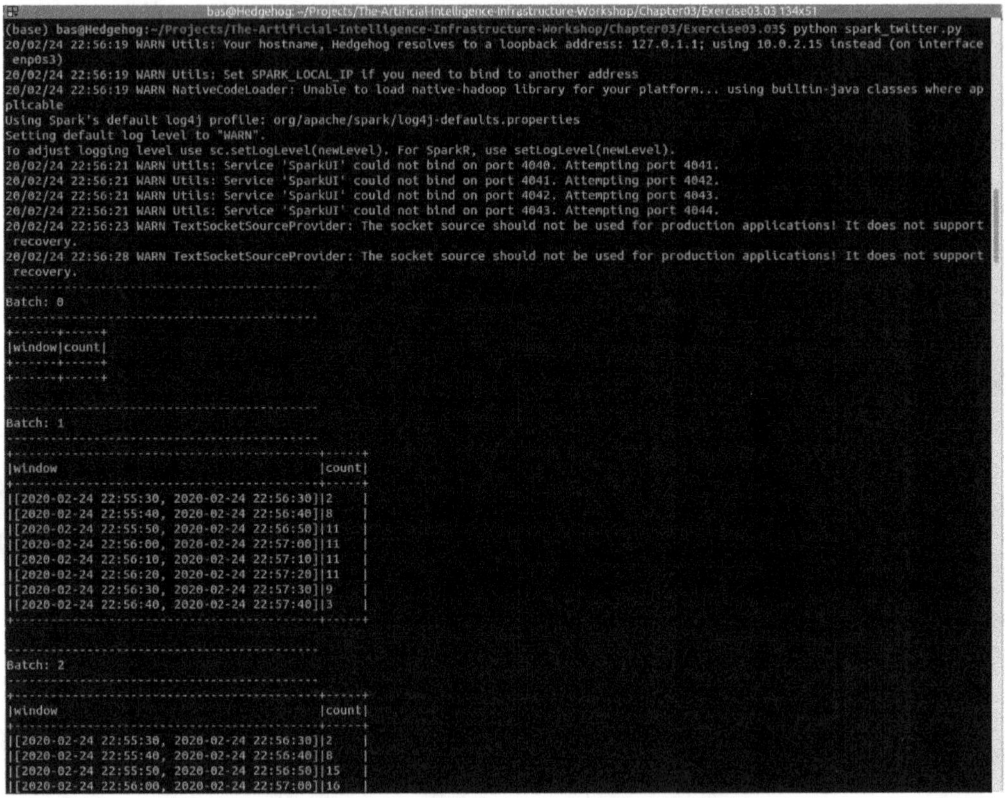

Figure 3.21: Output for the spark_twitter.py file

As you are experiencing now, your code runs in an infinite loop. The stream never ends. But, of course, the processing of the stream on your local machine can be stopped. To stop the stream processor, simply press *Ctrl + C*.

> **NOTE**
>
> To access the source code for this specific section, please refer to https://packt.live/2BYZEXE.

In this exercise, you have used Spark Structured Streaming to analyze a live Twitter feed. In the next activity, you'll continue to work with this framework to analyze the tweets and count the words in a certain time window.

ACTIVITY 3.02: COUNTING THE WORDS IN A TWITTER DATA STREAM TO DETERMINE THE TRENDING TOPICS

In this activity, you'll connect to Twitter in the same way as in *Exercise 3.03, Streaming Data Processing with Spark*. You can reuse the **get_tweets.py** or **get_tweets.ipynb** file that we have created. Only, this time, your goal is to group and count the words in the specified time window rather than the total amount of tweets. In this way, you can create an overview of the trending topics per time window.

> **NOTE**
>
> The code can be found here: https://packt.live/3iX0ODx.

Perform the following steps to complete the activity:

1. Create a new Python file and name it **spark_twitter.py**.

2. Write the Python code for the required imports and connect to a Spark cluster or a local instance with the **SparkSession.builder** object.

3. Connect to the socket on the localhost and create the raw data stream by calling the **readStream** function of a **SparkSession** object.

4. Convert the raw event strings from JSON to a useful format, namely a **text** field with the tweet text and a **timestamp** field that contains the event time. Do this by specifying the date-time format and the JSON schema and using the **from_json** PySpark function.

5. Convert the field that contains the event time to a timestamp with the **to_timestamp** function.

6. Split the text of the tweets into words by using the **explode** and **split** functions.

   Hint: the tutorial of Spark Structured Streaming contains an example of how to do this.

7. Create a tumbling window of **10 minutes** with **groupBy(window(…))**. Make sure to group the tweets in two fields: the window, and the words of the tweets.

8. Add a watermark that ensures that we have a slack of 1 minute before the window evaluates.

9. Specify the evaluation function of the window: a count of all the words in the window.

10. Send the output of the stream to the console and start executing the stream with the **awaitTermination** function.

You should get the following output:

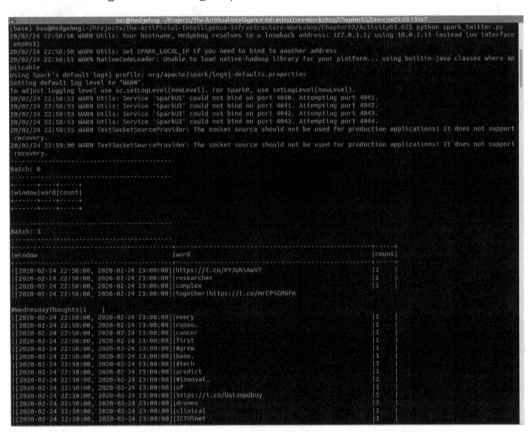

Figure 3.22: Spark structured streaming job that connects to Twitter

> **NOTE**
>
> The solution to this activity can be found on page 592.

SUMMARY

In this chapter, we have discussed many ways to prepare data for machine learning and other forms of AI. Raw data from source systems had to be transported across the data layers of a modern data lake, including a historical data archive, a set of (virtualized) analytics datasets, and a machine learning environment. There are several tools for creating such a data pipeline: simple scripts and traditional software, ETL tools, big data processing frameworks, and streaming data engines.

We have also introduced the concept of feature engineering. This is an important piece of work in any AI system, where data is prepared to be consumed by a machine learning model. Independent of the programming language and frameworks that are chosen for this, an AI team has to spend significant time writing the features and ensuring that the resulting code and binaries are well managed and deployed, together with the models themselves.

We have performed exercises and activities where we have worked with Bash scripts, Jupyter Notebooks, Spark, and finally, stream processing with live Twitter data.

In the next chapter, we will look into a less technical but very important topic for data engineering and machine learning: the ethics of AI.

4

THE ETHICS OF AI DATA STORAGE

OVERVIEW

In this chapter, you will learn about how ethics relate to **Artificial Intelligence** (**AI**) by looking at some of the largest industry scandals where AI ran up against morality. We will dive deep into several case studies, examining everything from AI being used to manipulate elections to AI displaying racial and sexist prejudices. We'll implement a simple sentiment classifier to differentiate between positive and negative words and sentences. We'll observe how this works in many cases and display the problematic biases and human stereotypes in the classifier. We'll gain hands-on experience with word embeddings and see how word embeddings can be used to represent how certain words relate to each other.

By the end of this chapter, you'll have learned how to evaluate the ethical aspects of the AI systems you build. You will know how to examine the potential consequences of using AI-based systems, and you'll know where to dive deeper to look for potential ethical problems. You'll be able to identify appropriate problems where word embeddings might be useful, and you'll be able to use Python's spaCy library for some common **natural language processing** (**NLP**) tasks.

INTRODUCTION

In the previous chapter, you learned how to prepare data using **extract**, **transform**, and **load** (**ETL**) pipelines to feed it efficiently into an AI system. In contrast, in this chapter, we'll take a break from looking at how things could be done, and we'll start asking whether they should be done. As with many new fields, AI has run up against ethical considerations. Ethics itself is always a topic that sparks controversy but combine that with a field that people often still associate with killer robots, and you're bound to find some very difficult and hotly debated topics.

Even outside of AI, robots can get into ethical trouble. For example, in 2014, the artistic group "!Mediengruppe Bitnik" created an automated trading bot that could buy random items from the so-called "dark web." The dark web is like the world wide web that most of us use every day, but you will not find its pages indexed on search engines such as Google. Instead, pages on the dark web hide behind specialized software, making sure that only specific people who are in the know can access them.

This dark web is often associated with illegal marketplaces, and people infamously trade illegal substances and weapons and run even darker marketplaces. The most widely publicized of these was known as "Silk Road," and it was shut down several times by authorities between 2013 and 2014.

!Mediengruppe Bitnik's bot was given a daily budget of $100, which was used to purchase random items and ship them to be exhibited as part of an art display. Of course, the bot's random purchases soon included an unsavory assortment of things, including illegal ecstasy tablets. In a highly publicized case, the world had to decide whether the bot itself had done something illegal or the creators of the bot had done something illegal.

After some legal debate, the police concluded that the purchases were in the name of "art" and were therefore not illegal, mainly because the artists never intended to sell or consume the illegal substance. The robot (the hardware running the AI) and its purchases were temporarily seized by the police but eventually returned. An exception was the MDMA tablets the robot had bought, which were destroyed. Overall, it was decided that it is acceptable to push boundaries in the name of art.

But what about the hundreds of companies using AI in similarly controversial ways for non-artistic purposes? There have been several high-profile cases over the past several years, and in this chapter, we will go through several case studies, examining some of the highest-profile cases where ethics and AI became a talking point.

Specifically, we will look at the following:

- **The Cambridge Analytica scandal**: This was where a private company helped a range of clients, from businesses to politicians, achieve their goals through AI and gained access to a vast trove of highly personal data via Facebook. It is widely believed that this was used to influence democratic elections, potentially bringing into question the validity of global democratic processes.

- **Amazon's AI recruiting tool**: This was where Amazon experimented with using AI to help it decide who to hire. It turned out that the algorithm displayed strong biases, specifically preferring male candidates over female ones. Amazon claims that it never used this tool for hiring purposes, but the fact that such a tool can exist makes us question whether other companies might have blindly depended on similar automated decisions without noticing built-in biases.

- **COMPAS software and racist AI**: This was the case of an algorithm that was used by US courts to predict whether prisoners are likely to commit further crimes being found to be biased against black people. A report in 2016 by ProPublica found that the algorithm made mistakes regarding black people twice as often as for whites, often incorrectly assessing black people as having a high "risk of re-offending."

The overall themes of these case studies are causing political harm at a country or global level via privacy invasion, causing social and financial harm through gender biases, and causing direct harm to individuals through the miscarriage of justice due to racial biases:

Abstract Political Harm   Social and Financial Harm   Miscarriage of Justice
Privacy   **Sexism**   **Racism**

Figure 4.1: AI can display political, sexist, and racist biases

Let's get started by looking at a large, recent scandal: Cambridge Analytica.

CASE STUDY 1: CAMBRIDGE ANALYTICA

Cambridge Analytica Ltd (**CA**) was in the headlines of all major news publications for months after it became apparent that they, as a company, had used personal data to drive mass targeted campaigns, aiming to influence political elections. But how exactly did they use data? How did they use AI? And how does this relate to ethics?

Let's take a look at how Cambridge Analytica broke trust, and probably the law, through the following actions:

- Mass storage of personal data

- Invading people's privacy through mass collection and aggregation of data that would have been harmless in isolation

- Using AI to categorize the resulting mass of personal data and to identify specific groups of people who could be influenced in specific ways

- Using AI further to target specific advertisements at these groups to influence the way that they voted in elections

There is still much controversy, debate, and investigation concerning exactly what Cambridge Analytica did, how much they did it, and how effective it was. Some people claim that the story was overblown and that targeted advertising does not work that well in any case.

However, it certainly seems probable that there was a large and concerted effort to use data and AI for what many people would regard as "evil"—subverting political systems for financial gain.

Let's take a look at exactly how this might have worked.

With over 2.5 billion monthly active users, Facebook is perhaps the largest centralized store of personal data ever collected. A lot of this data is innocuous enough: someone's gender, their date of birth, where they live, and what they are interested in. More interestingly, Facebook also stores who everyone is friends with, how often they interact with each other, what brands and organizations they "like," what events they go to, and much more besides.

Many third-party companies gain access to subsections of this data via Facebook games, quizzes, and applications. These are small pieces of software that interact with your Facebook account in some way and require some of the personal data to work. Due to Facebook's popularity, these pieces of software have sometimes been political, with quizzes titled "Which political party are you most aligned with?" and similar, and this has often been useful for statistical and scientific research.

The value of this data also presents a business opportunity, though. Politicians need votes, and they spend large amounts of time and money trying to convince people to vote for them. Any politician or political party can broadly divide a population up into three groups:

- Strong supporters who will almost definitely vote for them, no matter what

- Strong opposers, who will almost definitely not vote for them, no matter what

- Moderates, who will follow events and opinions closely in the months and weeks leading up to an election and choose whether or not to vote for them based on these factors:

Figure 4.2: Strong supporters, strong opposers, and moderates

It is the last group, the moderates, who are the most interesting for politicians to spend time and effort trying to persuade. And one way to do this is through targeted digital advertising. Unlike advertising via traditional media, in which a politician pays to show the same advert to all readers, digital advertising allows for targeted advertising. This means that politicians can, in theory, show adverts only to the moderates, or people who might vote either way.

Additionally, they can further subdivide the moderates' group into different categories and show them each advert that would be more meaningful to their own beliefs. For example, young women with an interest in feminism might be a subgroup that can be influenced by hearing more about the politicians' plans to close the gender wage gap. Older males who have previously served in the army might be influenced by hearing more about politicians' plans to alleviate the financial hardship that many veterans face. Instead of creating a single campaign to try to find common interests in all groups, politicians can easily put on a "different face" when targeting each group.

It is not proven, but many now accept that voters can be manipulated by being shown targeted adverts relating to topics about which they are specifically sensitive. If this is true, the consequences are worrying. Do we want to live within a political system where the politicians with a digital marketing team and the best access to personal data are those who are voted into power?

Cambridge Analytica, by most accounts, created their innocuous-looking political quiz. To take the quiz, Facebook users had to allow the quiz software access to their private information. On top of this, Cambridge Analytica used a specific weakness in Facebook's Graph API to retrieve not only more personal information than might be expected from people who took the quiz, but also the same information from friends of people who took the quiz.

Using this treasure trove of personal data, Cambridge Analytica was able to use AI to not only categorize people into specific political affiliation groups and personality subtypes but also to effectively predict what kind of advertisement would sway these people to vote for their clients.

Knowing someone's birthday would not have been helpful for this kind of influence, and people know that their birthdays are not really private information and willingly give them out.

Knowing someone's location would similarly not have been helpful, and again, people are not guarded about sharing where they live.

The same goes for other individual bits of information, such as political leanings, favorite brands, events, and acquaintances. Bits of data on their own are not that interesting, but with enough data, and enough understanding of how the bits of data are connected, we start to derive information:

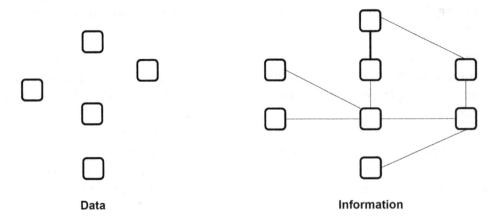

Data **Information**

Figure 4.3: Data is only valuable when it is connected to other data in a meaningful way

But gathering all of this information about one person starts to paint a very specific picture. Imagine we know the following facts about an anonymous person:

- He is a 30-year-old male Facebook user.

- He has been to a Trump rally and a lecture by Bernie Sanders.

- He lives in the US in a state known to be a "battleground" or "swing" state.

- He has a large network of friends, many of whom match a similar profile.

- He often posts statuses that seem to be persuasive or argumentative in nature.

- His posts get an above-average number of likes and comments.

This person might be a prime candidate for targeted advertising. He is probably politically undecided, looking at arguments from both sides; in a position to influence many other people to match his views; and is located in a state that has better than average chances of being influential in the final election outcome.

Knowing **a lot** of basic information about **some** people or knowing **some** information about **a lot** of people wouldn't help Cambridge Analytica or their political clients. But knowing **a lot** of basic information about **a lot** of people is powerful. By looking at these facts and how they connect, it is possible to build a huge "knowledge graph" of information. Such a graph links people to facts, beliefs, and each other, and it can be used to manipulate them and therefore exert control over the world. We can use this information in combination with AI systems to help target specific messages at specific people:

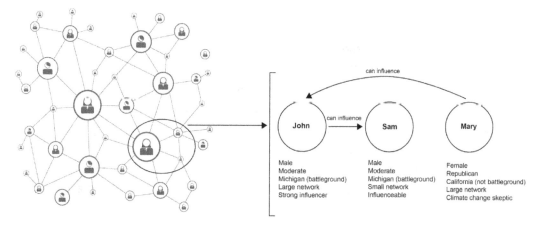

Figure 4.4: A knowledge graph – a huge collection of facts and connections
can be used to train AI

How much data did Cambridge Analytica collect? While exact figures are hard to prove, the CEO claimed in 2016 that they had "four or five thousand" data points on every adult in the United States. Related specifically to the Facebook scandal, Cambridge Analytica claimed that they collected data from only 30 million Facebook profiles, but it was later confirmed that this number was closer to 87 million, with a large majority of these being from the USA.

Facebook lost a collective $100 billion due to a drop in share price after the scandal emerged in 2018, and further received a $5 billion fine from the Federal Trade Commission.

These numbers are scary. Data privacy laws are still catching up with what is possible to do through global aggregation and analysis of this type of data, so where laws fall short, we have to look to ethics to keep people and corporations in check.

One major ethical problem with this kind of use of personal data is that it relies on ignorance. Many people who use Facebook and the third-party software add-ons, such as the Cambridge Analytica quiz, do not realize that this kind of use of their data is possible. Therefore, while they may have agreed to let the quiz app access their data, this is not an **informed choice**. Imagine the following scenario based on the preceding knowledge graph:

- John is an active user of Facebook. He does not care about privacy and does not understand why his data might need to be kept private. He installs many apps and quizzes on Facebook (including the Cambridge Analytica one) and clicks "agree" on the notifications about the terms and conditions and privacy policy without reading them.

- John is Facebook friends with Mary. Mary is very privacy-aware. All of the personal information associated with her Facebook profile is set so that only her friends can see it. She understands the implications of making personal data public, and she never installs any apps or quizzes on Facebook.

- Because John installed the Cambridge Analytica quiz, it can use his account on behalf of him and can see everything he can see. Therefore, Cambridge Analytica can now see all of Mary's data too, even though she took measures to prevent this. While in this case, Cambridge Analytica might have obtained legal consent (via John), many people would argue that this is an ethical privacy violation from Mary's point of view.

Perhaps a more serious ethical problem is that this kind of use of data has the potential to undermine (or, depending on who you talk to, has already undermined) our entire global political system. Most countries today rely on some form of democracy—a form of government that strongly depends on all citizens having **unbiased** access to accurate information. If the political side with the larger checkbook and the better marketing team can influence specific groups of people by overloading them with evocative one-sided information, democracy as a system ceases to work as it is intended to work.

SUMMARY AND TAKEAWAYS

In this case study, we saw how Cambridge Analytica built a system that used mass data storage and AI to collect personal information from a lot of people without their informed consent. It further allegedly used this information to present potential voters with biased information, specifically targeting certain groups and individuals with adverts that were more likely to influence them.

This has large implications not only for our right to personal privacy but also for our political systems, which rely on transparent practices and equal access to information.

What can we learn from this while building our own AI systems? The most important factor is to think carefully about each piece of personal data that we process or store. Ask yourself whether your system must store and use every piece of data. Ask yourself how the data might be used, not only by you but if distributed to third parties. And ask yourself how it might be combined with other information, and whether the combined data might have ethical implications.

In cases where personal data is necessary and cannot be anonymized, it is important to restrict access as far as possible. Make sure that only the people who need the data have access to it and that there are processes in place to prevent backups or other copies lying around or being distributed.

Let's take a look at another case, where AI was almost used to reinforce gender biases in the hiring processes for Amazon engineers.

CASE STUDY 2: AMAZON'S AI RECRUITING TOOL

Every day, millions of resumes are evaluated. Hiring managers and HR representatives spend an average of 6 seconds looking at a CV before throwing it out or advancing it to the next round.

This is not an accurate process. Many bad candidates are pushed through, and many good ones are discarded based on the CV. But the hiring process, from the company's perspective, doesn't have to be perfect. It only has to be good enough to get at least some qualified candidates to the final stage; it does not matter to the company how many good candidates are skipped along the way.

But companies do care about efficiency. Deciding which resumes represent good candidates is a classification task and, as we saw in *Chapter 1, Data Storage Fundamentals*, computers are very good at many classification tasks.

Could computers pick good job candidates over bad ones?

Amazon, in another infamous AI scandal, decided to try it. Anonymous leaks of early results painted a pretty grim picture, and Amazon, embarrassed, threw out the entire system.

But that's not to say that other companies are not automating decisions in their hiring pipelines, probably behind closed doors.

Let's take a look at how bias and stereotypes can make AI behave in unexpected ways, and how that can bring harm to people.

IMBALANCED TRAINING SETS

In *Chapter 1, Data Storage Fundamentals*, we trained our AI by giving it a dataset split into two: 50% of the dataset contained clickbait headlines and 50% contained normal ones. This is called a **balanced** dataset and having a classification problem where the classes are evenly split makes a lot of things easier in machine learning.

However, it's rare to be this lucky. Often, we find that some classes or attributes are much rarer than others. If we were to use the population of humans in the world as a dataset and try to predict something such as life expectancy, one attribute we might look at is the nationality of a person. We would soon realize that we had more people from China than we did from Luxembourg, and we would have to account for that in the results.

Unlike country populations, gender splits are more often balanced. For example, most countries have an approximately even split of women and men. So, can we assume balanced datasets when specifically looking at gender?

Unfortunately, it's not that simple. Despite overall balanced genders, only about 20% of software engineers at large technology companies such as Amazon and Google are women, and fewer women apply for job openings at these companies. Any machine learning algorithm that tries to meaningfully classify people and uses gender as an attribute must actively work to avoid any problems that this could cause.

Amazon has not published details on exactly how the hiring algorithm was designed to work, but we can make some assumptions. If we managed to get the resumes of 1,000 people who work at Amazon, we could take those as "good" resumes: the people who have them are the kind of people that Amazon is looking for. We could then take another 1,000 resumes at random that were rejected by human recruiters and call those "bad" resumes. We could use this dataset to train a classifier to differentiate between "good" and "bad" resumes.

Our dataset is balanced in the most basic sense: it has the same number of "good" and "bad" resumes:

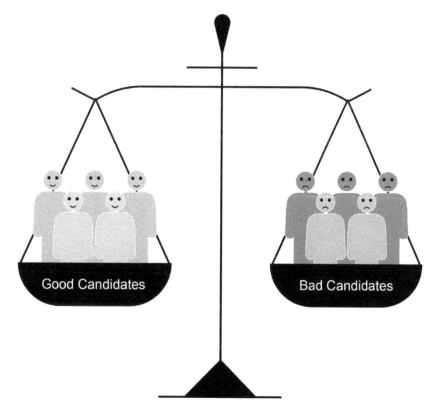

Figure 4.5: A balanced dataset – we have an equal number of good and bad candidates

However, it is highly imbalanced in a more complex sense: if we look within each class of **good** and **bad**, we might find a large difference in the gender of each class. Again, Amazon did not publish details of the algorithm or datasets, so we can only assume. But the distribution might have been something similar to the following:

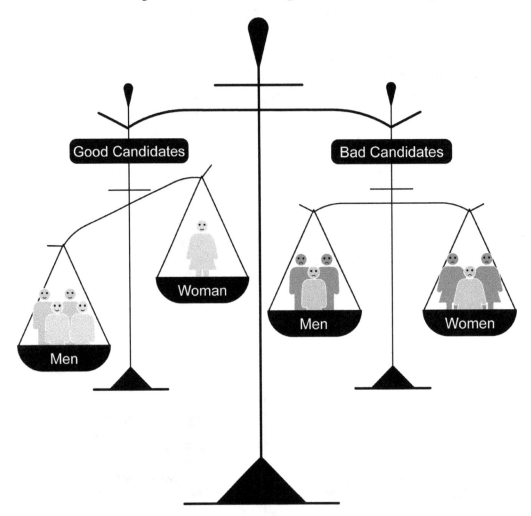

Figure 4.6: A partially balanced dataset – the quality of candidates is balanced, but gender is not balanced within the "good" class

Overall, the dataset was balanced, but there were more men in the "good candidates" class. As we already know, 80% of the "good" resumes are from male engineers while 20% are from female engineers who are working in Amazon. On the contrary, there was a more equal distribution in the "bad applicants" class.

If we explicitly use gender as one of the dimensions, the algorithm will quickly learn to favor male applicants, reinforcing biases that already plague Amazon's workforce.

It is unlikely that Amazon's algorithm would have explicitly used gender as an attribute to use in resume classification, but unfortunately, that does not stop a shrewd AI from picking up on the existing bias. For example, the resumes of many rejected candidates might have contained phrases such as "Women's chess club captain," or "Women's hockey coach," leading Amazon's algorithm to unfairly associate these words with bad candidates. On the other hand, men statistically use words such as "executed" and "captured" more often on their resumes, leading Amazon's algorithm to unfairly prefer candidates who use these words.

By internal accounts, phrases such as "Women's Chess Club Captain" were taken as a negative indicator, while words such as "executed" and "captured," which are found more often on men's resumes, were taken as positive.

No one explicitly asked the algorithm to do this, but they asked the algorithm to follow an example set by humans, and unfortunately, that example is often flawed.

In the previous case, there was a strong pattern, by many accounts, of Facebook and Cambridge Analytica claiming that they had not done anything wrong and refusing to accept accountability.

In this case, Amazon saw the problem and pulled the AI system at an experimental stage, before it could cause any harm in the real world, and this is already a key learning point for this.

SUMMARY AND TAKEAWAYS

In *Chapter 1, Data Storage Fundamentals*, we saw how good AI was at performing classification tasks—sorting data into two or more classes based on a deep analysis of many attributes. This works great for things such as spam classification and detecting cancer in X-rays, but can we use it to classify people? We can, but it's ethically very problematic. The way AI performs classification is by looking to exploit any bias in the data that it can, and many times the data is already affected by human biases, which AI is all too ready to take advantage of and reinforce.

Before building an AI system, we should ask whether it might be susceptible to bias through the following:

- Imbalances in the training dataset

- Existing human biases in the dataset

And we should also be wary of variables being modeled "by proxy." Sometimes, we do not want the algorithm to use specific aspects of the data such as gender, and so we remove it. However, it might still be possible for the algorithm to recalculate the missing pieces, based on what we left.

In the next case study, we will look at how AI is used in criminal justice.

CASE STUDY 3: COMPAS SOFTWARE

We have now seen two different ways that AI can cause harm if ethics is not considered. In the Cambridge Analytica case, the harm was abstract. No one was directly harmed, but we could argue that its results were a global harm through damage to our political systems.

In the second case, we saw that Amazon's hiring algorithm could have hurt people more directly: by rejecting them unfairly for jobs. But some people still might argue that this is not harmful enough to be worried. Some might argue that sexism is something that is a human problem and that the reinforcement of this is the price to pay for the benefits brought through AI efficiency.

In this case, we'll see how AI can cause direct and long-lasting harm to specific individuals by keeping them in prison. Being rejected from a job based on who you are is a terrible thing, but hopefully, you will be able to find another job. If you are kept in prison unfairly for 20 years, there is barely any compensation that can make up for that.

The US government uses predictive algorithms to decide who to keep in prison. These are based on trying to predict a criminal who might commit crimes again. It makes sense that we don't want to let people out of prison if they are likely to re-offend. But ethical considerations get complicated when we get a computer to help us make that decision.

The **Correctional Offender Management Profiling for Alternative Sanctions(COMPAS)** software used within the US judicial system is an example of exactly this. You give it all the details of a specific prisoner, and it spits out how likely that prisoner is to re-offend. It is not clear whether race is explicitly given as a variable to the COMPAS system, but this does not affect the result. As we saw in the case of Amazon, variables can be modeled by proxy. These are known as latent variables. For example, by looking at a combination of where a prisoner is originally from, who their relatives are, and which area they live in, the algorithm might be able to infer the prisoner's race. And, again, as we saw with the Amazon case, if a bias already exists in society, the algorithm could reinforce this bias.

The creators of COMPAS did try to control this in their software, and it seemed to predict equally that black and white defendants are likely to re-offend. At first glance, then, it looks like the algorithm has no racial prejudice. However, looking retrospectively at when the algorithm makes mistakes reveals a very different picture.

Classification algorithms can be wrong in two ways: they can have "false positives" and "false negatives." In COMPAS's case, they would be as follows:

- **False positive**: Predict that someone will re-offend, but they don't

- **False negative**: Predict that someone won't re-offend, but they do

Both cases are bad. In the first case, you keep someone in prison who shouldn't be there, and in the second case, you let someone out of prison who causes further harm to society:

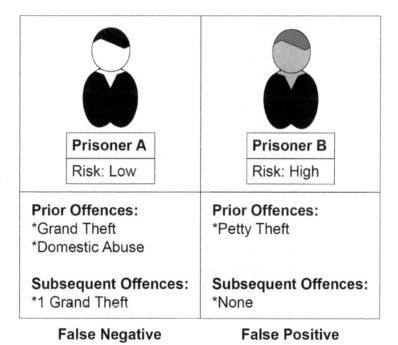

False Negative   **False Positive**

Figure 4.7: COMPAS predictions representation

Even though COMPAS predictions on who will re-offend are split equally between black and white prisoners, it predicts almost twice as many **false positives** for black prisoners than for white prisoners. Furthermore, it predicts almost twice as many **false negatives** for white prisoners than for black prisoners.

This means that the algorithm often recommends keeping low-risk black defendants behind bars, and often recommends that high-risk white defendants should be released. It is acceptable and expected that the algorithm will sometimes make mistakes. However, it is hugely problematic to continue using software for something as important as keeping people in prison if it allows miscarriage of justice. This is especially true if it appears that this miscarriage happens through racism.

SUMMARY AND TAKEAWAYS

Using biased algorithms to assess who should be kept in prison has the potential to cause (and quite possibly does cause) concrete and direct ethical harm. This is in contrast to algorithms that violate privacy, which cause harm on a more abstract level.

We have seen how AI can cause harm intentionally or unintentionally, and how existing human biases, along with imbalanced datasets, can contribute toward the latter.

In this case, we saw that the COMPAS algorithm specifically seemed to pass initial bias tests. However, with more time and with analysis from outside researchers, we noticed that a few points had not been initially considered, some of which are as follows:

- These tests were not deep enough. They looked only at summary statistics on when the algorithm made mistakes, but not at the details of these mistakes.

- It took time and expertise to expose the details around how false positives were biased against black people.

Even if the problem were now fixed, it would not be acceptable for the people who fell victim to the false positives in the interim, and it is unfortunately not clear whether COMPAS changed their algorithm to account for the research. If they did make changes, it is difficult to verify that these changes fixed the problem.

Next, let's look at some more practical examples of how bias can be found in machine learning models.

FINDING BUILT-IN BIAS IN MACHINE LEARNING MODELS

We've seen how large and complicated systems can create ethical harm, and we've seen a strong theme of prejudice. In 2017, Robyn Speer wrote the article *"How to make a racist AI without really trying."* In this article, Speer showed how word vectors, a form of vectorized text, could be combined with a basic sentiment classifier. In Speer's examples, phrases that should be evaluated with a similar sentiment such as "Mexican food" and "Italian food" are given very different sentiment ratings. Because the word vectors are built from large corpora of human-written text, there is a negative sentiment toward specific racial groups, and this is apparent even when looking at phrases concerning food instead of people.

Let's take a deeper look at how word vectors work.

While the TF-IDF vectors we looked at in *Chapter 1, Data Storage Fundamentals*, are one way to vectorize words, these are character-based vectors. Using that method, there is no relation between "cat" and "kitten," as the words look very different. Word embeddings rely on a more complicated algorithm to convert text into vectors, but it has some of the advantages of structures such as TF-IDF. Word embeddings are trained using huge amounts of text data, and they look at what words often occur in similar contexts. Each word is represented as a vector with n dimensions, which can be thought of as a point in multidimensional space.

Just as a vector [3,4] might represent a point in two-dimensional space along an x- and y-axis, a word represented by a vector with 96 dimensions can be thought of as a point in 96-dimensional space. The word "machine" might look as follows:

```
array([ 1.1362579 ,  0.6309743 , -0.3325829 ,  0.37675968,  1.6817904 ,
        0.23466611,  4.135811  ,  0.5422696 ,  2.2826018 ,  2.2008624 ,
        3.8998048 ,  0.859442  ,  2.7883368 , -0.21905822, -1.5583985 ,
       -2.5670037 , -1.5083698 , -0.11448759, -2.9931982 , -0.9069863 ,
        1.3111084 , -1.1638764 , -0.97565454, -1.5257355 ,  0.63322806,
       -2.5783947 ,  1.2363327 , -1.8938144 ,  1.5853525 ,  0.31022626,
        1.7970477 ,  0.47506928, -1.6084212 ,  1.684785  ,  1.5389957 ,
        0.1386419 , -0.7013227 ,  0.00961751, -1.7244251 , -0.7209445 ,
        3.187506  ,  2.1730027 , -1.9845129 , -3.85418   , -0.22612506,
       -0.01370215,  0.15269166, -0.6928604 , -0.63267624,  2.1143537 ,
       -0.8674346 , -1.9403771 , -1.6406736 , -2.6019177 , -1.0126046 ,
        1.6190633 ,  1.2659423 ,  0.05900469,  1.4160635 , -0.07090846,
        0.2659979 , -1.8755596 , -0.05048826,  0.6290233 , -2.322978  ,
       -1.944304  ,  0.9604355 , -2.9318776 , -3.2819483 ,  3.4064205 ,
       -2.6280713 ,  0.39027372,  1.645208  ,  4.37038   , -3.3575606 ,
       -0.42797962,  1.6369323 ,  0.28004026, -2.5933466 ,  3.7668483 ,
        0.22324765, -0.33607954, -1.2183836 , -2.4340582 ,  0.3160853 ,
        3.938485  , -0.5823619 ,  0.559906  , -0.7645585 , -2.1697512 ,
        1.4418402 ,  1.4624262 , -1.1901258 ,  2.5993197 , -0.20187286,
        0.22061825], dtype=float32)
```

Figure 4.8: The word "machine" as a word vector

It's hard thinking in terms of multidimensional space but remember that many of the rules that apply in two-dimensional space are valid in 300-dimensional space too. A recent discovery was that words that are semantically similar to "cat" and "kitten" can be close to each other in our multi-dimensional space, while words that are not related will be far away.

We will use the spaCy library to easily convert words into word vector representations and analyze these words by looking at their distances from other words.

As a very brief introduction, we can see that word embeddings, which are used in spaCy's pretrained models, are very powerful out of the box. In the following example, we're loading spaCy and comparing a few words using the **similarity** function, which simply looks at how close together words are in the multi-dimensional vector space described previously:

```
>>> import spacy
>>> nlp = spacy.load("en_core_web_lg")
>>> nlp("cat").similarity(nlp("kitten"))
0.821555342995594
>>> nlp("cat").similarity(nlp("dog"))
0.8016854705531046
>>> nlp("cat").similarity(nlp("cow"))
0.4746069221481621
>>> nlp("cat").similarity(nlp("coffee"))
0.2722319630373339
```

Cat and kitten are 82% similar, while cat and dog are 80% similar. Remember, word embeddings are based on context, so cat and dog might have a higher similarity than expected simply because the two words are often used in the same context; for example, "yesterday I fed my dog" and "yesterday I fed my cat" are both valid and common sentences.

There is some similarity between "cat" and "cow" (they are both animals), but very little similarity between "cat" and "coffee."

Therefore, even without teaching a computer what "good" or "bad" are (or "positive" or "negative"), we can build a very basic sentiment classifier by seeing how close words and phrases are to "good" and "bad".

spaCy is useful because we can use it, as humans, to operate directly on strings such as "cat." Internally, spaCy will convert this into a vectorized representation that we can inspect by using the `.vector` attribute, as shown in the following figure:

```
>>> nlp("cat").vector
array([-0.15067  , -0.024468 , -0.23368  , -0.23378  , -0.18382  ,
        0.32711  , -0.22084  , -0.28777  ,  0.12759  ,  1.1656   ,
       -0.64163  , -0.098455 , -0.62397  ,  0.010431 , -0.25653  ,
        0.31799  ,  0.037779 ,  1.1904   , -0.17714  , -0.2595   ,
       -0.31461  ,  0.038825 , -0.15713  , -0.13484  ,  0.36936  ,
       -0.30562  , -0.40619  , -0.38965  ,  0.3686   ,  0.013963 ,
```

Figure 4.9: Looking at the vector for "cat" using spaCy

We can even check the total vectors in which SpaCy has converted the string using the **vector.size** attribute, as shown in the following figure:

```
>>> nlp("cat").vector.size
300
```

EXERCISE 4.01: OBSERVING PREJUDICES AND BIASES IN WORD EMBEDDINGS

In this exercise, we're going to build a prejudiced AI using sentiment analysis. You'll write some Python code, use the spaCy library, and get a hands-on feel for how word vectors (also known as word embeddings) work.

By the end of this exercise, you will have built a very crude form of sentiment classifier, but it will be enough to see some biases and prejudices that are to be expected whenever we start looking at large amounts of text data.

Before proceeding to the exercise, we need to install the spaCy library, along with large pre-trained models, in the local dev environment. Please follow the instructions in the *Preface* to install it.

Perform the following steps to complete this exercise:

1. Create a directory called **Chapter04** for all the exercises of this chapter. In the **Chapter04** directory, create the **Exercise04.01** directory to store the files for this exercise.

2. Open your Terminal (macOS or Linux) or Command Prompt (Windows), navigate to the **Chapter04** directory, and type **jupyter notebook**. The Jupyter Notebook should look as follows:

Figure 4.10: The Chapter04 directory in the Jupyter Notebook

3. Select the **Exercise04.01** directory and click **New** -> **Python3** to create a new Python 3 notebook:

Figure 4.11: The Exercise04.02 directory in the Jupyter Notebook

4. Import spaCy and load the pre-trained model:

```
import spacy
nlp = spacy.load('en_core_web_lg')
```

Note that the model file is large and that it will probably take quite a few seconds to run this, depending on your machine.

5. Parse **"good"** and **"bad"** into variables to see how close other words are to them, as shown in the following code:

```
good = nlp("good")
bad = nlp("bad")
```

6. Compare the word **awful** to the words **good** and **bad**, as shown in the following code:

```
print(nlp("awful").similarity(bad))
print(nlp("awful").similarity(good))
```

You should get the following output:

```
0.7721672894451931
0.551066389568291
```

The output shows that **awful** is 77% similar to **bad** and 55% similar to **good**. However, **awful** is fairly similar to both concepts, as it used in the same contexts. Let's try a more abstract concept: night and day. Traditionally, "day" is used in more positive settings (for example, "a new day") and night in more negative ones (for example, "For the night is dark and full of terrors").

7. Compare **day** to **good** and **bad**, as shown in the following code:

```
print(nlp("day").similarity(bad))
print(nlp("day").similarity(good))
```

You should get the following output:

```
0.4455448306231779
0.5082143964475753
```

8. Compare **night** to **good** and **bad**, as shown in the following code:

```
print(nlp("night").similarity(bad))
print(nlp("night").similarity(good))
```

You should get the following output:

```
0.45386439630840125
0.4425397729689741
```

Note that both words are less like both **good** and **bad** than the word **awful**. When we look at sentiment using this crude definition, we aren't interested in the absolute distances between these concepts, but rather we are interested in the relative distance. Is a particular word *closer* to **good** than it is to **bad** or vice versa? Let's define a basic function to allow us to calculate this.

9. Define a **polarity** function to calculate whether a word is closer to **good** or closer to **bad**, as shown in the following code:

```
def polarity_good_vs_bad(word):
    """Returns a positive number if a word
    is closer to good than it is to bad,
    or a negative number if vice versa
    IN: word (str): the word to compare
    OUT: diff (float): positive if the word
    is closer to good, otherwise negative
    """

    good = nlp("good")
    bad = nlp("bad")
    word = nlp(word)
    if word and word.vector_norm:
        sim_good = word.similarity(good)
        sim_bad = word.similarity(bad)
        diff = sim_good - sim_bad
        diff = round(diff * 100, 2)
        return diff
    else:
        return None
```

The preceding code takes in a string (word) as an argument. It creates vector representations based on the pre-trained vectors that ship with spaCy for three words: **good**, **bad**, and the passed-in word. It then looks at the distance in vector space between that passed-in word and the **good** and **bad** vectors and calculates the difference between these. Therefore, a closer (more similar) word to **good** will have a positive value and a word that is closer (more similar) to **bad** will have a negative value.

Note that the differences are often quite small, so we multiply by **100** to magnify them and make them more interpretable to humans.

10. Now, we need to create a function to display the polarity of words which have opposite meanings :

```
def contrast_pairs(pairs):
    for pair in pairs:
        pos_word = pair[0]
        neg_word = pair[1]
```

```
pos_score = polarity_good_vs_bad(pos_word)
neg_score = polarity_good_vs_bad(neg_word)
print(f"{pos_word}({pos_score}): \
{neg_word}({neg_score})")
```

11. And now, we need to test our polarity function with words of opposite meanings:

```
word_pairs_neutral = [
('day', 'night'),
('light', 'dark'),
('happy', 'sad'),
('love', 'hate'),
('strong', 'weak'),
('healthy', 'sick'),
('free','captive'),
('high', 'low')
]


contrast_pairs(word_pairs_neutral)
```

You should get the following output:

```
day(6.27): night(-1.13)
light(9.81): dark(-2.73)
happy(12.52): sad(-12.81)
love(11.95): hate(-19.3)
strong(18.09): weak(-10.16)
healthy(15.04): sick(-14.86)
peace(10.31): war(-9.9)
free(5.13): captive(-4.27)
high(8.43): low(-0.3)
```

Figure 4.12: Comparing the scores of positive and negative words

As expected, some opposites such as **healthy** and **sick** show a very strong difference in polarity, while others such as **day** and **night** or **high** and **low** are less obvious, but still have some sentiment attached to them. Let's see what happens if we look at word pairs that are more closely associated with current and historical prejudices.

12. Compare polarity scores for words often associated with **prejudice**, as shown in the following code:

```
word_pairs_prejudice = [('white', 'black'),
('christian', 'jew'), ('christian', 'muslim'),
('christian', 'atheist'), ('man', 'woman'),
('king', 'queen'), ('citizen', 'immigrant'),
('resident', 'migrant'), ('rich', 'poor'),
('engineer', 'janitor'), ('young', 'old'),
('pizza', 'sprout'), ('native', 'foreigner'),
('italian', 'iranian'), ('swiss', 'mexican'),
]


contrast_pairs(word_pairs_prejudice)
```

13. You should get the following output:

```
white(2.23): black(-1.9)
christian(0.15): jew(-8.32)
christian(0.15): muslim(-4.78)
christian(0.15): atheist(-3.86)
man(3.63): woman(-1.25)
king(0.81): queen(-5.11)
citizen(3.89): immigrant(-5.87)
resident(2.17): migrant(-0.19)
rich(11.43): poor(-10.06)
engineer(4.23): janitor(-7.97)
young(5.25): old(-2.84)
native(4.86): foreigner(-3.7)
italian(4.07): iranian(-5.07)
swiss(4.55): mexican(-3.53)
```

Figure 4.13: Comparison of scores for word pairs associated with prejudices

NOTE

To access the source code for this specific section, please refer to https://packt.live/3elL7Ch.

While this is a small set of examples and does not prove anything, we can see that there is likely some kind of latent prejudice in our pre-trained model. We can see some strong elements of sexism (**man** is better than the **woman**), racism (**white** is better than **black**), classism (**rich** is better than **poor**), and religious prejudice (**Christian** is better than **jew**, **Muslim**, and **atheist**).

Remember that the word vectors are created by using large amounts of human-created training data. Although spaCy changes the data sources from time to time, in the examples shown here, we used the default built-in spaCy model, which is trained on the OntoNotes dataset. This is a collection built from telephone conversations, newswire, newsgroups, broadcast news, broadcast conversation, and weblogs, and we can easily imagine how existing human prejudices might be represented in these datasets.

With that, we have seen how our sentiment classifier performed in this toy example, where we made up some data. We'll try it with a proper dataset in the next exercise.

EXERCISE 4.02: TESTING OUR SENTIMENT CLASSIFIER ON MOVIE REVIEWS

Sentiment analysis is a huge field on its own, and it is actively evolving. There are many very sophisticated sentiment models, but the one we built in the previous exercise is very crude indeed. In this exercise, we'll see how the classifier performs on the common task of discriminating between movie reviews. Reviewers leave positive or negative text associated with a star rating between 1 and 10 on internet sites such as IMDb. If our algorithm knows what kind of words are "good" and which ones are "bad," it should be able to tell the difference between good and bad movie reviews without using the star rating.

We'll use the **aclImdb** dataset of 100k movie reviews from IMDb, 50k each for training and testing. Each dataset has 25k positive reviews and 25k negative ones, so this is a larger dataset than our headlines one. The dataset can be found in our GitHub repository at the following location: https://packt.live/2C72sBN.

You need to download the **aclImdb** folder from the GitHub repository.

Dataset Citation: Andrew L. Maas, Raymond E. Daly, Peter T. Pham, Dan Huang, Andrew Y. Ng, and Christopher Potts. (2011). Learning Word Vectors for Sentiment Analysis. The 49th Annual Meeting of the Association for Computational Linguistics (ACL 2011).

Perform the following steps to complete this exercise:

1. Create the **Data** and **Exercise04.02** directories in the **Chapter04** directory to store the files for this exercise.

2. Move the **aclImdb** folder to the **Data** directory.

3. Open your Terminal (macOS or Linux) or Command Prompt (Windows), navigate to the **Chapter04** directory, and type **jupyter notebook**.

4. In the Jupyter Notebook, click the **Exercise04.02** directory and create a new notebook file with the **Python3** kernel.

5. Import the same libraries as before, along with **matplotlib**, so that we can do some plotting, as shown in the following code:

```
%matplotlib inline

import spacy
import os
import numpy
from matplotlib import pyplot as plt
from statistics import mean, median
```

In this code, we imported the libraries that we need. Note that we use **%matplotlib inline** right at the top of the notebook file to make our plots display correctly.

6. Load **spacy** and define our **polarity** function, as shown in the following code:

```
nlp = spacy.load('en_core_web_lg')

def polarity_good_vs_bad(word):
    """Returns a positive number if a word
is closer to good then it is too bad,
or a negative number if vice versa
    IN: word (str): the word to compare
    OUT: diff (float): positive if the
word is closer to good, otherwise negative
    """

    good = nlp("good")
```

```
    bad = nlp("bad")
    word = nlp(word)
    if word and word.vector_norm:
        sim_good = word.similarity(good)
        sim_bad = word.similarity(bad)
        diff = sim_good - sim_bad
        diff = round(diff * 100, 2)
        return diff
    else:
        return None
```

As in the previous exercise, the preceding code takes in a string (**word**) as an argument. It creates vector representations based on the pre-trained vectors that ship with spaCy for three words: **good**, **bad**, and the passed-in word. It then looks at the distance in vector space between the passed-in word and the **good** and **bad** vectors and calculates the difference between them. Therefore, a closer (more similar) word to **good** will have a positive value and a word that is closer (more similar) to **bad** will have a negative value.

7. Now, loop through each review in the **pos** and **neg** subdirectories of the **train** folder and calculate the sentiment score for each, keeping two distinct arrays of positive and negative reviews:

> **NOTE**
>
> Make sure you change the path to the train folder based on its location on your system.

```
review_dataset_dir = "../Data/aclImdb/train"

pos_scores = []
neg_scores = []

LIMIT = 2000

for pol in ("pos", "neg"):
    review_files = os.listdir(os.path.join(\
                    review_dataset_dir, pol))
    review_files = review_files[:LIMIT]
```

```
    print("Processing {} review files"\
    .format(len(review_files)))
    for i, rf in enumerate(review_files):
        with open(os.path.join(review_dataset_dir,\
        os.path.join(pol,rf)), encoding ="utf-8") as f:
            s = f.read()
            score = polarity_good_vs_bad(s)
            if pol == "pos":
                pos_scores.append(score)
            elif pol == "neg":
                neg_scores.append(score)
```

You should get the following output:

```
Processing 2000 review files
Processing 2000 review files
```

We grabbed some files from each of the **neg** (negative) and **pos** (positive) training folders, calculated the **polarity** score for each using our crude sentiment analyzer from *Exercise 4.01, Observing Prejudices and Biases in Word Embeddings*, and keep a record of these scores. If the sentiment analyzer is good, it will give high scores to the **pos** reviews and low scores to the **neg** reviews.

8. Calculate the **mean** and **median** of each set of scores:

```
mean_pos = mean(pos_scores)
mean_neg = mean(neg_scores)
med_pos = median(pos_scores)
med_neg = median(neg_scores)

print(f"Mean polarity score of positive reviews: \
{mean_pos}")

print(f"Mean polarity score of negative reviews: \
{mean_neg}")
print(f"Median polarity score of positive reviews: \
{med_pos}")
print(f"Median polarity score of negative reviews: \
{med_neg}")
```

You should get the following output:

```
Mean polarity score of positive reviews: 4.88497
Mean polarity score of negative reviews: 3.220575
Median polarity score of positive reviews: 4.84
Median polarity score of negative reviews: 3.29
```

We can see that there is some difference in the direction that we expect, with the positive reviews, on average, having a higher polarity score than the negative ones.

Note that we used two average functions: **mean** and **median**. The **median** is a useful average when you expect some outliers in your distribution to have undue influence. In our case, the two averages are quite similar, so there are probably not many outliers. Interestingly, the negative reviews have an on-average *positive* polarity score. Let's look at the distribution of both classes.

9. Plot histograms of the scores for positive and negative reviews:

```
bins = numpy.linspace(-10.0, 10.0, 50)

plt.hist(pos_scores, bins, alpha=0.9, label='pos')
plt.hist(neg_scores, bins, alpha=0.9, label='neg')
plt.legend(loc='upper right')

plt.show()
```

You should get the following output:

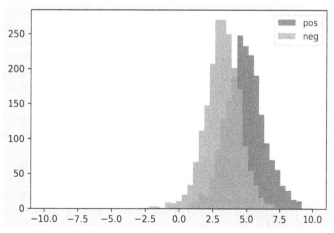

Figure 4.14: The distribution of polarity scores for positive and negative movie reviews

The preceding chart shows a histogram of polarity scores for positive and negative reviews. The *x*-axis shows the polarity scores, while the *y*-axis shows the number of reviews with that score. For example, we can see that over **250** negative reviews have a polarity score of **2.5**, while only about 50 positive reviews have this score. The range of polarity scores is between **−10** and **10**, with nearly all reviews falling between **−1** and **9**. These exact numbers don't mean anything on their own, but they show us relatively well which reviews are regarded as more positive.

We can see that the trend is in the right direction but that the scores are not enough to fully differentiate between positive and negative movie reviews. One important limitation of our classifier is that it is based on single words and not phrases. Language is complicated. Let's look at how words such as "not" can make sentiment analysis more difficult.

10. Calculate the polarity scores of some phrases relating to how good or bad a movie was:

```
phrases = [
    "the movie was good",
    "the movie was not good",
    "good",
    "not",
    "the movie was very good",
    "the movie was very very good",
    "the movie was bad",
    "the movie was very very very bad"
]

for phrase in phrases:
    print(phrase, polarity_good_vs_bad(phrase))
```

You should get the following output:

```
the movie was good 8.14
the movie was not good 6.78
good 26.45
not 0.33
the movie was very good 11.35
the movie was very very good 13.27
the movie was bad -8.47
the movie was very very very bad 5.31
```

Figure 4.15: The polarity scores of short, invented move reviews

Because our crude polarity calculator averages together the "meanings" of words by using their word vectors, the phrase **not good** is the average of the words **not** and **good**. Now, **Good** has a strongly positive score, and **not** has a neutral score, so **not good** is still seen as positive overall. On the other hand, **very** is closer to **good** in our vector space than it is to **bad**, so with enough occurrences of **very** in a phrase, the negative score from **bad** can be canceled out, leaving us with an overall positive score.

> **NOTE**
>
> To access the source code for this specific section, please refer to https://packt.live/38QwJAV.

In this exercise, we built a polarity classifier. We fed movie reviews into it and we saw that it worked as expected. On average, it gave higher scores for positive movie reviews and lower scores for negative movie reviews. The fact that we can verify that our classifier does work for a normal sentiment classifier task makes it more concerning that it displays some prejudices.

In the next activity, you will use the classifier for a third time and see what other prejudices it might display.

ACTIVITY 4.01: FINDING MORE LATENT PREJUDICES

Imagine that you have built the preceding classifier and have noted potential ethical concerns. You need to further validate your assumption that the model displays problematic prejudices.

You aim to further test the classifier with a wider range of inputs and to see whether you can predict what it will output.

You should build your word list of concepts that might have a prejudiced association and try to predict whether the classifier will classify them as positive or negative. You should then test your assumptions by running the words through the classifier.

> **NOTE**
>
> The code and the resulting output for this exercise have been loaded into a Jupyter Notebook that can be found here: https://packt.live/2OxwZeZ.

Perform the following steps to complete this activity:

1. Create a list of at least 16 words that you think might have a positive or negative prejudice. We are using the following 16 words for this activity:

```
sporty
nerdy
employed
unemployed
clever
stupid
latino
asian
caucasian
disabled
pregnant
introvert
extrovert
politician
florist
```

2. Define the same classification model that we used in previous exercises.

3. Before running the code, guess whether each of the words you chose would be classified as a positive or negative word. We have guessed the words (sporty, employed, clever, caucasian, extrovert, and florist) as positive and the remaining ones as negative.

4. Run the classifier on the word list and see how close your predictions were.

   You should get the following output:

```
sporty 13.26
nerdy -6.96
employed 4.46
unemployed -9.36
clever 7.27
stupid -24.37
latino -5.41
asian -4.76
caucasian 1.08
disabled -8.82
pregnant -8.14
introvert -0.36
extrovert 3.92
politician -5.98
florist 6.96
CEO -7.16
```

Figure 4.16: The polarity scores for each word in our new list

In our activity, we gained further evidence that our classifier might display prejudices. Let's move on and close this chapter.

> **NOTE**
>
> The solution to this activity can be found on page 595.

SUMMARY

In this chapter, we saw examples of prejudice in AI systems through several case studies. We saw how Cambridge Analytica developed an AI based on stolen personal data, how Amazon developed an AI that displayed sexist traits, and how the US justice system, to some extent, relies on AI that displays racist traits.

We built our own AI system that displayed some elements of prejudice and discussed how important it is to be aware of in-built biases, especially when using pre-trained models. We gained experience with the Python library spaCy and saw how word embeddings work. We verified that our sentiment analyzer worked on movie reviews, and then tested it further with some more words associated with prejudices.

In the next chapter, we will be studying the fundamentals of SQL and NoSQL databases by taking a practical approach. We will be learning and performing queries in MySQL, MongoDB, and Cassandra. Don't forget to consider the ethical considerations of any data that you store.

5

DATA STORES: SQL AND NOSQL DATABASES

OVERVIEW

In this chapter, we will study the SQL and NoSQL databases. First, you will learn how to decide on the ideal database to be used based on the format of the data at hand. Then, we will implement storing and querying data in different databases such as MySQL, MongoDB, and Cassandra. Furthermore, to improve efficiency, we will optimize the databases for maximum performance. This chapter includes hands-on experiences so that you can learn how to deal with practical real-world problems related to databases. By the end of this chapter, you will be able to develop and implement data models on MySQL, MongoDB, and Cassandra databases.

INTRODUCTION

In the previous chapter, we covered case studies on how AI has been controversial, along with various industry-standard use cases and hands-on information about the tools and frameworks related to storage. In a continuation of earlier discussions, we will be diving into detailed discussions on the kinds of databases that are available, along with their evolution, internal structures, and best possible use cases and examples.

Understanding database management is quite important as it gives you a clear idea about how data flows, that is, the transaction of data between the application and the database. For businesses and organizations, multiple types of data can be handled quite effectively using the knowledge of database management. Nevertheless, there are several challenges associated with data storage, such as infrastructure, cost, compatibility, and security.

In practical scenarios, databases enable us to accelerate our predictive analytics power by deploying **Artificial Intelligence** (**AI**) to get rid of several real-world problems such as credit card fraud detection, stock market prediction, and early stage disease detection.

Currently, various kinds of databases are available on the market, such as Relational (SQL), Document store (MongoDB), Columnar (Cassandra), Key-Value store (NoSQL), and Graph (ArangoDB) databases. Each of these databases has its characteristics and advantages and disadvantages. These databases are broadly divided into two groups: SQL and NoSQL databases.

SQL databases are **relational database systems** (**RDBMS**), which are structured/relational databases. The main language that's used to deal with the data is **Structured Query Language** (**SQL**). It is used to form a query and access the data.

NoSQL databases are **Not Only SQL** databases, wherein the data is in a semi-structured format. The main language that deals with data can vary for each database. This type of database has triggered the evolution of large distributed data stores running over a bunch of commodity hardware called **nodes**. Without the ability to use distributed data stores and having our data stored on a cluster of inexpensive nodes, we would have to rely on expensive single machines with very large capacities. This is not only expensive, but it also presents a single point of failure infrastructure. Generally, these databases are used where the data is huge and needs to be accessed at the lowest latency possible.

In this chapter, we are going to learn about the relational, document store, and columnar databases by going through examples and case studies. Lastly, we will learn how these databases are used in a hybrid architecture. In the next section, we will discuss database components to understand concepts including the query engine and interface, tables, schemas, and buffer pool.

DATABASE COMPONENTS

The typical database components can be seen in the following diagram. These components include the following:

- **Query Engine & Interface**: A translation unit that translates the query into a machine-readable language.

- **Tables**: Collection of rows where the actual data is stored. Tables can be accessed via the query engine.

- **Schemas**: Collection of tables where each schema represents a functional group of tables; for example, HR and Payroll.

- **Buffer pool**: A component that holds recent or intermediate transaction data in memory for optimized operation execution:

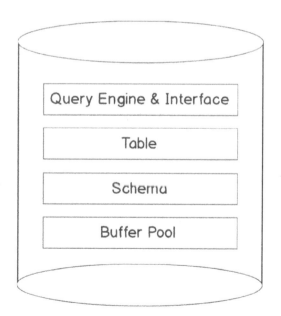

Figure 5.1: Typical database components

SQL DATABASES

These are the row-oriented type of databases. Here, each row of the database is a record. A collection of columns forms a record and is stored in a structure called a table. These collections of tables form the database.

These databases are accessed using SQL. It was developed by two IBM software researchers in the 1970s and it became more and more dominant in the field after that due to its straightforward, understandable syntax, ease of learning, and fewer lines of code being required to use complex functionalities:

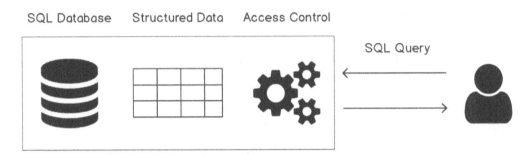

Figure 5.2: User database interaction through SQL

In the preceding diagram, user database interaction through SQL is demonstrated, where the efficacy of the SQL architecture to deal with "structured data" is quite visible.

SQL databases work on **Atomicity**, **Consistency**, **Isolation**, and **Durability** (**ACID**) properties, which guarantees strong consistencies and isolated transaction behavior. These are crucial while designing a data storage system.

Also, when we think about AI, having just one type of database is not sufficient due to the various forms of data. Therefore, there is a need for a hybrid data architecture in various cases, as we covered in the last section, *Exploring the Collective Knowledge of Databases,* of the chapter. In some hybrid architectures, we use the properties of the consistency and availability of MySQL and Mongo DB, respectively.

Undoubtedly, MySQL and Oracle are the most popular choices among RDBMS, as per their efficacies, but we will restrict our discussions to just MySQL in the next section as it is open source.

MYSQL

MySQL is an open source relational database system that's used by websites such as Facebook, Flickr, Twitter, and YouTube because it supports stored procedures, cursors, triggers, and all the SQL database features.

ADVANTAGES OF MYSQL

- Simple and easy to use

- Open-source

- Good performance with a wide range of queries and primary keys

- Effectively processes data that's GBs to a few TBs in size, with a limited level schema complexity

- Follows the ACID methodology accurately

- Follows the schema-on-write methodology where it pre-validates the records during insertion

DISADVANTAGES OF MYSQL

- Difficult to scale MySQL after a certain level.

- Designed to run on a single machine, which means a single point of failure and lesser availability.

- Being a part of the relational database system focuses on normalization and data duplication, which results in multiple joins. After a while, when data grows exponentially, this leads to a massive problem, especially for use cases such as user activity logging.

- Not appropriate in the case of read-heavy systems with high complexity schemas.

QUERY LANGUAGE

As you already know, MySQL uses the SQL language to deal with data. Let's take a look at the syntax and semantics.

TERMINOLOGY

SQL is a structured query language; it is composed of predicates or clauses:

- A column is a single field in the data; it's a single unit of data.

- A row is a collection of columns.

- A table is a collection of rows.

- A database is the collection of tables.

DATA DEFINITION LANGUAGE (DDL)

The following SQL commands are used to deal with the structure of the database or table:

- **CREATE DATABASE**

  This command creates a database:

  ```
  CREATE DATABASE fashionmart;
  ```

- **CREATE TABLE**

  This clause is used to create a table. Here, you have to specify the columns and their data types and constraints:

  ```
  CREATE TABLE products (p_id INT,
                  p_name TEXT,
                  p_buy_price FLOAT,
                  p_manufacturer TEXT,
                  p_created_at TIMESTAMP,
                  PRIMARY KEY (p_id)
                  );
  ```

  This command will create a table product that has five columns. **p_id** will have **PRIMARY KEY** as a constraint. We will discuss the primary key later.

- **DESC**

  The table structure can be displayed using this command:

  ```
  DESC products;
  ```

- **ALTER TABLE**

  This clause is used to modify the table structure to, for example, add, rename, or drop a column:

  ```
  ALTER TABLE products ADD COLUMN p_modified_at TIMESTAMP NOT NULL;
  ```

  > ### NOTE
  >
  > In case you run into an error while running the preceding code, you can set the **SQL_MODE** in MYSQL to **Allow Invalid Dates** using the following command: **SET SQL_MODE='ALLOW_INVALID_DATES';**

  In this command, we have added the **p_modified_at** column with the **TIMESTAMP** data type and the **NOT NULL** constraint.

- **DROP TABLE**

  This command drops the table and removes all the records:

  ```
  DROP TABLE products;
  ```

DATA MANIPULATION LANGUAGE (DML)

The following SQL commands are used to deal with the actual data in the table:

- **INSERT**

  This command is used to insert records into the table:

  ```
  INSERT INTO products (p_id, p_name, p_buy_price,
                        p_manufacturer, p_created_at)

  VALUES (1, 'Z-1 Running shoe', 34, 'Z-1', now());
  ```

  In this command, columns can be specified after the table name and the values inside the **VALUES** clause.

Adding single rows for multiple queries is exhausting and time-consuming. We can add multiple rows in a single query like so:

```
INSERT INTO products(p_id, p_name, p_buy_price,
                     p_manufacturer, p_created_at)
VALUES
(2, 'XIMO Trek shirt', 15, 'XIMO', now()),
(3, 'XIMO Trek shorts', 18, 'XIMO', now()),
(4, 'NY cap', 18, 'NY', now());
```

- **UPDATE**

  This command is used to update existing records inside the table:

```
UPDATE products SET p_buy_price=40
WHERE p_id=1;
```

  In this command, the **SET** clause is used to set a value for the column, while the **WHERE** clause filters the rows for the matching condition. If the condition is missing, it will update the value for all the records. In this example, we are updating the buy price to **40** for the product whose ID is **1** in the **products** table.

- **DELETE**

  This command is used to delete records from the table:

```
DELETE FROM products WHERE p_id=1;
```

  Here, if the condition is not given, then it will delete all the records from the table. As we have given the condition where the product ID is **1**, it will only delete that record.

DATA CONTROL LANGUAGE (DCL)

The following SQL commands are used to deal with providing control access to data stored in a database:

- **GRANT**

  This command allows the owner of the object with **GRANT** access to accord privileges to other users. The syntax for the **GRANT** command is as follows:

```
GRANT privilege_name ON object_name
TO {user_name | PUBLIC | role_name} [with GRANT option];
```

An example of the **GRANT** command is as follows:

```
GRANT SELECT ON student TO user5
```

This command grants a **SELECT** permission on the **student** table to **user5**.

- **REVOKE**

  This command cancels previously granted or denied permissions. The syntax for the **REVOKE** command is as follows:

```
REVOKE privilege_name ON object_name
FROM {User_name | PUBLIC | Role_name}
```

An example of the **REVOKE** command is as follows:

```
REVOKE SELECT ON student TO user5
```

This command revokes the **SELECT** permission on the **student** table from **user5**.

TRANSACTION CONTROL LANGUAGE (TCL)

The following SQL commands are used to deal with SQL's logical transactions:

- **COMMIT**

  This command permanently saves any transaction in the database. The syntax is as follows:

```
COMMIT;
```

- **ROLLBACK**

  This command restores the database to the last committed state. The syntax is as follows:

```
ROLLBACK;
```

- **SAVEPOINT**

  This command temporarily saves a transaction so that we can roll back to that point if required. The syntax is as follows:

```
SAVEPOINT savepoint_name;
```

To retrieve data in SQL, we often use the **SELECT** statement as it plays a significant role. Now, let's discuss data retrieval in terms of the appropriate commands and examples of how to use them.

DATA RETRIEVAL

This section focuses on data access:

- **SELECT**

  This command is used to access the data from the table:

  ```
  SELECT * FROM products;
  ```

  This command will display all the records from the **products** table. Here, the **FROM** clause specifies the table name.

- **JOINS**

  To compare and combine in SQL, joins are used. As output joins return specific rows of data from two or more tables in a database. Hence, this is not a command but a concept, where we can join more than one condition based on another condition. Let's understand this better by going through an example. Consider we have two tables, sales, and products, with the following data:

```
+------+------+--------------+----------+---------------------+
| s_id | p_id | s_sale_price | s_profit | s_created_at        |
+------+------+--------------+----------+---------------------+
|    1 |    2 |           18 |        3 | 2019-12-15 17:02:28 |
|    2 |    3 |           20 |        2 | 2019-12-15 17:02:28 |
|    3 |    3 |           19 |        1 | 2019-12-15 17:02:28 |
|    4 |    1 |           40 |        6 | 2019-12-15 17:02:28 |
|    5 |    1 |           34 |        0 | 2019-12-15 17:02:28 |
+------+------+--------------+----------+---------------------+
5 rows in set (0.00 sec)
```

sales table

```
+------+------------------+-------------+--------------+---------------------+
| p_id | p_name           | p_buy_price | p_manufacturer | p_created_at      |
+------+------------------+-------------+--------------+---------------------+
|    1 | Z-1 Running shoe |          34 | Z-1          | 2019-12-15 16:53:05 |
|    2 | XIMO Trek shirt  |          15 | XIMO         | 2019-12-15 16:55:42 |
|    3 | XIMO Trek shorts |          18 | XIMO         | 2019-12-15 16:55:42 |
|    4 | NY cap           |          18 | NY           | 2019-12-15 16:55:42 |
+------+------------------+-------------+--------------+---------------------+
```

products table

Figure 5.3: Data of the sales and products tables

Using the following code, we will join the **sales** table with **p_id** as the
FOREIGN KEY pointing to the products table:

```
SELECT
    products.p_name AS product_name,
    products.p_manufacturer AS manufacturer,
    sales.s_profit AS profit
FROM products
JOIN sales
ON products.p_id=sales.p_id;
```

This should give us the following output:

```
+---------------------+--------------+--------+
| product_name        | manufacturer | profit |
+---------------------+--------------+--------+
| Z-1 Running shoe    | Z-1          |      6 |
| Z-1 Running shoe    | Z-1          |      0 |
| XIMO Trek shirt     | XIMO         |      3 |
| XIMO Trek shorts    | XIMO         |      2 |
| XIMO Trek shorts    | XIMO         |      1 |
+---------------------+--------------+--------+
5 rows in set (0.00 sec)
```

Figure 5.4: Joined data on the sales and products tables

This command joins **products** and **sales** using the **JOIN** and **ON** clauses. The
ON clause holds the condition to join.

There are various types of joins available. The one we observed in the previous
example was an inner join. Conceptually, inner joins find and return matching
data from tables, whereas outer joins find and return matching data, as well
as some dissimilar data from tables. Let's discuss outer joins by going through
several examples.

LEFT OUTER JOINS

This join, in the preceding scenario, will show non-matching data from the left table, **products**; that is, data from the **products** table that does not have sales data associated with it:

```
SELECT
    products.p_name AS product_name,
    products.p_manufacturer AS manufacturer,
    sales.s_profit AS profit
FROM products
LEFT OUTER JOIN sales
ON products.p_id=sales.p_id;
```

This should give us the following output:

```
+----------------------+----------------+---------+
| product_name         | manufacturer   | profit  |
+----------------------+----------------+---------+
| Z-1 Running shoe     | Z-1            |       6 |
| Z-1 Running shoe     | Z-1            |       0 |
| XIMO Trek shirt      | XIMO           |       3 |
| XIMO Trek shorts     | XIMO           |       2 |
| XIMO Trek shorts     | XIMO           |       1 |
| NY cap               | NY             |    NULL |
+----------------------+----------------+---------+
```

Figure 5.5: Left outer join over the sales and product tables

RIGHT OUTER JOINS

This join, in alternate scenarios, will show non-matching data from the right table, **sales**; that is, data from the **sales** table that does not have products table data associated with it:

```
SELECT
    products.p_name AS product_name,
    products.p_manufacturer AS manufacturer,
    sales.s_profit AS profit
FROM products
RIGHT OUTER JOIN sales
ON products.p_id=sales.p_id;
```

This should give us the following output:

Figure 5.6: Right outer join over the sales and product tables

- Aggregate functions and **GROUP BY**

The **GROUP BY** clause is used to group records on one or more columns and then perform all the operations associated with it. One thing to take care of is that the column not used in aggregate functions should be included in the **GROUP BY** clause.

Aggregate functions include **SUM**, **MIN**, **MAX**, **COUNT**, and **AVG**. Let's look at an example:

```
SELECT
    sales.p_id AS p_id,
    SUM(sales.s_profit) AS profit
FROM sales
GROUP BY sales.p_id;
```

This command is grouping the table on the **p_id** column and performing addition (using the SUM function) over the **profit** column to find out the total profit for each **p_id**. You can use any aggregate function with the **GROUP BY** clause.

This should give us the following output:

Figure 5.7: SUM (Aggregate) function over profit in the sales table

- **ORDER BY**

  This clause is used to sort the result on a column. **ORDER BY** executes last in query lineage. You can sort the column in ascending or descending order:

  ```
  SELECT
      sales.p_id AS p_id,
      SUM(sales.s_profit) AS profit
  FROM sales
  GROUP BY sales.p_id
  ORDER BY profit ASC;
  ```

  This should give us the following output:

Figure 5.8: SUM (aggregate) function over profit in the sales table, ordered by profit

SQL CONSTRAINTS

This section is dedicated to the constraints/rules in the table. These constraints help MySQL pre-validate the record before inserting it into the table. These constraints can be created by using the **CREATE TABLE** statement. We will discuss a few constraints here.

The following are the main constraints:

- **PRIMARY KEY**: This specifies the main ID column of the table. This should be unique and not null.

- **FOREIGN KEY**: This allows us to ensure the referential integrity of data. This specifies the relationship between two tables. While declaring it, you can use **ON DELETE CASCADE**, which implies that if the parent data is deleted, the child data record is automatically deleted.

- **UNIQUE**: This constraint specifies that non-duplicate records are accepted.

- **NOT NULL**: This constraint specifies that any value of the column should not be null.

To understand the importance of SQL databases, let's take an example of a store called FashionMart that sells products such as clothing and jewelry to its customers. We have been given the task of making an online e-commerce website for the store. Initially, we want to achieve the following tasks with this website:

- Store the product information in the inventory

- Store the selling prices and calculate profits

Let's assume that we saved this information in file storage such as **Products.csv** and **Sales.csv** files. Now, we can easily add a new product to **Products.csv** and also add new sales data to **Sales.csv**; all you need to do is append one record to the end of the file:

Products.csv

```
p_name,p_type,p_price,p_manufacturer
T-Shirt,clothing,10,Adidas
```

Sales.csv

```
s_date,s_name,s_price
10/10/2012,T-Shirt,10
```

Figure 5.9: Products and sales data representation in the file

At this point, everything is working fine, but suddenly, the store owner finds out that one of the product's prices and names needs to be changed. So, to apply these changes, the product entry record needs to be found in **Products.csv** and then modified. Even after doing this, there will be records with duplicate product names in **Sales.csv** that need to be changed. This will lead to inconsistency in the data and its management becoming tedious. So, it's clear that file storage is not a good option for this use case. There are some more points to consider, such as a lock on the file when the file is being read/modified simultaneously. Also, if there are a million products in the store, how efficient will the search be?

Considering the previous criteria, we need to build the data system so that users can digitally run the FashionMart store. The following exercise focuses on designing a data model for the database and converts this CSV-based non-maintainable data system into a relational database system.

Let's design a basic building block of the database, that is, a data model.

A data model is a way of representing entities and their inter-relationships. We will discuss data modeling in detail later in this chapter:

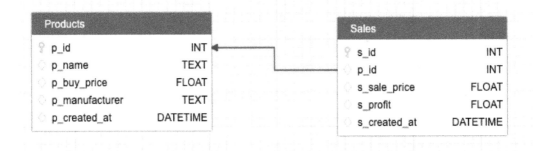

Figure 5.10: Data model for FashionMart

This data model gives us an overview of a relational database consisting of two tables, **Products** , and **Sales**. **Products** contains fields such as **p_id**, **p_name**, **p_buy_ price**, **p_manufacturer** (we could also have a manufacturer table and refer to it here with a foreign key), and **p_created_at**, while the **Sales** table has fields such as **s_id**, **p_id** (foreign key of the **Products** table), **s_sale_price**, **s_profit**, and **s_created_at**. This data model prevents the duplicity of product name data and overcomes problems we saw in the file storage system.

Since the physical data model is ready, we'll implement it in the next exercise.

EXERCISE 5.01: BUILDING A RELATIONAL DATABASE FOR THE FASHIONMART STORE

In this exercise, we need to implement the relational database for the FashionMart store.

We will be creating the **products** and **sales** tables and then performing basic SQL operations on them. We will also join the tables and find the overall profits.

Before proceeding with this exercise, we need to set up the MySQL database. Please follow the instructions on the *Preface* to install it.

Perform the following steps to complete this exercise:

1. Open a Terminal and run the MySQL client using the following command in the terminal based on your OS:

    Windows:

    ```
    mysql
    ```

    Linux:

    ```
    sudo mysql
    ```

    macOS:

    ```
    mysql
    ```

2. Once in MySQL, Create and select a database (**fashionmart**, in this case) using the following commands:

    ```
    create database fashionmart;
    use fashionmart;
    ```

    You should get the following output:

    ```
    mysql> create database fashionmart;
    Query OK, 1 row affected (0.14 sec)

    mysql> use fashionmart;
    Database changed
    mysql>
    ```

    Figure 5.11: Created and selected database for operation

    We have used a **USE** statement to select a database. This particular database remains as the default until the end of the session or until we deploy another **USE** statement to select another database.

3. Create the **products** table based on the data model of FashionMart, as shown in the following query:

```
CREATE TABLE products (p_id INT,
                       p_name TEXT,
                       p_buy_price FLOAT,
                       p_manufacturer TEXT,
                       p_created_at TIMESTAMP,
                       PRIMARY KEY (p_id)
                       );
```

The **products** table is created with the required columns and **p_id** is declared as the primary key.

4. Create the **sales** table based on the data model of FashionMart, as shown in the following query:

```
CREATE TABLE sales (s_id INT,
p_id INT,
s_sale_price FLOAT,
s_profit FLOAT,
s_created_at TIMESTAMP,
PRIMARY KEY (s_id),
FOREIGN KEY (p_id)
    REFERENCES products(p_id)
    ON DELETE CASCADE
);
```

The **sales** table is created with **s_id** as a primary key and **p_id** as a foreign key to the **products** table's **p_id** column.

5. Check the structure of the **products** table using the **desc** query:

```
desc products;
```

You should get the following output:

```
+------------------+-----------+------+-----+---------+-------+
| Field            | Type      | Null | Key | Default | Extra |
+------------------+-----------+------+-----+---------+-------+
| p_id             | int(11)   | NO   | PRI | NULL    |       |
| p_name           | text      | YES  |     | NULL    |       |
| p_buy_price      | float     | YES  |     | NULL    |       |
| p_manufacturer   | text      | YES  |     | NULL    |       |
| p_created_at     | timestamp | YES  |     | NULL    |       |
+------------------+-----------+------+-----+---------+-------+
5 rows in set (0.01 sec)
```

Figure 5.12: Products table structure

6. Check the structure of the **sales** table using the **desc** query:

```
desc sales;
```

You should get the following output:

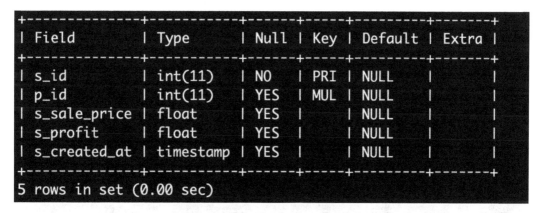

```
+---------------+-----------+------+-----+---------+-------+
| Field         | Type      | Null | Key | Default | Extra |
+---------------+-----------+------+-----+---------+-------+
| s_id          | int(11)   | NO   | PRI | NULL    |       |
| p_id          | int(11)   | YES  | MUL | NULL    |       |
| s_sale_price  | float     | YES  |     | NULL    |       |
| s_profit      | float     | YES  |     | NULL    |       |
| s_created_at  | timestamp | YES  |     | NULL    |       |
+---------------+-----------+------+-----+---------+-------+
5 rows in set (0.00 sec)
```

Figure 5.13: Sales table structure

With that, we have created our tables and database for the FashionMart store successfully. Now, let's insert the data into it.

7. Insert the data into the **products** table using the **INSERT INTO** command, as shown in the following query:

```
INSERT INTO products(p_id, p_name, p_buy_price,
                     p_manufacturer, p_created_at)
VALUES
(1, 'Z-1 Running shoe', 34, 'Z-1', now()),
(2, 'XIMO Trek shirt', 15, 'XIMO', now()),
(3, 'XIMO Trek shorts', 18, 'XIMO', now()),
(4, 'NY cap', 18, 'NY', now());
```

8. Similarly, insert the necessary records into the **sales** table, as shown in the following query:

```
INSERT INTO sales(s_id, p_id, s_sale_price,
                  s_profit, s_created_at)
VALUES
(1,2,18,3,now()),
(2,3,20,2,now()),
(3,3,19,1,now()),
(4,1,40,6,now()),
(5,1,34,0,now());
```

The formula that's used for calculating profit is (sale price – buy price). In each record, we can see that the sale price is different for the same product. It is assumed that the selling prices are different based on various sale offers.

9. View the **product** table's data using the **SELECT** query:

```
SELECT * FROM products;
```

You should get the following output:

```
+------+-------------------+-------------+----------------+---------------------+
| p_id | p_name            | p_buy_price | p_manufacturer | p_created_at        |
+------+-------------------+-------------+----------------+---------------------+
|    1 | Z-1 Running shoe  |          34 | Z-1            | 2019-12-15 16:53:05 |
|    2 | XIMO Trek shirt   |          15 | XIMO           | 2019-12-15 16:55:42 |
|    3 | XIMO Trek shorts  |          18 | XIMO           | 2019-12-15 16:55:42 |
|    4 | NY cap            |          18 | NY             | 2019-12-15 16:55:42 |
+------+-------------------+-------------+----------------+---------------------+
```

Figure 5.14: Data from the products table

10. View the **sales** table's data using the **SELECT** query:

```
SELECT * FROM sales;
```

You should get the following output:

```
+------+------+--------------+----------+---------------------+
| s_id | p_id | s_sale_price | s_profit | s_created_at        |
+------+------+--------------+----------+---------------------+
|    1 |    2 |           18 |        3 | 2019-12-15 17:02:28 |
|    2 |    3 |           20 |        2 | 2019-12-15 17:02:28 |
|    3 |    3 |           19 |        1 | 2019-12-15 17:02:28 |
|    4 |    1 |           40 |        6 | 2019-12-15 17:02:28 |
|    5 |    1 |           34 |        0 | 2019-12-15 17:02:28 |
+------+------+--------------+----------+---------------------+
5 rows in set (0.00 sec)
```

Figure 5.15: Data from the sales table

11. Join the two tables using the primary and foreign keys, as shown in the following query:

```
SELECT
    products.p_name AS product_name,
    products.p_manufacturer AS manufacturer,
    sales.s_profit AS profit
FROM products
JOIN sales
ON products.p_id=sales.p_id;
```

You should get the following output:

```
+-------------------+--------------+--------+
| product_name      | manufacturer | profit |
+-------------------+--------------+--------+
| Z-1 Running shoe  | Z-1          |      6 |
| Z-1 Running shoe  | Z-1          |      0 |
| XIMO Trek shirt   | XIMO         |      3 |
| XIMO Trek shorts  | XIMO         |      2 |
| XIMO Trek shorts  | XIMO         |      1 |
+-------------------+--------------+--------+
5 rows in set (0.00 sec)
```

Figure 5.16: Joined data for the sales and products tables

This result is achieved using the **JOIN** clause and the **ON** clause, but with a condition. Here, in the joined table, the column name is renamed using the **AS** query and a join operation is executed based on the **p_id** column. There are multiple entries for the same product, so we will group the products.

12. Group the products and show the overall profit using **GROUP BY**, as shown in the following query:

```
SELECT
  products.p_name AS product_name,
  products.p_manufacturer AS manufacturer,
  sales_subq.profit AS total_profit
FROM products
JOIN
  (SELECT
    sales.p_id AS p_id,
    SUM(sales.s_profit) AS profit
  FROM sales
  GROUP BY sales.p_id) AS sales_subq
ON products.p_id=sales_subq.p_id;
```

You should get the following output:

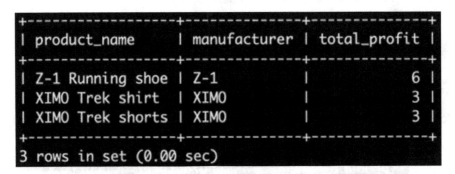

Figure 5.17: Showing the total profits for all the products

In this query, the aggregate operator, **SUM**, is used to sum the **profit** column among grouped records. We used a **SELECT** subquery to create an output of the total profit for each **p_id** by applying the **SUM** operator over the **s_profit** column. Furthermore, we used the **GROUP BY** command over the **p_id** column and assigned an alias of **sales_subq**. After that, a join operation was performed on the **products** and **sales_subq** tables to map the **p_id** column, resulting in the outcome of three columns, that is, **product_name**, **manufacturer**, and **total_profit**.

Notice that the **NY Cap** product is inside the store but that not a single unit was sold and is not shown. But still, we need to display it with profit as **0**.

13. Display the profit of all the products using a **LEFT OUTER JOIN**, as shown in the following query:

```
SELECT
    products.p_name AS product_name,
    products.p_manufacturer AS manufacturer,
    IFNULL(sales_subq.profit,0) AS total_profit
FROM products
LEFT OUTER JOIN
    (SELECT
        sales.p_id AS p_id,
        SUM(sales.s_profit) AS profit
    FROM sales
    GROUP BY sales.p_id) AS sales_subq
ON products.p_id=sales_subq.p_id;
```

You should get the following output:

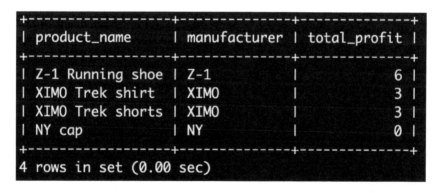

Figure 5.18: Showing all the records, regardless of whether or not they have sales data

This result is achieved with the **LEFT OUTER JOIN** clause. The **IF NULL** operator is used to display the record if the joining condition does not match.

> **NOTE**
>
> To access the source code for this specific section, please refer to https://packt.live/3gUYcUR.

By completing this exercise, you have successfully implemented a **fashionmart** database where the deployment of SQL queries helped you keep track of records on sales-related issues. Now, let's learn how to optimize our data model.

DATA MODELING

Data modeling is the technique of deciding on how the data should be stored so that inserts, updates, and data retrieval happens with the lowest latency possible. If the data modeling is done efficiently, then we can utilize the data store to its fullest, irrespective of the database being used.

There are three types of data models:

- **Conceptual data model**

  The terminology that's used in this data model includes Entities, Attributes, and Relationships. This data model is independent of the database and is very much understandable by a business user and they can relate to it very easily. When you want to explain data storage to non-technical audiences, use the conceptual data model.

- **Logical data model**

  This is the prerequisite for a physical data model. Here, you specify the fields and their data types. Although this is database-independent, the model has to abide by normalization rules to go to the next stage.

- **Physical data model**

  This is the final stage where database-specific details are presented. Everything from columns, data types, constraints, indices, views, and keys should be mentioned here.

To design a data model, you should know the structure of the data and how a user sees the application.

NORMALIZATION

As per ACID properties, it is very much required in the relational model that there should not be any duplicate records. The dependencies need to be defined clearly as well.

To standardize these, the process of normalization is used to structure the database. The following are the main forms of normalization:

1. 1st Normalization form

   This form says that the columns should be atomic and should have unique names. The column should not have multiple values for multiple purposes.

   Invalid schema example

   Here, `emp_id` with a value of **1** has got two departments, which is not a valid form. So, to accommodate and maintain the validity, we will store the data like so:

emp_id	name	department
1	John Doe	IT, Pre-sales
2	Max	HR
3	Claire Yates	IT

   Figure 5.19: Invalid schema – table displaying the department column with multiple values

   Valid schema example

   In the following example of a valid schema, we have created a separate record for **John Doe** by splitting the **department** column values, such as **IT** and **Pre-sales**, into two different rows:

emp_id	name	department
1	John Doe	IT
1	John Doe	Pre-sales
2	Max	HR
3	Claire Yates	IT

   Figure 5.20: Valid schema in 1st Normalization form

2. 2nd Normalization form

This form tells us that the table should be in 1st Normalization form and that all the columns should be fully dependent on the primary key.

Let's take an example of the performance of employees.

Invalid schema example

In the current example, the primary key is **emp_id** but the location is dependent on the department. This schema holds the information in the **location** column, which does not depend on the key column, **emp_id**, so there is a partial dependency here. Each department has a specific location, so **emp_id** is partially dependent on **location**. If we maintain **department** and **location** with employee ID in the same table, it will create redundancy of data:

emp_id	name	department	location
1	John Doe	IT	London
1	John Doe	Pre-sales	Luxembourg
2	Max	HR	New York
3	Claire Yates	IT	London

Figure 5.21: Invalid schema – partial dependency

To achieve the necessary validity and remove partial dependency, we will store the data as shown in *Figure 5.22* and *Figure 5.23*, respectively.

Valid schema example

The schema creates a department table to remove the partial dependency. Here, the location is dependent on **department_id**, which can be seen in the following table:

department_id	name	location_id
1	IT	London
2	Pre-sales	Luxemburg
3	HR	New York

Figure 5.22: Department table to remove the partial dependency

So, now, we have two tables, **department** and **employee**. **department_id** and **emp_id** stand as the primary keys for the **department** and **employee** tables, respectively. With the previously mentioned approach, we removed the partial dependency by only having **department_id** in the **employee** table:

emp_id	name	department_id
1	John Doe	1
1	John Doe	2
2	Max	3
3	Claire Yates	1

Figure 5.23: Valid schema in 2nd Normalization form

3. 3rd Normalization form:

This form says that the schema should be in 2nd Normalization form and that transitive dependencies should not exist in the single table. A transitive dependency is if A determines B (A-B) and B determines C (B-C) so that A determines C (A-C).

Invalid schema example

This schema shows that a transitive dependency such as **emp_id** determines **delivery_status** and that **delivery_status** determines **performance_label**, which is a transitive dependency and is not valid for the 3rd Normalization form. Let's elaborate more by explaining the components of the table. If **John Doe** delivers the package on time, he is meeting his set expectations. But if he delivers it before the given time, this infers that he has exceeded his expectations. The same scenarios are taking place for **Max** and **Claire Yates** as well. So, based on the delivery statuses of **John**, **Max**, and **Claire**, we can understand their performance labels. Since this is a transitive dependency, it is not a valid 3rd Normalization form:

emp_id	name	delivery_status	performance_label
1	John Doe	Before time	Exceeds expectations
1	John Doe	On-time	Matches expectations
2	Max	Delayed	Does not meet expectations
3	Claire Yates	On-time	Matches expectations

Figure 5.24: Invalid schema – transitive dependency

Valid schema example

This schema creates a delivery table to measure performance, eliminate transitive dependency, and remove duplicity, as shown in the following table:

delivery_status_id	delivery_status	performance_title
1	On-time	Meets expectations
2	Before time	Exceeds expectations
3	Delayed	Does not meet expectations
4	Not delivered	Failed

Figure 5.25: Delivery table to remove the transitive dependency

Now, **emp_id** only determines **delivery_status_id. delivery_status** is stored in the delivery table, altogether forming a valid schema, as shown in the following table:

emp_id	name	delivery_status_id
1	John Doe	2
1	John Doe	1
2	Max	3
3	Claire Yates	1

Figure 5.26: Valid schema in 3rd Normalization form

DIMENSIONAL DATA MODELING

This technique is used while building a data warehouse. It consists of facts and dimensions. Dimensions are categories of the data and their information; for example, Time, Place, and Company. These dimensions are used to build the fact tables that contain the metrics.

There are two types of data models/schemas in dimensional data modeling:

1. **Star schema**

   In this, there is a fact table at the center of the star, followed by several associated dimension tables. As per its similarity with the star structure, it is referred to as a star schema. It is the simplest type of data warehouse schema and is optimized for querying large datasets. In this schema, the dimension tables do not have interrelationships, so they only interact through the Fact table:

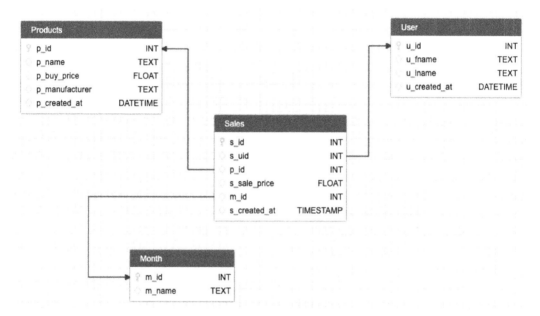

Figure 5.27: Star schema

2. **Snowflake schema**

In this schema, the dimensions can be interconnected to form the Fact table. It looks just like a snowflake, which is complex looking. This is used where complex schemas are integrated to form a bigger family:

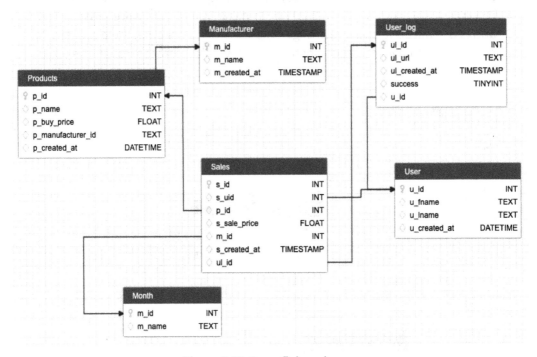

Figure 5.28: Snowflake schema

These popular modeling techniques are key to successful and future-proof database design, although these tricks won't solve problems such as having a huge number of records. We will discuss this later in this chapter.

The star schema is an appropriate choice when only a single join creates a relationship between the fact table and any dimension tables, whereas the snowflake schema is an appropriate choice when there is a requirement to have many joins to fetch the data. To simplify this, if we want to design a simple database, then the star schema is appropriate, whereas the snowflake schema is an appropriate choice for designing a complex database.

Now that you know about relational database design, modeling, and its implementation, let's talk about performance tuning.

PERFORMANCE TUNING AND BEST PRACTICES

Companies such as Facebook use MySQL to deal with petabytes of data, but it's difficult to achieve such performance with default settings and configuration. So far, we have set up a model using SQL queries. However, it is quite understandable that by getting more products, the entries in tables of e-commerce sites such as FashionMart will also be increasing. In such a scenario, the expectation from the databases would be to perform at the same durability and speed. To ensure that, we need to tune the performance of our commands. Some of the best practices to ensure performance tuning are as follows:

- Try to avoid using **SELECT ***:

  Instead of selecting all the columns unless required, try to mention column names in the **SELECT** clause. This will lower the unnecessary overhead on the MySQL query engine.

- Limited use of the **ORDER BY** clause:

  ORDER BY orders the results in ascending or descending order, but this also impacts the performance of MySQL since before showing the results, they need to be sorted.

- Use **LIMIT OFFSET** for pagination:

  Many web applications use pagination to show records batch by batch. You can use **LIMIT** and **OFFSET** to skip and limit the records you want to show. This will improve query performance since you don't need to bring in all the records from the result.

- Sub-queries are expansive:

  Sub-queries put extra load on the query engine. Instead, use joins if possible, which are treated as part of the same query.

- Use **UNION ALL** instead of **UNION**:

  UNION combines two query results and removes duplicates, which are classed as overhead for the query engine. **UNION ALL** does not remove duplicates and instead leads to better performance.

- Join on numerical columns:

  Always try to perform a join on a numerical column instead of a text column as this will save a lot of time.

- Don't use too much indexing:

  A database index is a data structure that improves the speed of data retrieval operations on a database table by taking a search key as input. Indexing returns a collection of matching records efficiently. It is used to quickly locate and access the data in the database. Indexes provide good performance, but too many of them will spoil the show as every index needs to be maintained and modified for every change in the data.

These tips and practices can be used to implement production-ready MySQL databases, where you can deal with a large number of queries. We'll implement some MySQL queries in the following activity.

ACTIVITY 5.01: MANAGING THE INVENTORY OF AN E-COMMERCE WEBSITE USING A MYSQL QUERY

Imagine we have a globally successful e-commerce site for the PacktFashion store. PacktFashion is facing tough challenges to manage the products in its stock. As products are being sold out quickly, PacktFashion is unable to maintain the inventory status. One of the customers who is staying in Prague needs a quick delivery and would like to know whether their product is available in the nearby warehouse. PacktFashion is facing trouble resolving the customer's query related to product availability in stock.

This activity aims to implement a data model so that we manage the inventory of the site. We will be listing those products that are in and out of stock. Furthermore, we need to find the list of available products in the warehouse of Prague.

You need to implement the data model as shown in the following screenshot. This model will help us manage the products, sales, inventory, manufacturer, the status of the product, and the warehouse location:

> **NOTE**
>
> We have created a data model based on the snowflake schema. You can use another approach if you prefer. It should be in its 3rd Normalization form at a minimum.

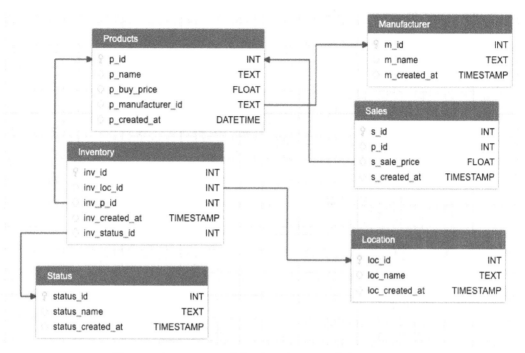

Figure 5.29: Data model using the snowflake schema

NOTE

The code for this activity can be found here: https://packt.live/2Wvf9xB.

Perform the following steps to complete this activity:

1. Create and use the PacktFashion database.

2. Create the necessary tables based on the data model using the **PRIMARY KEY**, **FOREIGN KEY**, and **ON DELETE CASCADE** queries.

3. Add at least three records for each table.

NOTE

The **location** table and **products** table must have **Prague** and **XIMO Trek shirt** as one of their records. Also, the **status** table should have values of **IN** or **OUT** only.

4. View the data of all the tables using the **SELECT** query.

The data in the tables should be similar to the following outputs:

products table

```
+------+----------------+--------------+------------------+---------------------+
| p_id | p_name         | p_buy_price  | p_manufacturer_id | p_created_at       |
+------+----------------+--------------+------------------+---------------------+
|    1 | Z-1 Running shoe |         34 |                1 | 2020-03-30 09:00:49 |
|    2 | XIMO Trek shirt  |         15 |                2 | 2020-03-30 09:00:49 |
|    3 | XIMO Trek shorts |         18 |                2 | 2020-03-30 09:00:49 |
|    4 | NY cap           |         18 |                3 | 2020-03-30 09:00:49 |
+------+----------------+--------------+------------------+---------------------+
```

sales table

```
+------+------+--------------+----------+---------------------+
| s_id | p_id | s_sale_price | s_profit | s_created_at        |
+------+------+--------------+----------+---------------------+
|    1 |    2 |           18 |        3 | 2020-03-30 09:00:49 |
|    2 |    3 |           20 |        2 | 2020-03-30 09:00:49 |
|    3 |    3 |           19 |        1 | 2020-03-30 09:00:49 |
|    4 |    1 |           40 |        6 | 2020-03-30 09:00:49 |
|    5 |    1 |           34 |        0 | 2020-03-30 09:00:49 |
+------+------+--------------+----------+---------------------+
```

manufacturer table

```
+------+--------+---------------------+
| m_id | m_name | m_created_at        |
+------+--------+---------------------+
|    1 | Z-1    | 2020-03-30 09:00:48 |
|    2 | XIMO   | 2020-03-30 09:00:48 |
|    3 | NY     | 2020-03-30 09:00:48 |
+------+--------+---------------------+
```

location table

```
+--------+------------+---------------------+
| loc_id | loc_name   | loc_created_at      |
+--------+------------+---------------------+
|      1 | California | 2020-03-30 09:00:49 |
|      2 | London     | 2020-03-30 09:00:49 |
|      3 | Prague     | 2020-03-30 09:00:49 |
+--------+------------+---------------------+
```

status table

```
+-----------+-------------+---------------------+
| status_id | status_name | status_created_at   |
+-----------+-------------+---------------------+
|         1 | IN          | 2020-03-30 09:00:50 |
|         2 | OUT         | 2020-03-30 09:00:50 |
+-----------+-------------+---------------------+
```

inventory table

```
+--------+------------+----------+---------------+---------------------+
| inv_id | inv_loc_id | inv_p_id | inv_status_id | inv_created_at      |
+--------+------------+----------+---------------+---------------------+
|      1 |          1 |        3 |             1 | 2020-03-30 09:00:53 |
|      2 |          3 |        4 |             1 | 2020-03-30 09:00:53 |
|      3 |          2 |        2 |             2 | 2020-03-30 09:00:53 |
|      4 |          3 |        2 |             2 | 2020-03-30 09:00:53 |
|      5 |          1 |        1 |             2 | 2020-03-30 09:00:53 |
+--------+------------+----------+---------------+---------------------+
```

Figure 5.30: Table data for the fashionmart data model

5. Find the total number of products in the inventory using the **JOIN** clause, where **status_name ='IN'**.

   You should get an output similar to the following:

```
+------------------------+
| total_in_stock_products |
+------------------------+
|                      2 |
+------------------------+
1 row in set (0.00 sec)
```

Figure 5.31: Table showing the total of in-stock products

6. Find out the total number of products that are not in the inventory using the **JOIN** clause, where **status_name ='OUT'**.

   You should get an output similar to the following:

```
+----------------------------+
| total_out_of_stock_products |
+----------------------------+
|                          3 |
+----------------------------+
1 row in set (0.00 sec)
```

Figure 5.32: Table showing the total of out-stock products

7. Find the status of the **XIMO Trek shirt** product for the Prague location.

   You should get an output similar to the following:

```
+--------+
| status |
+--------+
| OUT    |
+--------+
1 row in set (0.00 sec)
```

Figure 5.33: Table showing the status of the product

We have completed the SQL databases section. Now, designing and implementing relational data models and querying them should not feel like a difficult task.

We'll learn about NoSQL databases in the next section.

> **NOTE**
>
> The solution to this activity can be found on page 598.

NOSQL DATABASES

In the previous section, we looked at how SQL databases are used in the industry, along with some examples. However, with all the great things SQL databases can do, there is still significant room for improvement when it comes to dealing with complex and huge data with quick retrieval requirements. The times before and after NoSQL databases were created can be seen in the following diagram:

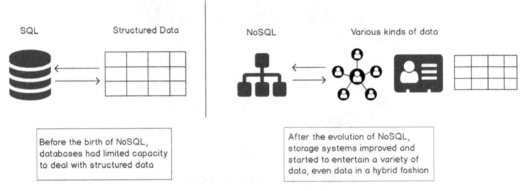

Figure 5.34: NoSQL database formats

These NoSQL databases evolved because of the fundamental limitations of SQL databases. The data in almost every sector has grown exponentially over time and the variety of usage has also expanded. Earlier, databases were limited to enterprise resource systems, research projects, telecommunications, and so on, but today, new domains have emerged with huge data in a variety of formats, such as social media and profiling millions of users, oil and gas, with billions of binary records, e-commerce, and life sciences. Because of the single server system, complex schemas, and the expense of vertical scaling, it became difficult for SQL databases to cope with high-performance requirements.

SQL databases such as RDBMS have a fixed, static, or predefined schema and these databases are not suited for hierarchical data storage, which is why these databases are best suited for complex queries. On the other hand, NoSQL databases are non-relational or distributed database systems. They have a dynamic schema. These databases are best suited for hierarchical data storage and are not very good for complex queries.

NEED FOR NOSQL

NoSQL databases are distributed databases built and supported in various languages.

There are various kinds of NoSQL databases available. Some of the most popular include document store, columnar, key-value store, and graph databases. Let's take a look at them:

- All of these databases can handle terabytes to petabytes of data in a distributed fashion, which helps in achieving a latency of a few milliseconds, such as the data we see in Amazon, Twitter, Facebook, and Google.

- There is a developer-friendly abstraction layer built by the communities of respected databases to deal with data smoothly. A lot of companies contribute to the community to come up with the abstraction layer. For example, Facebook invented Hive to deal with Hadoop data using a SQL-like language.

- The distributed database makes it easy to scale horizontally if data grows and also makes it highly fault-tolerant.

- The majority of databases are open source, which is why they are picked by the industry.

- The decision of which database to use is taken based on the use case at hand; for example, if you want to store/analyze the logs or profile products or users, then MongoDB is the best choice since it treats each record as a document with a flexlble schema.

We will discuss document store and columnar databases and provide examples in the following sections. Before we dive into NoSQL in more detail, let's have a careful look at one of the most important big data theorems: CAP theorem. This theorem is used to set the agenda for the data storage system.

CONSISTENCY AVAILABILITY PARTITIONING (CAP) THEOREM

This theorem is the *de facto* standard for deciding on the most appropriate NoSQL data storage:

- **Consistency**: This is the same concept we studied in ACID properties; it is about maintaining the same state of the data at any given point in time.

- **Availability**: This concept is about the system or, specifically, speaking data being available, even during node/cluster failures so that successful reads and writes can take place.

- **Partitioning**: According to this concept, the data is partitioned across the network as a secondary source in case of network congestion or node failures:

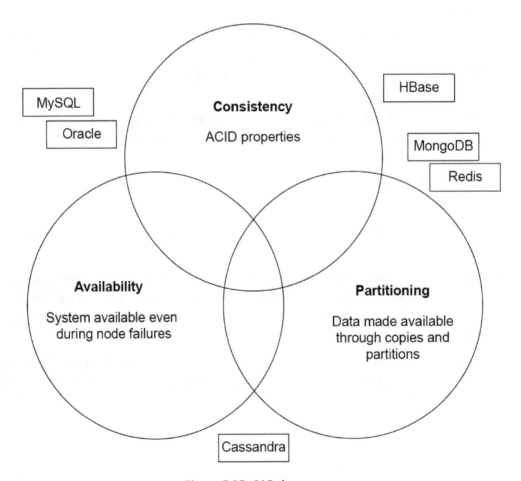

Figure 5.35: CAP theorem

As shown in the preceding diagram, the databases that give good consistency and fair availability are relational databases/SQL databases. One of the databases that can guarantee partitioning and availability is Cassandra. Some of the databases that have a good mixture of consistency and partitioning include HBase, MongoDB, and Redis.

So, with this theorem, the following is clear:

1. If you partition the data, then it's difficult to sync the state of data (that is, Consistency), but the data is available all the time. This is known as AP (Availability and Partitioning).

2. If you sync the data between its copies (joins), then there is a chance of a single point failure occurring, but the state of data can be maintained easily. This is known as CP (Consistency and Partitioning).

So, ideally, only two designs (we call this an agenda) can be chosen (either CP or AP) while considering data storage. Let's start with CP and learn how it's implemented through the MongoDB database.

MONGODB

MongoDB is a document-oriented database and is a generalized form of the NoSQL database. This database is popular among the community, especially for use cases such as logging, user profiling, search engines, geolocation data, and configuration storage.

It solves the scalability limitation of relational databases with its distributed architecture. So, it is considered an alternative to SQL databases. The installation and implementation of MongoDB is quite easy. For instance, Coinbase (digital currency exchange) uses MongoDB to build faster applications, diversify data type handling, and for efficient application management purposes. Let's now discuss the advantages and disadvantages of MongoDB.

ADVANTAGES OF MONGODB

- Since MongoDB stores, the data in consecutive memory locations, querying for one of the fields of the whole document can be retrieved from memory quite easily. Hence, MongoDB does not need to look through the whole disc.

- It supports complex data types, such as Map and List.

- It is a distributed database, which means we can set up a cluster of MongoDB with the Master-Slave paradigm. This cluster partition replicates the data to provide low latency and high availability.

- It's a schema-less database, which means that the collection adopts the schema based on the data that's inserted. So, when you are not sure about the schema, MongoDB is the right choice for such a scenario.

- It comes with a powerful query engine for NoSQL queries using JavaScript.

DISADVANTAGES OF MONGODB

- MongoDB does not support joins to ensure high availability.

- Relational data modeling approaches are not useful with MongoDB.

- Partitioning and denormalization lead to data duplication and data inconsistency.

- No default transaction support, so users need to handle these issues by themselves.

QUERY LANGUAGE

The query language that's used for MongoDB is JavaScript. The MongoDB team has provided a variety of functions to deal with collections and data.

TERMINOLOGY

As we discussed previously, you must be aware of SQL terminology. We will try to align the terminology of MongoDB with the former to aid our understanding:

MySQL	MongoDB
Column	Field
Row/Record	Document
Table	Collection
Database	Database
Index	Index

Figure 5.36: Comparison of MySQL and MongoDB terminology

Now, let's talk about the functions of the newly created sales MongoDB collection:

- **Use**

  Databases are selected through the **use** command:

  ```
  use fashionmart
  show dbs
  ```

  If you create the collection here, only the database is created.

- **Create Collection**

  ```
  db.createCollection("Sales")
  ```

  This function creates the **collection** with the **sales** name in the **fashionmart** database.

- **Insert**

  You can create a JavaScript object and insert it into the collection:

  ```
  db.Sales.insert(<JavaScript object or JSON object)
  ```

  Multiple records can be combined in this object:

  ```
  db.Sales.insert(
    [
    {
    "ShirtID":"1",
    "ShirtColor":"2",
    "ShirtSize":"XL",
    "ShirtDetails": [
        {
            "styleID": "4",
            "designerID": "6",
        }],
    },
    {
    "ShirtID":"2",
    "ShirtColor":"Red",
    "ShirtSize":"XXL",
    "ShirtDetails": [
        {
  ```

```
            "styleID": "5",
            "designerID": "6",
      }],
  }
  ]
  );
```

• **Find**

This function can be used to find the records from the collection:

```
db.Sales.find({ShirtID:"2"}).pretty()
```

It should give us the following output:

Figure 5.37: Finding a query on the records of a collection

Here, we are querying the sales collection with the **ShirtID** criteria (the **where** clause in MySQL) as **2**. The **pretty** function at the end displays the output JSON in a structured way. Here, the **pretty()** method is used mainly to display the result in an easier-to-read format.

Let's take another example of searching for an element value in an array and implement it using the Find query and then, alternatively, with the **$elemMatch** operator:

```
db.Sales.find({"ShirtDetails.styleID":"4"}).pretty()
```

It should give us the following output:

```
{
        "_id" : ObjectId("5e60bb8adf7688ca2692fb03"),
        "ShirtId" : "1",
        "ShirtColor" : "2",
        "ShirtSize" : "XL",
        "ShirtDetails" : [
                {
                        "styleID" : "4",
                        "designerID" : "6"
                }
        ]
}
```

Figure 5.38: Finding a query on the records of a collection

Alternately, you can use the **$elemMatch** operator, as follows:

```
db.Sales.find({ShirtDetails: {$elemMatch :{ styleID:"4"}}}).pretty()
```

It should give us the following output:

```
{
        "_id" : ObjectId("5e60bb8adf7688ca2692fb03"),
        "ShirtId" : "1",
        "ShirtColor" : "2",
        "ShirtSize" : "XL",
        "ShirtDetails" : [
                {
                        "styleID" : "4",
                        "designerID" : "6"
                }
        ]
}
```

Figure 5.39: The $elemMatch operator as an alternative to the Find query

Both commands will retrieve the whole matching document.

- **Aggregate**

  In MongoDB, aggregation pipelines are a breakthrough. Using these, you can chain various operations, such as grouping, aggregation, and exploding the embedded document. Here, operations such as "exploding the embedded document" imply the deconstruction of an array field from the input documents to the output document for each element.

  It is not a standard practice to use hardcoded values such as column name, alias name, and even the column that we want to perform the count on. To solve this, we can create JavaScript functions that are dynamic and return an object depending on the arguments so that we don't have to rely on the hardcoded values.

  Let's understand this through an example:

  ```
  db.Sales.aggregate([{$unwind: "$ShirtID"}]).pretty();
  ```

  It should give us the following output:

  ```
  {
          "_id" : ObjectId("5e60bd8bdf7688ca2692fb05"),
          "ShirtID" : "1",
          "ShirtColor" : "2",
          "ShirtSize" : "XL",
          "ShirtDetails" : [
                  {
                          "styleID" : "4",
                          "designerID" : "6"
                  }
          ]
  }
  {
          "_id" : ObjectId("5e60bd8bdf7688ca2692fb06"),
          "ShirtID" : "2",
          "ShirtColor" : "Red",
          "ShirtSize" : "XXL",
          "ShirtDetails" : [
                  {
                          "styleID" : "5",
                          "designerID" : "6"
                  }
  ```

  Figure 5.40: Aggregate query on the Sales collection

- **Pipeline concept**

  Here, we are using a JavaScript function that returns the aggregate pipeline object:

```
function sales_designer(match_field, match_value) {
var matchObject = {};
var eq_object = {};
eq_object["$eq"] = match_value
matchObject[match_field] = eq_object

return [

    {
      "$unwind":"$ShirtDetails"
    },
    {
      "$match": matchObject
    },
    {
      "$project": {
        "StyleID":"$ShirtDetails.styleID"
        }
      }
    ]
  };
```

  The JavaScript function executes in the following sequence:

- **$unwind** explodes/deconstructs the **ShirtDetails** array to each individual element and then uses them.

- **$match** filters the result over **match_field** equal to **match_value**, which is given as a parameter in the function.

- **$project** is just like the **SELECT** clause in relational databases, returning just **styleID** from **ShirtDetails**.

> **NOTE**
>
> Other criteria can also be parameterized and used as input in the pipeline function.

In this example, the **$unwind** operator in MongoDB has been used to deconstruct an array field from the input documents to the output document for each element:

```
db.Sales.aggregate(sales_designer("ShirtDetails.designerID","6")).
pretty()
```

It should give us the following output:

```
{ "id" : ObjectID("5e60bd8bdf7688ca2692fb05"), "StyleID" : "4" }
{ "id" : ObjectID("5e60bd8bdf7688ca2692fb06"), "StyleID" : "5" }
```

This command runs this pipeline and executes the commands in a sequence of declaration.

To understand MongoDB better, we will look at a simple example. For ease of understanding, we will be using the FashionMart example.

Imagine that you want to build a report for each product and that there are thousands of products in the inventory. Each product has millions of sales. If we need to build the solution using a relational database, then it will be a combination of the product's dimension and its sales, and the query will include a join between these two. This will result in extensively scanning and mapping each product to its sales data. In the case of a billion sales records for over a hundred thousand products, this join would be complex, which makes it an expensive solution. Another major drawback would be the query being performed on a single database server.

Now, let's think about this problem using MongoDB. As we know, joins are not encouraged, so the data will probably be in a denormalized manner. So, if we want reports for each product, the sales entity will be embedded inside the product entity. This way, we avoid joins and put data in a single collection. Also, using the aggregations in MongoDB, we can get interesting analytics out of the data.

Let's create a data model for this example. We will call the collection **Product**.

This collection will hold the product and sales information inside the same collection. It should look as follows:

```
{
"_id": "5498da1bf83a61f58ef6c6d5",
"p_name": "Z-1 Running shoe",
"p_manufacturer": "Z-1",
"p_buy_price": 34,
"p_created_at": Date(),

  "sales": [
    {
"s_sale_price": 40,
      "s_profit": 6,
      "p_created_at": Date(),
    },
    {
      "s_sale_price": 41,
      "s_profit": 7,
      "p_created_at": Date()
    }
  ]
}
```

In the data model, the example document is shown with an autogenerated mongo ID. Then, the **sales** object is added to the Sales array field. If you query for a product of **Z-1** manufacturer, then without joining the collection, all the necessary information will be fetched.

We will implement this example in the next exercise using the MongoDB database.

EXERCISE 5.02: MANAGING THE INVENTORY OF AN E-COMMERCE WEBSITE USING A MONGODB QUERY

In this exercise, we will implement the data model for FashionMart in MongoDB. Then, we'll display the report of a specific product, along with its sales data. Then, we will find the total products in the inventory and the bestselling product whose profit is greater than 6.

Before proceeding with this exercise, we need to set up a MongoDB database. Please follow the instructions on the *Preface* to install it.

Perform the following steps to complete this exercise:

1. Open the MongoDB shell:

    ```
    $mongo
    ```

    > **NOTEC**
    >
    > For additional help on starting the MongoDB shell, refer to the following link: https://docs.mongodb.com/manual/mongo/#start-the-mongo-shell-and-connect-to-mongodb. For a quick reference of all the important MongoDB shell commands, visit: https://docs.mongodb.com/manual/reference/mongo-shell/.

2. Use the **show** query to see the list of existing databases:

    ```
    show dbs
    ```

    You should get the following output:

    ```
    admin          0.000GB
    config         0.000GB
    demo           0.000GB
    local          0.000GB
    ```

3. Create a database called **fashionmart** through the **use** query:

```
use fashionmart
```

You should get the following output:

```
switched to db fashionmart
```

Now, you can directly start using the database without creating it. It will be created when you create a collection in it.

4. Create a products collection, as shown in the following query:

```
db.createCollection("products");
```

You should get the following output:

```
{
        "ok" : 1,
        "$clusterTime" : {
                "clusterTime" : Timestamp(1593812618, 7),
                "signature" : {
                        "hash" :
BinData(0,"XLov2h+xWNKbYrcWIqsCOROMuKY="),
                        "keyId" : NumberLong("6808867179686002691")
                }
        },
        "operationTime" : Timestamp(1593812618, 7)
}
```

In MongoDB, we do not have a structure for the collection, so we will start with data insertion in the **products** collection.

5. Insert the necessary data into the **products** collection. For the record, we will need today's date:

```
todayDate=new Date()
```

You should get the following output:

```
ISODate("2019-12-21T16:05:48.765Z")
```

We can also create a variable for time efficiency. Since the query language to access MongoDB is JavaScript, the way we access the interface is a mix of procedural and declarative.

6. Now, we'll create the product's JavaScript object so that it can be inserted into a collection:

```
products={
"p_name": "Z-1 Running shoe",
"p_manufacturer": "Z-1",
"p_buy_price": 34,
"p_created_at": todayDate,
"sales": [
    {
        "s_sale_price": 40,
        "s_profit": 6,
        "p_created_at": todayDate,
    },
    {
        "s_sale_price": 41,
        "s_profit": 7,
        "p_created_at": todayDate
    }
    ]
}
```

You should get the following output:

```
{
    "p_name" : "Z-1 Running shoe",
    "p_manufacturer" : "Z-1",
    "p_buy_price" : 34,
    "p_created_at" : ISODate("2019-12-21T16:38:27.274Z"),
    "sales" : [
            {
                "s_sale_price" : 40,
                "s_profit" : 6,
                "p_created_at" : ISODate("2019-12-21T16:38:27.274Z")
            },
            {
                "s_sale_price" : 41,
                "s_profit" : 7,
                "p_created_at" : ISODate("2019-12-21T16:38:27.274Z")
            }
    ]
}
```

Figure 5.41: Result

7. Lastly, insert the **products** object as a document in the collection:

```
db.products.insert(products);
```

You should get the following output:

```
WriteResult({ "nInserted" : 1 })
```

8. Create a JavaScript script object called **total_sales_for_product** for **"Z-1 Running shoe"** to find the total units sold, as shown in the following query:

```
total_sales_for_product_pipeline=[
  {
    "$match": {
      "p_name": "Z-1 Running shoe"
    }
  },
  {
    "$project": {
      "product":"$p_name",
      "total_sales": {
```

```
            "$size": "$sales"
        }
      }
   }
];
```

With this, we have created a JavaScript array object that will hold the **total_sales** pipeline operations.

9. Generate a report for using the JavaScript array object with the aggregate function, as shown in the following query:

```
db.products.aggregate(total_sales_for_product_pipeline).pretty()
```

You should get the following output:

```
{
        "_id" : ObjectId("5dfe4c0b211649ccef9a7426"),
        "product" : "Z-1 Running shoe",
        "total_sales" : 2
}
```

Here, we have filtered the records for **shoe** and then performed a count over the **sales** object field. We used the **aggregate** pipelining function to perform the functions. This way, various pipelines can be built as JavaScript objects and then executed to get the results.

10. Create a function called **total_product_sales** to parameterize the pipeline object's creation, as shown in the following query:

```
function total_product_sales(match_product,
projection_field,
count_field) {

var matchObject = {};
var countObject = {};
matchObject[projection_field] = match_product
countObject["$size"] = "$"+count_field;
return [
    {
      "$match": matchObject
    },
    {
      "$project": {
```

```
        "product":"$"+projection_field,

    "total_sales": countObject
      }
     }
    ]
  };
```

Here, we have created a **total_product_sales** function that will return the desired pipeline object. You can modify this function to suit your requirements. Now, it is time to call this function inside the aggregate function.

11. Generate the report using the **total_product_sales** and aggregate function, as shown in the following query:

```
db.products.aggregate(total_product_sales("Z-1 Running shoe", "p_
name","sales")).pretty();
```

You should get the following output:

```
{
        "_id" : ObjectId("5dfe4c0b211649ccef9a7426"),
        "product" : "Z-1 Running shoe",
        "total_sales" : 2
}
```

12. Find the total number of products in the inventory using the **count** function, as shown in the following query:

```
db.products.count();
```

You should get the following output:

```
1
```

13. Create the **gt_product_sales()** function to find the products with a profit greater than 6, as shown in the following query:

```
function gt_product_sales(gt_value, compare_field) {
      var gt_object = {};
      var match_object = {};
      gt_object["$gt"] = gt_value
      match_object[compare_field] = gt_object

    return [
        {
```

```
        "$unwind":"$sales"
    },
    {
      "$match": match_object
    },
    {
      "$project": {
          "product":"$p_name",
          "sales":"$sales"

      }

    }
  ]
};
```

Here, we created a JavaScript function called **gt_product_sales()**. This function will take **gt_value** and **compare_field** and return the JavaScript object that acts as input to the MongoDB **aggregate** function. Also, we used the **$unwind** operator to flatten the document and create one document for each element in the **sales** array field.

14. Find the product with a profit greater than 6 using the **aggregate** function, as shown in the following query:

```
db.products.aggregate(gt_product_sales(6, "sales.s_profit"));
```

You should get the following output:

```
{
    "_id" : ObjectId("5dfe4c0b211649ccef9a7426"),
    "product" : "Z-1 Running shoe",
    "sales" : {
            "s_sale_price" : 41,
            "s_profit" : 7,
            "p_created_at" : ISODate("2019-12-21T16:38:27.274Z")
    }
}
```

Figure 5.42: Result of aggregate with the unwind operator

This command executes the **aggregate** function to get the products that have a profit greater than the given **gt_value**. In our case, we have passed a value of 6.

> **NOTE**
>
> To access the source code for this specific section, please refer to https://packt.live/3epAaQ0.

With that, you have addressed a problem statement in MongoDB. Now, you can start thinking of various ways to perform similar operations and try to implement that on your local instance. By implementing this exercise, we have understood the importance of MongoDB to overcome complexities while dealing with large databases. Unlike SQL, since it's a schema-less NoSQL database, there are flexibilities while storing JSON documents. So far, we've learned how MongoDB reduces the complexity of deployment by speeding up application development.

Now, let's move on to data modeling.

DATA MODELING

In this section, we will discuss how to model the data for MongoDB. NoSQL data modeling is a new paradigm and MongoDB, being a document store database, saves data in **BSON (Binary JSON)** format. So, to make the most of MongoDB, the modeler has to think in JSON.

Let's go over the techniques of data modeling in MongoDB.

LACK OF JOINS

The reason why a record is called a document is that MongoDB is expected to find all the required information in one place. Imagine a real-world scenario where you are reading an important document. It's obvious that you expect to find all the information in the same document. It would be annoying if you could only read pieces of information through various documents.

To make reading over all the information as low latency as possible, MongoDB does not encourage joins in the collections. Here, we are talking about data duplication, which is a violation of normalization rules, but it is the key to quick reads from storage discs.

Let's take an example from the FashionMart use case. We created two tables, **Products**, and **Sales**. Right now, you can get sales data for a product by joining these two tables, but what if there are thousands of products and billions of sales records? Especially if you think in terms of distributed computing, the **Sales** table data is partitioned and stored in various machines in the cluster, so when you join the records of **Products** and **Sales**, then there will be a large amount of data transfer, as shown in the following diagram:

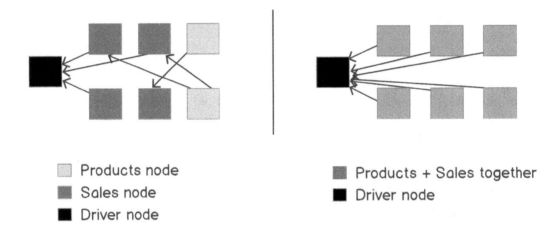

Products node
Sales node
Driver node

Products + Sales together
Driver node

Figure 5.43: Two scenarios performed with and without joins in collections

On the left-hand side, let's assume that the **Products** and **Sales** data is stored in separate collections, that is, separate nodes. As you can see, there is a lot of data exchange over the network because of the joins. However, on the right-hand side, the **Products** and **Sales** data is together, so there are no joins. This means that less data exchange happens over the network.

The resulting data model will be of the embedded type, where the Sales data will be embedded in the **Products** collection, as shown in the following query:

```
{
    "_id":ObjectId("5dfe4c0b211649ccef9a7426"),
    "p_name":"Z-1 Running shoe",
    "p_manufacturer":"Z-1",
    "p_buy_price":34,
    "p_created_at":ISODate("2019-12-21T16:38:27.274Z"),
    "sales":[
        {
```

```
            "s_sale_price":40,
            "s_profit":6,
            "p_created_at":ISODate("2019-12-21T16:38:27.274Z")
        },
        {

            "s_sale_price": 41,
            "s_profit":7,
            "p_created_at":ISODate("2019-12-21T16:38:27.274Z")
        }
    ]
}
```

JOINS

MongoDB is designed not to have joins, which is why there are two workarounds for this application: side joins and out-of-the-box joins. These are new concepts where you maintain SQL-like relationships between MongoDB collections. Joins have to be performed by the application layer. Let's have a look at one such example:

```
products
{

    "_id":ObjectId("5dfe4c0b211649ccef9a7426"),
    "p_name":"Z-1 Running shoe",
    "p_manufacturer":"Z-1",
    "p_buy_price":34,
    "p_created_at":ISODate("2019-12-21T16:38:27.274Z"),
}


Sales
{

    "_id":ObjectId("4dfe4c0b475649ccef9a7426"),
    "s_sale_price":40,
    "s_profit":6,
    "s_created_at":ISODate("2019-12-21T16:38:27.274Z")
    "p_id":ObjectId("5dfe4c0b211649ccef9a7426")
},
{

    "_id":ObjectId("2dfe4c0b475649jfkf9a7426"),
    "s_sale_price": 41,
    "s_profit":7,
```

```
    "s_created_at":ISODate("2019-12-21T16:38:27.274Z")
    "p_id":ObjectId("5dfe4c0b211649ccef9a7426")
}
```

It should give us the following output:

```
{
    "p_name" : "Z-1 Running shoe",
    "p_manufacturer" : "Z-1",
    "p_buy_price" : 34,
    "p_created_at" : ISODate("2020-03-30T04:42:47.845Z"),
    "sales" : [
            {
                "s_sale_price" : 40,
                "s_profit" : 6,
                "p_created_at" : ISODate("2020-03-30T04:42:47.845Z")
            },
            {
                "s_sale_price" : 41,
                "s_profit" : 7,
                "p_created_at" : ISODate("2020-03-30T04:42:47.845Z")
            }
    ]
}
```

Figure 5.44: Joins in MongoDB

The procedure to join these two collections is to take the product's ID and query the **sales** collection against the matching **p_id** field. In this case, we can create an index for the **p_id** column so that it can be retrieved and matching can be made simpler. This can be made possible at the application layer in a programming language by taking an ID from a collection and searching the records of the target collection based on the ID. You can also use the **$lookup** operator with the aggregate pipeline to join the two collections.

With that, we have covered the various techniques of data modeling in MongoDB. There's lots of domain-specific variety that can be used for the respective domains.

PERFORMANCE TUNING AND BEST PRACTICES

To make the database production-ready, there are some tricks to consider:

- Use the aggregation pipeline as much as possible since it's an in-built function that utilizes MongoDB functionality properly.

- Use less embedding by flattening the records if possible. This results in a major drop in latency.

- Use indexes wisely; indexing on textual columns occupies lots of storage.

- Use bulk operation where possible, such as bulk insertion and reading through code, to reduce network traffic.

- Whenever possible, perform lookups following the limit operation in the aggregation pipeline.

- Embedding hampers the insertion of records. For write-heavy applications, consider using the join method referenced to improve write latency.

- MongoDB is a very good fit for production-scale applications with huge data (that is, petabytes).

Using these techniques, you can build a high-performing MongoDB instance. How you use your resources is more important than what resources you use.

ACTIVITY 5.02: DATA MODEL TO CAPTURE USER INFORMATION

Let's take the example of a globally successful e-commerce site for the PacktFashion store. PacktFashion is facing tough challenges with managing the user logs on their site. They want to store users who are either visitors or buyers on their site so that they can send them sale offers in the future.

This activity aims to add product information and store information about users and their logs on our website. Furthermore, we will generate a report to show the user, along with the products they view, and the action (bought or not bought) that was performed by them.

You need to implement the data model shown in the following screenshot. The **user_logs** collection will store the events performed by the users, that is, whether they bought or returned a product. The many-to-many relationships between products and users are shown in this collection:

users

- _id,
- u_name
- u_created_at

user_logs

- _id
- user_id
- product_id
- action
- ul_crated_at

Figure 5.45: Data model for UserLogs

> **NOTE**
>
> The code for this activity can be found here: https://packt.live/2ZlIZGy.

Perform the following steps to complete this activity:

1. Create and use the **PacktFashion** database.

2. Create a **Products** collection and add data to it.

The data in the collection should be similar to the following output:

```
> db.products.find()
{ "_id" : ObjectId("5e817aecb292090c5cf31399"), "p_name" : "XIMO Trek shirt", "p_manufacturer" : "XIMO", "p_buy_price"
: 15, "p_created_at" : ISODate("2020-03-30T04:50:40.659Z"), "sales" : [ { "s_sale_price" : 30, "s_profit" : 15, "p_cr
eated_at" : ISODate("2020-03-30T04:50:40.659Z") }, { "s_sale_price" : 18, "s_profit" : 3, "p_created_at" : ISODate("20
20-03-30T04:50:40.659Z") }, { "s_sale_price" : 20, "s_profit" : 5, "p_created_at" : ISODate("2020-03-30T04:50:40.659Z"
) }, { "s_sale_price" : 15, "s_profit" : 0, "p_created_at" : ISODate("2020-03-30T04:50:40.659Z") } ] }
{ "_id" : ObjectId("5e817aecb292090c5cf3139a"), "p_name" : "XIMO Trek shorts", "p_manufacturer" : "XIMO", "p_buy_price
" : 18, "p_created_at" : ISODate("2020-03-30T04:50:40.659Z"), "sales" : [ { "s_sale_price" : 22, "s_profit" : 4, "p_cr
eated_at" : ISODate("2020-03-30T04:50:40.659Z") }, { "s_sale_price" : 18, "s_profit" : 0, "p_created_at" : ISODate("20
20-03-30T04:50:40.659Z") }, { "s_sale_price" : 20, "s_profit" : 2, "p_created_at" : ISODate("2020-03-30T04:50:40.659Z"
) } ] }
{ "_id" : ObjectId("5e817aecb292090c5cf3139b"), "p_name" : "NY cap", "p_manufacturer" : "NY", "p_buy_price" : 18, "p_c
reated_at" : ISODate("2020-03-30T04:50:40.659Z"), "sales" : [ { "s_sale_price" : 20, "s_profit" : 2, "p_created_at" :
ISODate("2020-03-30T04:50:40.659Z") }, { "s_sale_price" : 21, "s_profit" : 3, "p_created_at" : ISODate("2020-03-30T04:
50:40.659Z") }, { "s_sale_price" : 19, "s_profit" : 1, "p_created_at" : ISODate("2020-03-30T04:50:40.659Z") } ] }
>
```

Figure 5.46: Products collection

3. Insert an object for **Products**.

4. Create **users** and **user_logs** collections based on the data model.

    The data in the collections should be similar to the following outputs:

```
> db.users.find().pretty()
{
        "_id" : ObjectId("5e817b1ab292090c5cf3139c"),
        "name" : "Max",
        "u_created_at" : ISODate("2020-03-30T04:50:40.659Z")
}
{
        "_id" : ObjectId("5e817b1ab292090c5cf3139d"),
        "name" : "John Doe",
        "u_created_at" : ISODate("2020-03-30T04:50:40.659Z")
}
{
        "_id" : ObjectId("5e817b1ab292090c5cf3139e"),
        "name" : "Roger smith",
        "u_created_at" : ISODate("2020-03-30T04:50:40.659Z")
}
```

Figure 5.47: The users collection

```
> db.user_logs.find().pretty()
{
        "_id" : ObjectId("5e817b22b292090c5cf3139f"),
        "user_id" : "Max",
        "product_id" : "XIMO Trek shirt",
        "action" : "bought",
        "ul_crated_at" : ISODate("2020-03-30T04:50:40.659Z")
}
{

        "_id" : ObjectId("5e817b22b292090c5cf313a0"),
        "user_id" : "John Doe",
        "product_id" : "NY cap",
        "action" : "bought",
        "ul_crated_at" : ISODate("2020-03-30T04:50:40.659Z")
}
{

        "_id" : ObjectId("5e817b22b292090c5cf313a1"),
        "user_id" : "Roger smith",
        "product_id" : "XIMO Trek shorts",
        "action" : "bought",
        "ul_crated_at" : ISODate("2020-03-30T04:50:40.659Z")
}
```

Figure 5.48: The user_logs collection

5. Insert objects for the **user** and **user_logs** collections.

6. Join the **user** and **user_logs** collections using an aggregate function.

7. Generate the report of user logs using **$lookup**.

You should get the following output:

```
> db.user_logs.aggregate(user_logs_aggregate_pipeline).pretty()
{
        "_id" : ObjectId("5e817b22b292090c5cf3139f"),
        "user" : "Max",
        "product" : [
                "XIMO Trek shirt"
        ],
        "action" : "bought"
}
{

        "_id" : ObjectId("5e817b22b292090c5cf313a0"),
        "user" : "John Doe",
        "product" : [
                "NY cap"
        ],
        "action" : "bought"
}
{
        "_id" : ObjectId("5e817b22b292090c5cf313a1"),
        "user" : "Roger smith",
        "product" : [
                "XIMO Trek shorts"
        ],
        "action" : "bought"
}
```

Figure 5.49: Using $lookup to join the collections

With this activity, we've completed the MongoDB section. We have successfully learned about MongoDB's design, its implementation, and how it differs from relational databases. Now, let's move on to the next NoSQL database, Cassandra.

> **NOTE**
>
> The solution to this activity can be found on page 606.

CASSANDRA

The topic of NoSQL isn't complete without discussing columnar storage. In Cassandra, the data is stored and read in columns instead of rows. Each column is stored separately with the same row offset related to the table. We will study offset in the *Data Modeling* subsection:

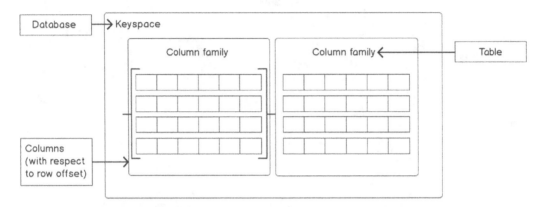

Figure 5.50: Keyspace and column family internal structure

Cassandra is a mixture of structured data, consistency, high scalability, and no single point of failure, and also has a powerful column family design. This particular database was developed by Facebook in 2008 and is horizontally scalable. Imagine you run a multinational company. You have thousands of employees, but you want the information of all the employees at the London location. In such a scenario, a traditional relational database will scan each row, parse it, and compare the `u_location` column value against the query. This is extremely time-consuming.

On the other hand, the columnar store only requires columns to operate. It not only saves resources but reduces the query processing time as well.

ADVANTAGES OF CASSANDRA

- Saves metadata for each column due to its columnar format design.
- Fetches the required columns into memory, avoiding full table scans.
- Efficient for micro writes and analysis queries.
- Highly scalable.
- No **single point of failure** (**SPOF**) due to a replication functionality of two or more.
- Open-source database.

- Capable of grouping columns into a family. This means that you can retrieve specific groups simultaneously rather than individual columns.

- Uses languages with SQL-like semantics to deal with data. This is known as the **Cassandra Query Language** (**CQL**). It was developed to make learning easy since the majority of people only know SQL.

DISADVANTAGES OF CASSANDRA

- Source replication means bad data is also replicated across clusters, which needs to be corrected.

- Data duplication due to denormalization.

- No validation, such as constraint checks and unique values.

- Steep learning curve to making it production-ready.

DEALING WITH DENORMALIZATIONS IN CASSANDRA

Let's look at an example where denormalized user data is stored in a user table and then converted into columnar format, as shown in the following diagram:

u_id	u_fname	u_lname	u_age	u_location	u_visits	u_created_at	U_last_visited
1	Derek	Gates	18	San Francisco	1	2019-12-24 17:58:18.606000+0000	2019-10-30 12:05:00+0000
2	Max	Frost	15	London	21	2019-12-24 17:58:18.606000+0000	2019-12-24 17:59:29.738000+0000
3	Roger	Moore	68	Amsterdam	4	2019-12-24 17:58:18.606000+0000	2019-11-07 07:30+0000
4	Isabela	McGuire	29	Seattle	18	2019-12-24 17:58:18.606000+0000	2019-12-02 14:50+0000

u_id	u_fname	u_lname	u_age	u_location	u_visits	u_created_at	u_last_visited
1	Derek	Gates	18	San Francisco	1	2019-12-24 17:58:18.606000+0000	2019-10-30 12:05:00+0000
2	Max	Frost	15	London	21	2019-12-24 17:58:18.606000+0000	2019-12-24 17:59:29.738000+0000
3	Roger	Moore	68	Amsterdam	4	2019-12-24 17:58:18.606000+0000	2019-11-07 07:30+0000
4	Isabela	McGuire	29	Seattle	18	2019-12-24 17:58:18.606000+0000	2019-12-02 14:50+0000

Figure 5.51: Columnar format representation

Now, each column is stored separately with metadata and offset the same as a row. The table user is converted into a columnar format where each column is separate, where, if queried with **u_fname**, then only **u_id** and **u_fname** will be brought into memory.

QUERY LANGUAGE

The query language that's used for Cassandra is CQL. Let's go over the terminology associated with it.

TERMINOLOGY

Just like how we understood the terminology of MongoDB, we will do the same to find out more about the key concepts of Cassandra. Before we dive into an example, we should know the basic terminology and concepts:

MySQL	Cassandra
Column	Column
Row/Record	Columns stored separately within the same offset
Table	Column family; any group of columns is called a column family
Database	Keyspace
Index	Index

Figure 5.52: Comparison of MySQL and Cassandra terminology

The Cassandra terminology is similar to that of the relational database:

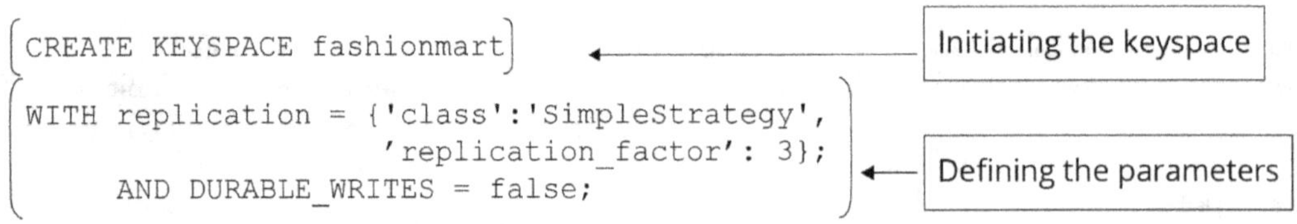

```
CREATE KEYSPACE fashionmart                          ← Initiating the keyspace
WITH replication = {'class':'SimpleStrategy',
                    'replication_factor': 3};         ← Defining the parameters
    AND DURABLE_WRITES = false;
```

Figure 5.53: Cassandra terminology

- **Keyspace**

    This is the collection of tables and column families as it stores all of its columns separately. There is one keyspace per node in a cluster.

    Keyspace has two attributes, Replication, and Durable Writes, which gives us the freedom to configure a particular keyspace. Let's understand replication first.

Replication has two options, as shown in the following table:

Options	Description
class	This is the option where we can specify the strategy of replication. The main classes are as follows:
	Simple strategy: Simple replication factor for the cluster.
	Network topology strategy: Each data center could have its replication factor.
	Old network topology strategy: Classic replication strategy.
replication_factor	We can specify the number of copies of a data unit in the cluster. The default value is set to 3.

Figure 5.54: Replication properties table

Durable write

If this is set to **TRUE**, then Cassandra maintains an update log for every data modification so that if the system crashes, we have the logs of the data to recover. The default value is **TRUE**; it should be kept as the default if possible. Just to show an example, we have added some syntax to change it to **false**.

Let's create a keyspace called **fashionmart** to understand replication and durable write better, as shown in the following code:

```
CREATE KEYSPACE fashionmart
WITH replication = {'class':'SimpleStrategy',
                    'replication_factor': 3}
                AND DURABLE_WRITES = false;

DESC keyspaces;
USE fashionmart;
```

Here, **'SimpleStrategy'** stands for "Strategy name" and describes a simple replication factor for the cluster. The replication factor value is 3.

- **COLUMNFAMILY**

  This is like a table in that it is a collection of columns. You can create a **ColumnFamily** like so:

  ```
  CREATE COLUMNFAMILY user(
  u_id int PRIMARY KEY,
  u_fname text,
  u_lname text,
  u_age int,
  u_location text,
  u_created_at timestamp,
  u_last_visited timestamp);
  ```

  It should give us the following output:

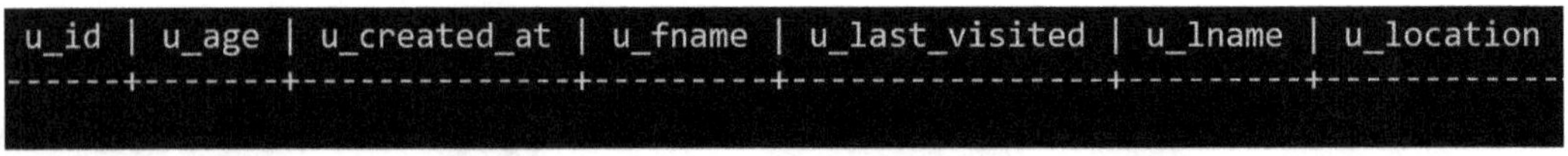

  Figure 5.55: Stdout to the CREATE KEYSPACE command

  Just like a relational database, you will have to specify column names and data types.

- **DATA RETRIEVAL**

  You can use the **SELECT** clause to retrieve records from **COLUMNFAMILY**. The Cassandra query engine locates the column with a byte offset and brings only the specified column into memory. This saves a lot of overhead while filtering the records:

  ```
  SELECT u_fname, u_lname, u_age FROM user;
  ```

  The **SELECT** query retrieves **u_fname**, **u_lname**, and **u_age** from the **user COLUMNFAMILY**.

Just like relational databases, you can also use aggregate functions, such as **SUM**, **AVG**, **MIN**, and **MAX**. This way, you can deal with data inside Cassandra using CQL.

Let's understand Cassandra in more detail using an example.

We have been discussing the FashionMart use case in this chapter and it has evolved from being a simple products and sales model to a SQL normalized model that has a searchable NoSQL design. Now, we will discuss the user side of it. We'll assume that there will be millions of users accessing this website, so we need to store their information to understand our audience better.

In this example, we need to store the user table in such a way that it can store the data of millions of users and give us access to the data with very low latency and high availability. Normally, with a relational database, this kind of table requires an index, but when querying an unindexed column that has this many records, it can take a lot of time to retrieve the data.

Instead of this, we will utilize the Cassandra columnar storage format, where each table will be stored separately in a column family. Let's implement this in the next exercise.

EXERCISE 5.03: MANAGING VISITORS OF AN E-COMMERCE SITE USING CASSANDRA

Let's take the previous problem statement where we need to store user information in a Cassandra column family called **user**. This exercise aims to store the number of visits that users have made to the site and perform queries on it. We will find the maximum and the minimum number of visits, along with the average age of the user.

Before proceeding with this exercise, we need to set up the Cassandra database. Please follow the instructions on the *Preface* to install it.

Perform the following steps to complete this exercise:

1. Launch the Cassandra CLI based on your OS, as shown here:

   Windows:

   ```
   Open Cassandra CLI application
   ```

   Linux:

   ```
   root@ubuntu: -$ cqlsh
   ```

   macOS:

   ```
   MyMac:~ root$ cqlsh
   ```

You should get the following output:

```
cqlsh>
```

> **NOTE**
>
> If you are having issues starting Cassandra CLI, you may try starting
> it with the **cassandra -f** command. For installation instructions
> for Cassandra on Windows 10, you can refer this link: https://medium.
> com/@sushantgautam_930/simple-way-to-install-cassandra-in-windows-10-
> 6497e93989e6. For installation instructions on Linux, you can visit https://
> cassandra.apache.org/doc/latest/getting_started/installing.html.

2. Create and select the **fashionmart** keyspace, as shown in the following query:

```
CREATE KEYSPACE fashionmart
WITH replication = {'class':'SimpleStrategy',
                    'replication_factor' : 3};

use fashionmart;
```

3. Create a **COLUMNFAMILY** called **user**, as shown in the following query:

```
CREATE COLUMNFAMILY user(
u_id int PRIMARY KEY,
u_fname text,
u_lname text,
u_age int,
u_location text,
u_created_at timestamp,
u_last_visited timestamp);
```

4. Check whether the **user** column family was created in the **fashionmart**
   keyspace using the following query:

```
DESC tables;
```

You should get the following output:

```
user
```

5. Check the **user** column family using the **SELECT** clause:

```
SELECT * FROM user;
```

You should get the following output:

```
u_id | u_age | u_created_at | u_fname | u_last_visited | u_lname | u_location
------+-------+--------------+---------+----------------+---------+-----------
```

Figure 5.56: User column family results

Column families in Cassandra are any group of columns, be they a table or an actual family. Now, let's add data to it.

6. Insert the sample data into the **user** column family, as shown in the following query:

```
INSERT INTO user(u_id, u_fname, u_lname, u_age,
                 u_location, u_created_at,
                 u_last_visited)
VALUES(1,'Derek', 'Gates', 18, 'San Francisco',
       toTimestamp(now()), '2019-10-30 12:05:00+0000');
```

7. Now, insert the records in batches, as shown in the following query:

```
BEGIN BATCH
  INSERT INTO user(u_id, u_fname, u_lname, u_age,
                   u_location, u_created_at,
                   u_last_visited)
  VALUES(2, 'Max', 'Frost', 15, 'London', toTimestamp(now()),
       toTimestamp(now()));
  INSERT INTO user(u_id, u_fname, u_lname, u_age,
                   u_location, u_created_at, u_last_visited)
  VALUES(3,'Isabela', 'McGuire', 29, 'Seattle',
       toTimestamp(now()), '2019-11-07 07:30+0000');
  INSERT INTO user(u_id, u_fname, u_lname, u_age,
                   u_location, u_created_at,
                   u_last_visited)
  VALUES(4,'Roger', 'Moore', 68, 'Amsterdam',
       toTimestamp(now()), '2019-12-02 14:50+0000');
APPLY BATCH;
```

8. View the **user** column family's data using the **SELECT** query:

```
SELECT * FROM user;
```

You should get the following output:

```
u_id | u_age | u_created_at                  | u_fname | u_last_visited                | u_lname | u_location
-----+-------+-------------------------------+---------+-------------------------------+---------+--------------
   1 |    18 | 2020-04-09 18:02:46.931000+0000 |   Derek | 2019-10-30 12:05:00.000000+0000 |   Gates | San Francisco
   2 |    15 | 2020-04-09 18:03:05.702000+0000 |     Max | 2020-04-09 18:03:05.702000+0000 |   Frost |        London
   4 |    68 | 2020-04-09 18:03:05.703000+0000 |   Roger | 2019-12-02 14:50:00.000000+0000 |   Moore |     Amsterdam
   3 |    29 | 2020-04-09 18:03:05.702000+0000 | Isabela | 2019-11-07 07:30:00.000000+0000 | McGuire |       Seattle
```

Figure 5.57: User column family data

9. Add the **u_visits** column to the **user** column family, as shown in the following query:

```
ALTER COLUMNFAMILY user ADD u_visits int;
```

By default, it will assign a **null** value to each record. Let's add the number of visits for each record with batch update statements.

10. Add some data to the **u_visits** column, as shown in the following query:

```
BEGIN BATCH
    UPDATE user
    SET u_visits=1
    WHERE u_id=1;
    UPDATE user
    SET u_visits=21
    WHERE u_id=2;
    UPDATE user
    SET u_visits=14
    WHERE u_id=3;
    UPDATE user
    SET u_visits=18
    WHERE u_id=4;
APPLY BATCH;
```

11. Confirm the data insertion of **u_visits** by viewing the **user** column family's data using the **SELECT** query:

```
SELECT * FROM user;
```

You should get the following output:

u_id	u_age	u_created_at	u_fname	u_last_visited	u_lname	u_location	u_visits
1	18	2020-04-09 18:02:46.931000+0000	Derek	2019-10-30 12:05:00.000000+0000	Gates	San Francisco	1
2	15	2020-04-09 18:03:05.702000+0000	Max	2020-04-09 18:03:05.702000+0000	Frost	London	21
4	68	2020-04-09 18:03:05.703000+0000	Roger	2019-12-02 14:50:00.000000+0000	Moore	Amsterdam	18
3	29	2020-04-09 18:03:05.702000+0000	Isabela	2019-11-07 07:30:00.000000+0000	McGuire	Seattle	14

Figure 5.58: User column family following the update operation

12. Find the maximum visits using aggregate functions, as shown in the following query:

```
SELECT MAX(u_visits) AS max_visits FROM user;
```

You should get the following output:

Figure 5.59: Showing max visits

The **MAX** function will find the highest value in the **u_visits** column and show it within the **max_visits** header.

13. Find the minimum visits using aggregate functions, as shown in the following query:

```
SELECT MIN(u_visits) AS min_visits FROM user;
```

You should get the following output:

Figure 5.60: Showing min visits

The **MIN** function will find the lowest value in the **u_visits** column and show it within the **min_visits** header.

14. Calculate the total number of visits, as shown in the following query:

```
SELECT SUM(u_visits) as total_visits FROM user;
```

You should get the following output:

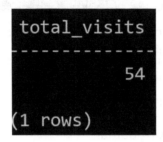

Figure 5.61: Showing total visits

Here, we have used the **SUM** aggregation function to get the result of total visits, that is, **54** in this query.

15. Calculate the average age group of the visitors, as shown in the following query:

```
SELECT AVG(u_age) as average_age FROM user;
```

You should get the following output:

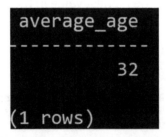

Figure 5.62: Showing average age

Here, we have used the **AVG** aggregation function to calculate the average age on the **u_age** column. We got an average age of **32** in this query.

> **NOTE**
>
> To access the source code for this specific section, please refer to https://packt.live/3eocOKI.

In this exercise, we have successfully implemented the inventory management of an e-commerce site using Cassandra. The column family, most visits, least visits, and total visits, along with the average age group, have been executed. We'll now move on to the data modeling techniques that can be performed in Cassandra in the next section.

DATA MODELING

We will discuss the structure and benefits of columnar storage in this section. Let's start with keyspace. A keyspace is a collection of column families, similar to the database as a collection of tables in RDBMS:

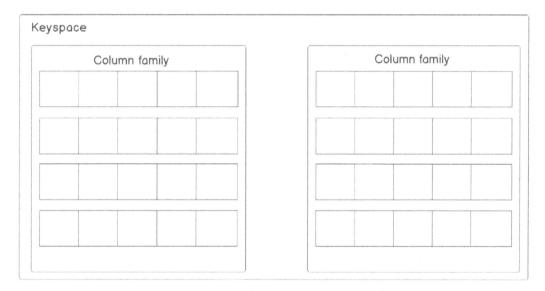

Figure 5.63: Keyspace and column family internal structure

Each column is stored on the same row offset as other column values. The internal structure of a column family is shown in the following diagram:

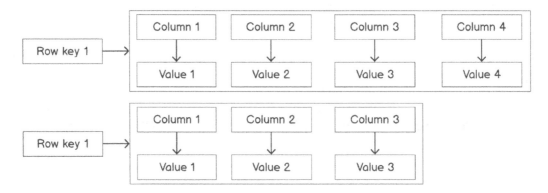

Figure 5.64: Column family, with all the columns mapped to a Row key

Here, each row has one key that is offset. Internally, each Cassandra table is a list of nested key-value pairs, that is, a map of maps. The column is a unit of storage and the row is used to replicate the data across the cluster. A column family contains columns and the table is a super column family.

COLUMN FAMILY DESIGN

We know that Cassandra is a columnar database and has the functionality to group columns to form a column family. The efficiency of this can be improvised if we create the column family based on query patterns. So, the columns that are accessed in a single query are stored in consecutive memory locations and are faster to retrieve. We can decide on the columns that need to be grouped based on the following questions:

- Which column is **GROUP BY** to be performed on?

- Which column is used for ordering the column family?

- Which column is required to filter the columns?

Once the column families have been designed properly, Cassandra will use the row key as the offset and read the required data. This is a good strategy in the case of read-heavy designs.

DISTRIBUTING DATA EVENLY ACROSS CLUSTERS

Data is distributed in various partitions and depends on the **partition_key**. The result of the hash function over the partition key is used to assign a slot/node to the data. Sometimes, the data is not distributed across clusters due to the presence of similar IDs or similar data in the partition column. So, our job is to evenly distribute the data by making sure that the **partition_key** is sufficiently unique. Also, the partition column we choose should have uniformly distributed values.

Let's understand it better through an example.

If you choose country column for partitioning the data and if 70% of the records are from the United States, then it's pretty obvious that 70% of data will end up on a single partition. So, the values should be distributed uniformly.

Apart from storing large quantities of data, Cassandra is capable of scaling out in keeping with the growth of data. This helps us pick out the right columns to avoid unbalanced cluster scenarios such as "Hot Spots" (one node ends up with higher tokens than others) and "Wide Rows" (a partition row grows larger than the other rows). To adjust to data growth and maintain even data distributions, you can use the **allocate_tokens_for_keyspace** setting.

CONSIDERING WRITE-HEAVY SCENARIOS

Cassandra is very good at micro writes where the data is pushed to map of maps to ensure that the performance has a constant lookup time. If you are looking for a data store that can perform micro writes with high availability, Cassandra would be a good choice.

In this section, we discussed the Cassandra storage strategy and the key things we need to consider while designing a successful data model. A better design model can give you acceptable performance. In the next section, we will dive deep into performance tuning and best practices.

PERFORMANCE TUNING AND BEST PRACTICES

Cassandra, is a distributed database, has been very popular among the NoSQL domain, but still, some tips need to be considered before running Cassandra in production:

- Concurrent reads and writes are an important part of Cassandra features as the ideal concurrent reads per processor core are 4 and writes are usually faster in Cassandra. If required, the concurrent writes value can be considered as equal to or higher than the concurrent reads. Hence, users do not need to worry about the optimum performance of the system. But raising these values beyond certain limits may have an impact on Cassandra's performance. The term "Certain Limit" varies from system to system as it depends on the CPU cores of a particular system. This can be configured in the **cassandra.yaml** file.

- Set the appropriate Java heap size as per the system's configuration. The rule of thumb is to use half of the system's memory for the Java heap.

- The complete content of the row is stored in the row cache. If the average size of data is huge, then disabling the row cache will turn into an overhead instead of a benefit.

- Commit logs should be stored at a separate location to the underlying **Sorted-String (SS)** table so that I/O can be shared across the disk.

These tips are crucial to getting the most out of your Cassandra setup as it can prove to be a really good form of data storage for your use cases.

ACTIVITY 5.03: MANAGING CUSTOMER FEEDBACK USING CASSANDRA

Now that we've had an overview of Cassandra, we'll try to go through the following problem statement and apply what we have learned. Let's assume you are heading the FashionMart project and, at this stage, the company has asked you to have feedback from customers to have a continuous evaluation of their activities. With millions of customers and their feedback, we have to create and maintain the column family for feedback and establish the total amount of feedback.

You need to implement the data model shown in the following screenshot. The **feedback_logs** column family will store the feedback and its details, such as user id, location, and creation date:

feedback_logs (COLUMNFAMILY)

- fl_id
- fl_feedback
- fl_location
- user_id
- fl_created_at

Figure 5.65: Data model for feedback logs

> **NOTE**
>
> The code for this activity can be found here: https://packt.live/3gTafC1.

Perform the following steps to complete this activity:

1. Create a **feedback_logs** column family based on the data model.

2. Insert the sample data into the **feedback_logs** column family.

   The data in **feedback_logs** should be similar to the following output:

```
 fl_id | fl_created_at                  | fl_feedback         | fl_location | user_id
-------+--------------------------------+---------------------+-------------+--------
     5 | 2019-10-30 12:05:00.000000+0000 |       Great website |      London |       1
     1 | 2019-10-30 12:05:00.000000+0000 |       Great website |      London |       3
     2 | 2019-10-03 12:05:00.000000+0000 |           Good work |     Seattle |       2
     4 | 2019-11-04 11:05:00.000000+0000 |         Not so good |   Hong Kong |       1
     7 | 2018-11-04 11:05:00.000000+0000 | Informative website |    Shanghai |       2
     6 | 2019-10-03 12:05:00.000000+0000 |           Good work |     Seattle |       4
     3 | 2019-11-04 11:05:00.000000+0000 |             Amazing |     Seattle |       2

(7 rows)
```

Figure 5.66: The feedback_logs column family data

3. Find the total feedback using an aggregate function.

You should get the following output:

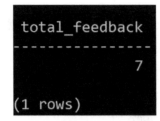

Figure 5.67: Showing total feedback

With this activity, we've completed the Cassandra section. We have successfully learned about Cassandra's design, its implementation, and how it differs from relational databases. Now, let's move on and combine all the databases for various business use cases.

> **NOTE**
>
> The solution to this activity can be found on page 613.

EXPLORING THE COLLECTIVE KNOWLEDGE OF DATABASES

This section is all about the implementation of what we've learned in this chapter. So far, we have gone through the nitty-gritty of databases and discussed MySQL for RDBMS and NoSQL databases such as MongoDB and Cassandra. Now, let's take the FashionMart store example and think about a scenario where we have the following schema:

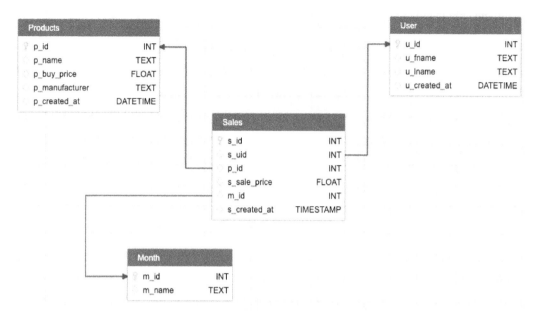

Figure 5.68: Star schema of FashionMart

In a case where the number of daily active users is 1 million and the Sales data is around 5 billion, then performing joins using MySQL between the dimensions will cause serious overhead. But here, the benefit is consistency, so if you want to update a Sales record, then it will be referenced in your next query.

Now, think about solving this use case using MongoDB. If we embedded a Sales entity inside Products by denormalizing the sales data, then it will drastically reduce latency as no data is shuffling or joins over the network. So, if you want to update sales data, then you will have to update it everywhere it is present due to data duplicity, otherwise we will lose consistency:

Figure 5.69: MongoDB embedded model

However, if we put both things together, then we could use them as follows:

1. MySQL for writing purpose, since it gives us good consistency (you can also use Cassandra in case you have a high volume of writes).

2. Once the data has been written, we could sync it to MongoDB by converting data objects into a collection-oriented format. You can use any programming language to perform this sync job.

3. Once the sync is done, we can use MongoDB as a read-only copy of the database because of its high-speed data retrieval.

The following diagram is a pretty good example of a MongoDB and MySQL-based hybrid approach in a real-world project scenario where users can achieve eventual consistency:

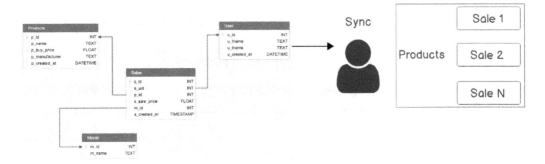

Figure 5.70: MongoDB and MySQL hybrid approach

Likewise, we use the hybrid (SQL and NoSQL) approach in the majority of projects in the real world. So, whenever there is a trade-off between consistency and availability, always remember this example.

SUMMARY

In this chapter, we had a hands-on overview of SQL and NoSQL databases. We looked into their usage, their best practices, and even various strategies to achieve near-perfect latency.

We learned about how to create a table, insert the data into a table, and how to select data from a table in various databases. Furthermore, we have explored complex aggregation over data, along with the resulting visualizations through sample data analytics.

Understanding the different types of databases and their evolutions, along with the various types of query languages and data modeling from this chapter, will help you to get to grips with the essence of big data file formats in the coming chapters. To start with big data, you will need to learn about the use of file formats. Such concepts can be assimilated by understanding the nitty-gritty of databases.

In the next chapter, you will learn about big data file formatting, along with compression, partitioning, and read-write strategy and performance optimizations.

6

BIG DATA FILE FORMATS

OVERVIEW

This chapter introduces popular big data file formats and skims through their advantages and disadvantages. The file formats that are covered in the chapter are Avro, ORC, and Parquet. It will walk through the code snippets required to implement their transformation and conversion to the desired file format. It will also educate you on attributes such as compression and the read-write strategy and executing queries to highlight the operational performance.

By the end of the chapter, you will be able to select the optimum file format for any user-specific case. You will strengthen these concepts by applying them to a real-world situation and get first-hand experience of performing the necessary queries.

INTRODUCTION

In the previous chapter, we learned about SQL and NoSQL databases. Further, we introduced different databases, such as MySQL, MongoDB, and Cassandra, and implemented a hands-on experience to deal with real-world problems. Now we will extend our understanding of these databases and study the big data file formats.

Ever-growing competition in the industry has been reducing reaction times to nil and to keep up with this situation, businesses have to improvise their responses to the problems strategically. Businesses are continually facing challenges to improve the product offering, production, human resources, customer services, operations, and other facets of the business. To get a cutting edge over their competition, organizations have to apply many strategies, and analytics is one of the strategies that feeds on big data. A few companies that rely extensively on real-time big data applications are Facebook, Twitter, Apple, and Google.

Processing and analyzing this big data can become nasty within a short span with a snowball effect. To overcome it, data scientists and analysts have to put in place a reliable data storage strategy with the best of the tools available. We will discuss various attributes with which we can build the strong foundation needed to serve multiple business needs. We will begin by understanding the basic input file formats, after which we will learn about big data file formats such as Avro, Parquet, and ORC, along with their respective technical setups and format conversions.

We will be using Scala and Apache Spark for this course. Spark is a cluster-computing framework. It is an open-source tool. It can be used to process both batch and streaming data. The Scala programming language can be used for multiple application domains and so is a general-purpose language. Scala supports functional programming. It has a strong static type system where type checking occurs during compile time.

Let's begin by exploring the common file formats used for the exchange of data between systems. In the next section, we will cover the CSV and JSON file formats.

COMMON INPUT FILES

Let's learn about the types of input files that are commonly used to exchange data between systems and the ways to convert them into various big data file formats. This section will also provide you with the programming skills required to transform these input files for the big data environment.

CSV – COMMA-SEPARATED VALUES

A CSV is a text file used to store tabular data separated by a comma. CSV is row-based data storage where each row is separated by a new line. For the exchange of tabular data, CSV files are frequently used.

The first row or header row of CSV files contains the schema detail, that is, column names for the data but not the type of data. CSV files fail to represent relational data, which means that a common column in multiple files does not have any relationship or hierarchy. Foreign keys are stored in columns of one or more files, but the CSV format itself does not express the linkage between these files.

The following figure is a screenshot of the Iris flower dataset, which contains data on three iris flower species. The dataset has 6 columns, namely, **ID**, **SepalLengthCm**, **SepalWidthCm**, **PetalLengthCm**, **PetalWidthCm**, and **Species**, and 150 rows. The **Species** column has a string value and the remaining four columns have rational numbers:

```
 1  Id,SepalLengthCm,SepalWidthCm,PetalLengthCm,PetalWidthCm,Species
 2  1,5.1,3.5,1.4,0.2,Iris-setosa
 3  2,4.9,3.0,1.4,0.2,Iris-setosa
 4  3,4.7,3.2,1.3,0.2,Iris-setosa
 5  4,4.6,3.1,1.5,0.2,Iris-setosa
 6  5,5.0,3.6,1.4,0.2,Iris-setosa
 7  6,5.4,3.9,1.7,0.4,Iris-setosa
 8  7,4.6,3.4,1.4,0.3,Iris-setosa
 9  8,5.0,3.4,1.5,0.2,Iris-setosa
10  9,4.4,2.9,1.4,0.2,Iris-setosa
11  10,4.9,3.1,1.5,0.1,Iris-setosa
```

Figure 6.1: Comma-separated value format

The size of a CSV file can easily range from **kilobytes** (**KBs**) to **gigabytes** (**GBs**). Other reasons for the popularity of the CSV file format include the simplicity of the file structure and its portability.

JSON – JAVASCRIPT OBJECT NOTATION

JSON (JavaScript Object Notation) data is a partially structured plain text file in which the data rests as key-value pairs. JSON can store data in a hierarchical format signifying the parent-child relationship between different pieces of data. JSON documents are small compared to other file formats of the same class, such as XML. The file size is the reason why it is preferred for network communication, notably in REST-based web services.

Various data processing applications support JSON serialization and deserialization. JSON documents can be converted or stored in highly performance-optimized files, but JSON works as raw data and can be used for the transformation of data.

The following figure is a screenshot of a JSON file:

```
1   {"Id":1,"SepalLengthCm":5.1,"SepalWidthCm":3.5,"PetalLengthCm":1.4,"PetalWidthCm":0.2,"Species":"Iris-setosa"}
2   {"Id":2,"SepalLengthCm":4.9,"SepalWidthCm":3.0,"PetalLengthCm":1.4,"PetalWidthCm":0.2,"Species":"Iris-setosa"}
3   {"Id":3,"SepalLengthCm":4.7,"SepalWidthCm":3.2,"PetalLengthCm":1.3,"PetalWidthCm":0.2,"Species":"Iris-setosa"}
4   {"Id":4,"SepalLengthCm":4.6,"SepalWidthCm":3.1,"PetalLengthCm":1.5,"PetalWidthCm":0.2,"Species":"Iris-setosa"}
5   {"Id":5,"SepalLengthCm":5.0,"SepalWidthCm":3.6,"PetalLengthCm":1.4,"PetalWidthCm":0.2,"Species":"Iris-setosa"}
6   {"Id":6,"SepalLengthCm":5.4,"SepalWidthCm":3.9,"PetalLengthCm":1.7,"PetalWidthCm":0.4,"Species":"Iris-setosa"}
7   {"Id":7,"SepalLengthCm":4.6,"SepalWidthCm":3.4,"PetalLengthCm":1.4,"PetalWidthCm":0.3,"Species":"Iris-setosa"}
8   {"Id":8,"SepalLengthCm":5.0,"SepalWidthCm":3.4,"PetalLengthCm":1.5,"PetalWidthCm":0.2,"Species":"Iris-setosa"}
9   {"Id":9,"SepalLengthCm":4.4,"SepalWidthCm":2.9,"PetalLengthCm":1.4,"PetalWidthCm":0.2,"Species":"Iris-setosa"}
10  {"Id":10,"SepalLengthCm":4.9,"SepalWidthCm":3.1,"PetalLengthCm":1.5,"PetalWidthCm":0.1,"Species":"Iris-setosa"}
11  {"Id":11,"SepalLengthCm":5.4,"SepalWidthCm":3.7,"PetalLengthCm":1.5,"PetalWidthCm":0.2,"Species":"Iris-setosa"}
12  {"Id":12,"SepalLengthCm":4.8,"SepalWidthCm":3.4,"PetalLengthCm":1.6,"PetalWidthCm":0.2,"Species":"Iris-setosa"}
13  {"Id":13,"SepalLengthCm":4.8,"SepalWidthCm":3.0,"PetalLengthCm":1.4,"PetalWidthCm":0.1,"Species":"Iris-setosa"}
14  {"Id":14,"SepalLengthCm":4.3,"SepalWidthCm":3.0,"PetalLengthCm":1.1,"PetalWidthCm":0.1,"Species":"Iris-setosa"}
15  {"Id":15,"SepalLengthCm":5.8,"SepalWidthCm":4.0,"PetalLengthCm":1.2,"PetalWidthCm":0.2,"Species":"Iris-setosa"}
16  {"Id":16,"SepalLengthCm":5.7,"SepalWidthCm":4.4,"PetalLengthCm":1.5,"PetalWidthCm":0.4,"Species":"Iris-setosa"}
17  {"Id":17,"SepalLengthCm":5.4,"SepalWidthCm":3.9,"PetalLengthCm":1.3,"PetalWidthCm":0.4,"Species":"Iris-setosa"}
18  {"Id":18,"SepalLengthCm":5.1,"SepalWidthCm":3.5,"PetalLengthCm":1.4,"PetalWidthCm":0.3,"Species":"Iris-setosa"}
19  {"Id":19,"SepalLengthCm":5.7,"SepalWidthCm":3.8,"PetalLengthCm":1.7,"PetalWidthCm":0.3,"Species":"Iris-setosa"}
20  {"Id":20,"SepalLengthCm":5.1,"SepalWidthCm":3.8,"PetalLengthCm":1.5,"PetalWidthCm":0.3,"Species":"Iris-setosa"}
21  {"Id":21,"SepalLengthCm":5.4,"SepalWidthCm":3.4,"PetalLengthCm":1.7,"PetalWidthCm":0.2,"Species":"Iris-setosa"}
22  {"Id":22,"SepalLengthCm":5.1,"SepalWidthCm":3.7,"PetalLengthCm":1.5,"PetalWidthCm":0.4,"Species":"Iris-setosa"}
23  {"Id":23,"SepalLengthCm":4.6,"SepalWidthCm":3.6,"PetalLengthCm":1.0,"PetalWidthCm":0.2,"Species":"Iris-setosa"}
24  {"Id":24,"SepalLengthCm":5.1,"SepalWidthCm":3.3,"PetalLengthCm":1.7,"PetalWidthCm":0.5,"Species":"Iris-setosa"}
25  {"Id":25,"SepalLengthCm":4.8,"SepalWidthCm":3.4,"PetalLengthCm":1.9,"PetalWidthCm":0.2,"Species":"Iris-setosa"}
```

Figure 6.2: JavaScript Object Notation format

The size of JSON can vary from between a couple of **megabytes** (**MBs**) to tens of GBs. However, the JSON file format can support complex schema and occupy a larger memory space when compared to the CSV file format.

Now that we've learned about common input file formats, let's learn about the attributes based on which you can evaluate and select the appropriate big data file format for your application.

CHOOSING THE RIGHT FORMAT FOR YOUR DATA

In an environment where the data grows exponentially, it can take only days to reach **terabytes** (**TBs**), and in weeks, it can turn into **petabytes** (**PBs**). By then, it is not very easy to alter the chosen file format. Let's consider that a gaming company launches a multiplayer game every week, which generates a log of all the users' activities. The log size may grow exponentially for each game, depending on the number of users. So, selecting the appropriate big data file format is a challenging task.

Let's define the framework to evaluate different file formats and apply these attributes to the specific use case. This framework will not provide a directive approach. Nonetheless, it will attempt to be as far-reaching as possible.

Let's consider the basic framework for the selection of the format.

Selection Criteria:

- Orientation: row-based or column-based

- Partitions

- Schema evolution

- Compression

ORIENTATION – ROW-BASED OR COLUMN-BASED

Orientation is one of the most crucial factors to consider when selecting a big data format. It explains how data is stored inside the file. There are primarily two ways of storing data on the physical drive. Based on the usage of the application, you can select the file type. The columnar-based format can result in faster processing when performing analytical queries over a subset of the columns. For example, when we have to apply a sum function to a column, in that case, the values of all the cells in the rows that form a single column need to be processed.

Whereas, when we have to scan through all or the majority of the columns, then the row-based format can turn out to be effective. For example, imagine that you have to retrieve a user profile; in that case, the values of all the columns from a single row are required to be processed.

Let's examine these two types by extending this concept using the COVID-19 dataset, considering the following subset:

Id	Citizen	ForeignCountry	Long	Lat	AbroadVisit	Infected
1	Indian	Italy	19.082847	72.874307	Yes	Yes
2	Indian	China	19.082847	72.874307	Yes	No

Figure 6.3: COVID 19 dataset

ROW-BASED

In the row-based format, all the columns of the rows are stored consecutively and retrieved one row at a time for processing. Row-based storage is highly efficient for transaction processing, which means that working with data involves interaction with all the columns over a group or all of the rows. In this orientation, writes of the data are very time-efficient.

When the previous table data is stored in the row-based format, it looks as follows:

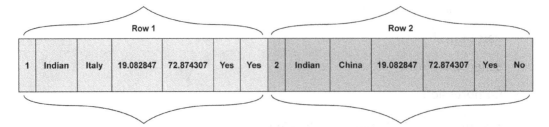

Figure 6.4: Data storage structure in the row-based format

COLUMN-BASED

In the columnar-based format, the data in a column is stored sequentially, that is, from the first row to the last row, and while retrieving the data, only selected columns can be processed. The column-based format is very performant when working with analytical query models that require only a subset of columns from large datasets. In such cases, using row-based orientation will lead to a need for a lot of memory, which is resource-expensive and doesn't perform satisfactorily.

Saving the file in the column-based format stores the same data type (that is, column data) in adjacent physical space on the disk, impacting the I/O performance. In the columnar-based format, the query for applying various analytical functions over a column is efficient because the data in a single column can be extracted quickly, so it can be productive when applying functions over a few columns of data.

When the previous table data is stored in the column-based format, it looks as follows:

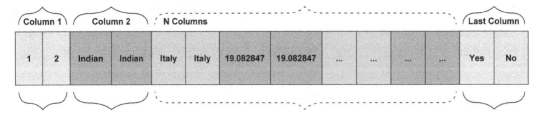

Figure 6.5: Data storage structure in the column-based format

The questions you need to ask when choosing the orientation are as follows:

1. Do you require more reads or writes?

    Try to identify the intent of the application, and consider whether the usage of the application will lead to more read commands, indicating a need for a data retrieval system, or more write commands, indicating a need for a data acquisition system.

2. Does your application have a transactional or analytical approach?

    For this question, you need to envision the volume of interaction with the data. For example, the user log for a multiplayer game will be analytically queried about first-time logins or how much time is spent on each level, and so on. These queries are applied across a single column.

    On the contrary, the gamer's profile will interact with the data limited to a single row. The gamer's profile may have attributes including name, age, level, batches, and so on. These attributes will be updated and retrieved in a predictable pattern accessing a single row from a table or only a few rows from multiple tables.

Let's learn about another important attribute, namely, partitions.

PARTITIONS

In a big data environment, processing data is challenging primarily because of the size of the files. These files have to be processed in chunks or small parts, which can be achieved with file partitioning. These splits are done as per the **Hadoop Distributed File System** (**HDFS**) block and impact the processing time.

The chunk that is created can be loaded in memory to be processed individually. The ability to partition the data and process it independently is the foundation of the scalable parallelization of processing. The parallel processing capability of the file is the key to its selection.

Let's learn about another important attribute, namely, schema evolution.

SCHEMA EVOLUTION

Choosing the optimum file format cannot be completed until schema evolution has been considered. In the case of continuous streaming data, the schema is bound to be changed over time and, most importantly, multiple times. The schema includes columns, data types, views, primary keys, relationships, and more.

Very rarely, it will happen that the schema present at the time of creating the file is perpetual. For other instances where the transition between the old schema and a new schema is required, a file format should have the provision to adopt these changes smoothly, such as if a gaming company decides to start a daily championship, which means that an additional leaderboard is required. This will require schema evolution.

There are a few questions to answer, without which your consideration of changing the schema will not be complete:

- Will there be difficulties in adopting the new schema?
- What is the impact on the disk storage of the file?
- Will there be compatibility issues between different versions?

We will be covering these questions with regard to each file type that we cover. Let's learn about another important attribute, namely, compression.

COMPRESSION

Data compression is a necessary evil that decreases the amount of data to be stored or transmitted. It helps in optimizing the use of time and money by shrinking data that otherwise would have used more space. It reduces the size of the file to be stored on the disk and speeds up data exchange throughout the network. This is attempted at the source of the data by using encoding on repeated data, which can result in a great reduction in size. A few encoding types are as follows:

- Dictionary encoding
- Plain encoding
- Delta encoding
- Bit-packing hybrid
- Delta-length byte array
- Delta strings (incremental encoding)

For illustration, let's look at dictionary encoding and consider the **Species** column from the Iris dataset.

Dictionary Encoding: This is a type of data compression technique that works by mapping the actual data to a token/symbol ("dictionary"). The tokens/symbols are smaller than the actual data, resulting in compression.

The unique values from the **Species** column are associated with corresponding values, and the dictionary will be stored separately. The new value will be stored in the file with other data, which is converted as follows:

Iris-setosa	0
Iris-versicolor	1
Iris-virginica	2

Figure 6.6: Dictionary encoding

It will create a dictionary of all the unique values from a column and link them to a corresponding unique value. This step reduces the storage of repeated values in a column, and overall reduces demand for storage space and the transmission of the data.

These attributes will provide a foundation based on which you can evaluate the file format and select the most suitable file format for your application. Let's learn about the various file formats in the next section.

INTRODUCTION TO FILE FORMATS

Now, let's understand the file structure in detail and distinguish between these file formats. This section will decompose the file formats and dive into the structure of files to elaborate on the efficiency of each file format.

PARQUET

Apache Parquet is an open-source column-oriented representation and stores data in an optimized columnar format. It is language-independent and framework-independent because the objective of creating this format was to optimize the operation and storage of data across Hadoop.

Shortly after its introduction, it acquired popularity in the industry. The reasons for its acceptance are primarily the fast retrieval and processing capabilities that it offers. However, writes are usually time-consuming and considerably expensive.

As it is a columnar-based format, homogenous data is stored together, resulting in better compression. The compression and encoding scheme can have a significant impact on performance.

Let's explore the elements of the Parquet file format and consider the following figure:

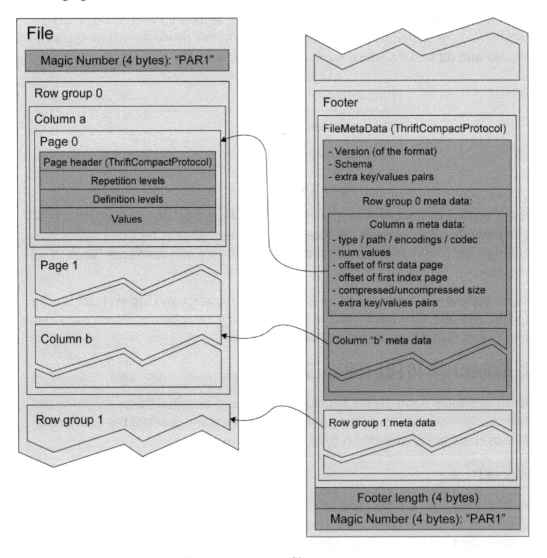

Figure 6.7: Parquet file structure

Parquet has nested data structures stored in a columnar-based format. A file can be decomposed into the following elements.

Firstly, the header, which is the apex of the file that contains the magic number (**PAR1**), communicating that the file type is **PARQUET** to the application.

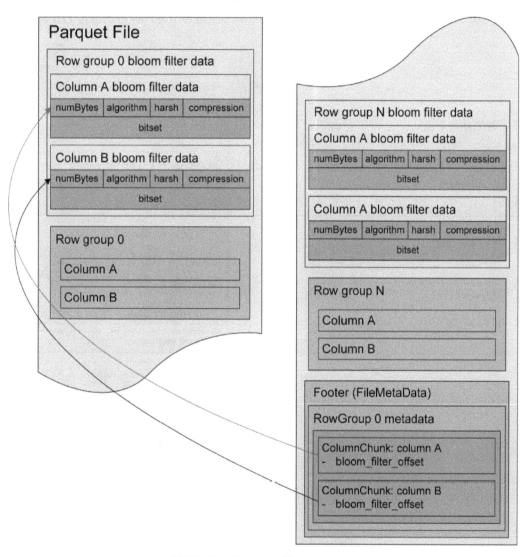

Figure 6.8: Parquet file structure

The second part is the data block, which is composed of the row groups, meaning a chunk of the actual data or a subset of the total number of rows. Each row group has multiple instances of column data that contains data pages for different chunks of column data. The page contains the metadata, repetition level, definition level, and the encoded column data. The metadata contains details about the chunk of column data, such as **min**, **max**, and **count**. The default value of the row group is 128 MB.

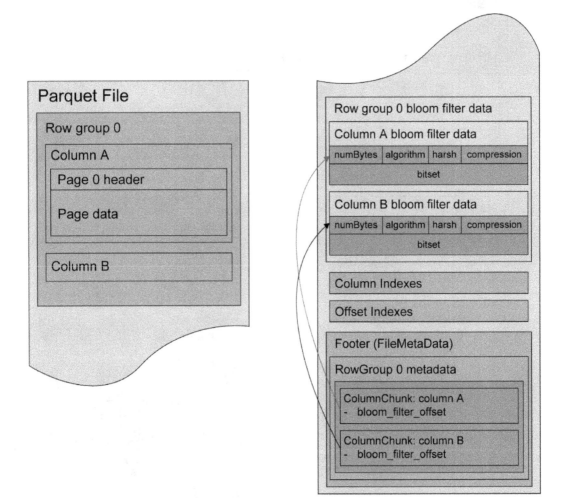

Figure 6.9: Parquet file structure

The last is the **Footer**, which includes the format version, the schema details, and the metadata of the columns in the file. It also includes an attribute for the footer length encoded as a **4-byte** field, terminated by a 4-byte magic number (**PAR1**), the same as the header.

Let's now learn about partitioning and compression in the Parquet file format.

Partitioning and Compression

The ability to partition Parquet files makes them a superior selection, and their columnar structure enables a comparatively faster scan than other file formats. The compression contributes to another significant attribute, which is that Parquet compression can significantly reduce the size. Therefore, the compression helps in enhancing performance, data ingestion, and data access.

Let's now learn about schema evolution for the Parquet file format.

Schema Evolution

Over the life of the application, there will be a need to alter the schema, and the incessant data flow may turn this into a nightmare. You can evolve the schema, which means adding new columns as and when needed. The schema evolution generates different Parquet files but they are mutually compatible schemas.

Platform Support

While taking a final decision on which file format is most suitable, taking into account the harmony between the file format and the platform is an important consideration. Spark has comprehensive support for the Parquet file format. Let's now look at a synopsis of the Parquet file format:

Parquet	
Partitioning	
Schema Evolution	
Compression	
Platform	Impala, Spark, Arrow Drill

Figure 6.10: Parquet attributes synopsis

The synopsis assesses the performance of each attribute of the Parquet file format and draws conclusions as a visually comparable scale. Parquet files perform moderately in schema evolution and above-moderately in attributes such as partitioning and compression.

Now we will proceed to the next exercise and create a Parquet file from common input file formats.

EXERCISE 6.01: CONVERTING CSV AND JSON FILES INTO THE PARQUET FORMAT

To start with a big data environment, you are required to import data into the environment. Usually, this data is available in CSV or JSON formats. This exercise aims to use a dataset in two input formats, that is, CSV and JSON files, and convert them into Parquet format. After converting the CSV file, we will read and display the contents of the output Parquet file to validate the conversion.

We will be working with a dataset containing census data that consists of 6 columns and 1,321,975 rows, taken from the New Zealand government's stats website (https://www.stats.govt.nz/large-datasets/csv-files-for-download/). The columns are **Year**, **Age**, **Ethnic**, **Sex**, **Area**, and **count**. This dataset is available in the CSV file format (41 MBs) and the JSON file format (106 MBs). The size difference is because the CSV file is compact, primarily due to its simpler schema structure. The dataset has some null values present as well.

The dataset can be found in our GitHub repository at the following location:

https://packt.live/2C72sBN.

You need to download and extract the **Census_csv.rar** and **Census_json.rar** files from the GitHub repository.

Before proceeding to the exercise, we need to set up a big data environment by installing Scala and Spark. Please follow the instructions in the *Preface* to install it.

Perform the following steps to complete the exercise:

1. Create a directory called **Chapter06** for all the exercises and activities of this chapter. In the **Chapter06** directory, create the **Exercise06.01** and **Data** directories to store the files for this exercise.

2. Move the extracted **CSV** and **JSON** files to the **Chapter06/Data** directory.

3. Open your Terminal (macOS or Linux) or Command Prompt window (Windows), move to the installation directory, and open the Spark shell in it using the following command:

```
spark-shell
```

You should get the following output:

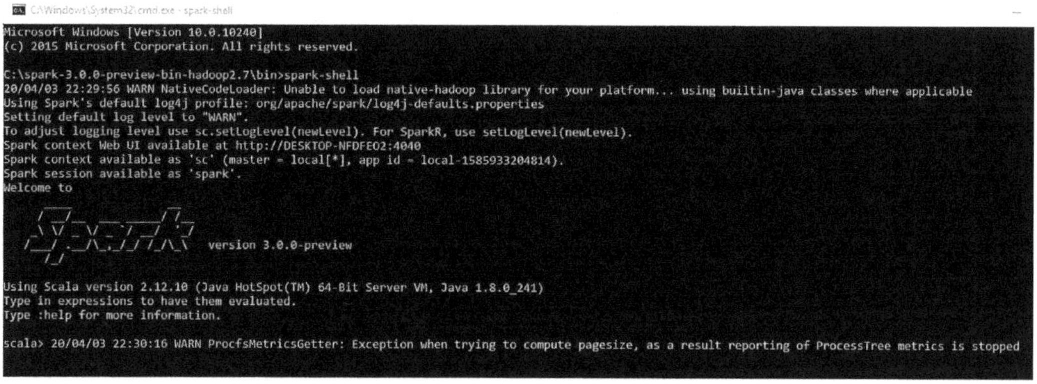

Figure 6.11: Spark shell

4. Read the CSV file using the following code:

```
var df_census_csv = spark.read.options(Map("inferSchema"-
    >"true","delimiter"->",","header"-
    >"true")).csv("F:/Chapter06/Data/Census.csv")
```

> **NOTE**
>
> Update the input path of the file according to your local file path throughout the exercises in this chapter.

You should get the following output:

```
df_census_csv: org.apache.spark.sql.DataFrame = [Year: int,
Age: int … 4 more fields]
```

This command creates a DataFrame variable (**df_census_csv**) and loads the contents of the CSV file into the variable.

5. Read the JSON file using the following code:

```
var df_census_json = spark.read.json(\
                    "F:/Chapter06/Data/Census.json")
```

> **NOTE**
>
> Owing to the size of the data, the commands in this exercise can take up to a few minutes to execute.

You should get the following output:

```
df_census_json: org.apache.spark.sql.DataFrame = [Age: bigint,
Area: string … 4 more fields]
```

6. Show the CSV file using the following command:

```
df_census_csv.show()
```

You should get the following output:

```
+----+---+--------+---+--------------------+-----+
|Year|Age|  Ethnic|Sex|                Area|count|
+----+---+--------+---+--------------------+-----+
|2018|  0|European|  1|Total - New Zeala...|  795|
|2018|  0|European|  1|Total - New Zeala...| 5067|
|2018|  0|European|  1|    Northland Region| 2229|
|2018|  0|European|  1|     Auckland Region| 1356|
|2018|  0|European|  1|      Waikato Region|  180|
|2018|  0|European|  1|Bay of Plenty Region|  738|
|2018|  0|European|  1|     Gisborne Region|  630|
|2018|  0|European|  1| Hawke's Bay Region| 1188|
|2018|  0|European|  1|     Taranaki Region| 2157|
|2018|  0|European|  1|       Tasman Region|  177|
|2018|  0|European|  1|       Nelson Region| 2823|
|2018|  0|European|  1|  Marlborough Region| 1020|
|2018|  0|European|  1|   West Coast Region|  516|
|2018|  0|European|  1|   Canterbury Region|  222|
|2018|  0|European|  1|        Otago Region|  219|
|2018|  0|European|  1|    Southland Region|  234|
|2018|  0|European|  1|    Waimate District|  ..C|
|2018|  0|European|  1|                null|19563|
|2018|  0|European|  1|Kapiti Coast Dist...|19557|
|2018|  0|European|  2|Total - New Zeala...|  768|
+----+---+--------+---+--------------------+-----+
only showing top 20 rows
```

Figure 6.12: Displaying the CSV DataFrame

This command will display the first 20 rows of the DataFrame created in the previous step.

7. Show the JSON file using the following command:

```
df_census_json.show()
```

You should get the following output:

```
+---+--------------------+--------+---+----+-----+
|Age|                Area|  Ethnic|Sex|Year|count|
+---+--------------------+--------+---+----+-----+
|  0|Total - New Zeala...|European|  1|2018|  795|
|  0|Total - New Zeala...|European|  1|2018| 5067|
|  0|    Northland Region|European|  1|2018| 2229|
|  0|     Auckland Region|European|  1|2018| 1356|
|  0|      Waikato Region|European|  1|2018|  180|
|  0|Bay of Plenty Region|European|  1|2018|  738|
|  0|     Gisborne Region|European|  1|2018|  630|
|  0|  Hawke's Bay Region|European|  1|2018| 1188|
|  0|     Taranaki Region|European|  1|2018| 2157|
|  0|       Tasman Region|European|  1|2018|  177|
|  0|       Nelson Region|European|  1|2018| 2823|
|  0|  Marlborough Region|European|  1|2018| 1020|
|  0|   West Coast Region|European|  1|2018|  516|
|  0|   Canterbury Region|European|  1|2018|  222|
|  0|        Otago Region|European|  1|2018|  219|
|  0|     Southland Region|European|  1|2018|  234|
|  0|     Waimate District|European|  1|2018|  ..C|
|  0|                null|European|  1|2018|19563|
|  0|Kapiti Coast Dist...|European|  1|2018|19557|
|  0|Total - New Zeala...|European|  2|2018|  768|
+---+--------------------+--------+---+----+-----+
only showing top 20 rows
```

Figure 6.13: Displaying the JSON DataFrame

This command will display the first 20 rows of the DataFrame created in the previous step.

8. Convert the **Census.csv** file to the Parquet format using the following code:

```
df_census_csv.write.parquet(\
        "F:/Chapter06/Data/Output/census_csv.parquet")
```

A file is created of the following size:

```
Output file size: 16.6 MB
```

This command will write the CSV DataFrame into the Parquet file format. The first argument is the path of the output file.

9. Convert the **Census.json** file to the Parquet format using the following code:

```
df_census_json.write.parquet(\
        "F:/Chapter06/Data/Output/census_json.parquet")
```

A file is created of the following size:

```
Output file size: 1.18 MB
```

This command will write the JSON DataFrame into the Parquet file format. The function argument is the path of the output file.

10. Verify the saved file using the following code:

```
val df_census_parquet = spark.read.parquet(\
        "F:/Chapter06/Data/Output/census_csv.parquet")
```

You should get the following output:

```
df_census_parquet: org.apache.spark.sql.DataFrame = [Year: int,
Age: int … 4 more fields]
```

This command will create a DataFrame (**df_census_parquet**) and load the Parquet file content into the DataFrame.

11. Open the Parquet file using the following code:

```
df_census_parquet.show()
```

You should get the following output:

```
+----+------+------+---+----+-----+
|Year|   Age|Ethnic|Sex|Area|count|
+----+------+------+---+----+-----+
|2013|999999|  null|  2|null|    9|
|2013|999999|  null|  2|null|  ..C|
|2013|999999|  null|  2|null|  339|
|2013|999999|  null|  2|null|  ..C|
|2013|999999|  null|  2|null| 1029|
|2013|999999|  null|  2|null|  264|
|2013|999999|  null|  2|null|  552|
|2013|999999|  null|  2|null|  ..C|
|2013|999999|  null|  2|null|  945|
|2013|999999|  null|  2|null|  699|
|2013|999999|  null|  2|null|  636|
|2013|999999|  null|  2|null|  585|
|2013|999999|  null|  2|null| 1038|
|2013|999999|  null|  2|null| 1329|
|2013|999999|  null|  2|null|  534|
|2013|999999|  null|  2|null| 1146|
|2013|999999|  null|  2|null|  561|
|2013|999999|  null|  2|null|  756|
|2013|999999|  null|  2|null|  546|
|2013|999999|  null|  2|null|  525|
+----+------+------+---+----+-----+
only showing top 20 rows
```

Figure 6.14: Displaying the Parquet DataFrame

This command will display the first 20 rows of the DataFrame created in the previous step.

NOTE

Similarly, the output Parquet file from the JSON DataFrame can be loaded and verified.

NOTE

To access the source code for this specific section, please refer to https://packt.live/32eGNT8.

By completing the exercise, you will be able to create a DataFrame and load data from CSV and JSON files into it. You will also be able to convert the created DataFrame into the Parquet file format and load the Parquet file in a new DataFrame. Let's learn about the Avro file format in the next section.

AVRO

Avro is a row-based file format that can be partitioned easily. It is a language-independent data serialization system that can be processed in multiple programming languages. The schema, which is saved as JSON format, is associated separately with the read and write operations. However, the actual data is converted to a binary format, resulting in reduced file size and improved efficiency.

The serialization converts the data into a highly compact binary format, which any application can deserialize.

It has a reliable provision for schema evolution that adds new fields or changes the existing fields. This feature makes the older file version compatible with the existing code with the least amount of modification.

Because the schema is written in JSON format, this makes it more comprehensive to understand the fields and their data types. In a big data environment, Avro is usually the first preference due to the extremely efficient write capability in a hefty workload.

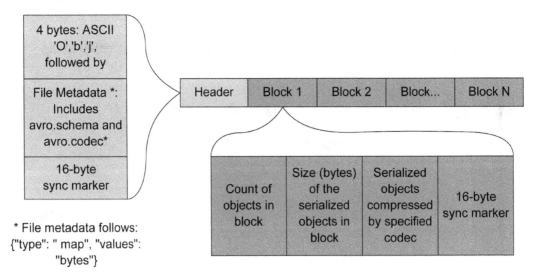

Figure 6.15: Avro file structure

The file structure of the Avro file contains a header and a data block. The header contains a 4-byte string, `'O'`, `'b'`, `'j'`, and **file metadata**. The data block is comprised of the count of the object in the block, a size 16-byte sync marker, and the serialized object of the actual data.

Platform Support

While making a final decision on which file format is most suitable, taking into account the potential harmony with the platform is an important consideration. The Avro file format works optimally with Kafka. Let's now look at a synopsis of the Avro file format:

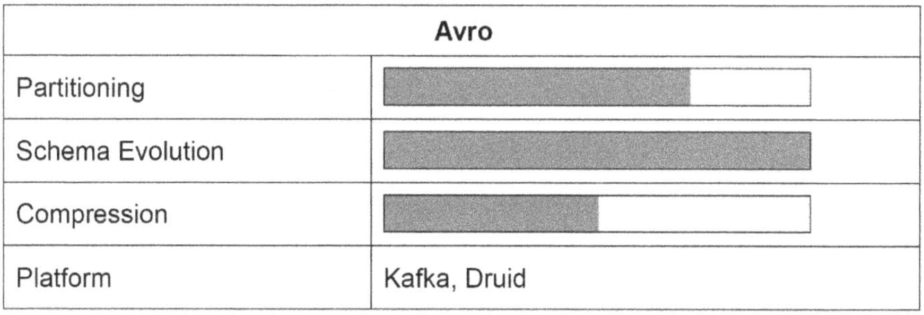

Avro	
Partitioning	
Schema Evolution	
Compression	
Platform	Kafka, Druid

Figure 6.16: Avro attribute synopsis

The synopsis assesses the Avro format's efficiency with regard to various attributes. The Avro file format performs moderately on compression, above-moderately on partitioning, and performs very well in the case of schema evolution.

Now we will proceed to the next exercise and create an Avro file from common input file formats.

EXERCISE 6.02: CONVERTING CSV AND JSON FILES INTO THE AVRO FORMAT

This exercise aims to use a dataset in both input formats, that is, CSV and JSON files, and convert it into the Avro format, which is another big data file format. We will use the same dataset from *Exercise 6.01, Converting CSV and JSON Files into the Parquet Format*. After converting the CSV file, we will read and display the content of the output Avro file to validate the conversion.

Perform the following steps to complete the exercise:

1. In the **Chapter06** directory, create an **Exercise06.02** directory to store the files for this exercise.

2. Open your Terminal (macOS or Linux) or Command Prompt window (Windows), move to the installation directory, and open the Spark shell in it using the following command:

```
spark-shell --packages org.apache.spark:spark-avro_2.11:2.4.5
```

> **NOTE**
>
> The **spark-avro** packages are not included by default and to launch the Spark shell with the **spark-avro** dependency, we will use the –**packages** argument.

You should get the following output:

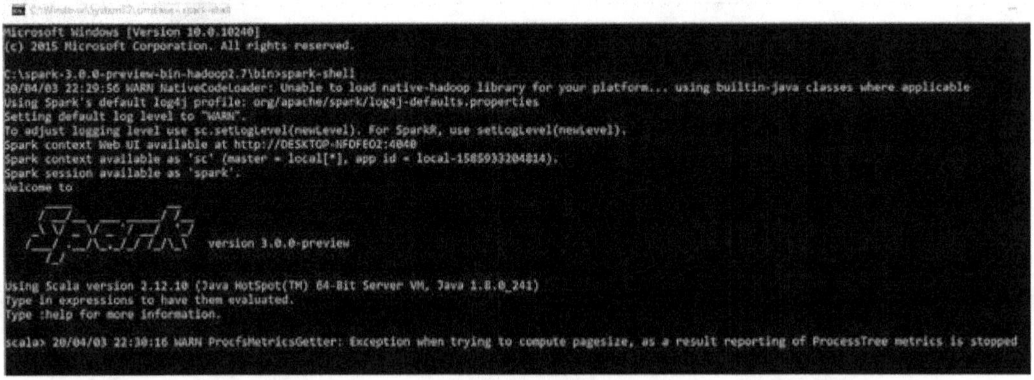

Figure 6.17: Spark shell

3. Read the CSV file using the following code:

```
var df_census_csv = spark.read.options(Map(\
    "inferSchema"->"true","delimiter"->",","header"->"true")\
    .csv("F:/Chapter06/Data/Census.csv")
```

> **NOTE**
>
> Update the input path of the file according to your local file path throughout the exercise.

You should get the following output:

```
df_census_csv: org.apache.spark.sql.DataFrame = [Year: int,
Age: int … 4 more fields]
```

This command creates a DataFrame variable (**df_census_csv**) and loads the contents of the CSV file into the variable.

4. Read the JSON file using the following code:

```
var df_census_json = spark.read.json("F:/Chapter06/Data/Census.json")
```

> **NOTE**
>
> Owing to the size of the data, the commands in this exercise can take up to a few minutes to execute.

You should get the following output:

```
df_census_json: org.apache.spark.sql.DataFrame = [Age: bigint,
Area: string … 4 more fields]
```

5. Show the CSV file using the following command:

```
df_census_csv.show()
```

You should get the following output:

```
+----+---+--------+---+--------------------+-----+
|Year|Age|  Ethnic|Sex|                Area|count|
+----+---+--------+---+--------------------+-----+
|2018|  0|European|  1|Total - New Zeala...|  795|
|2018|  0|European|  1|Total - New Zeala...| 5067|
|2018|  0|European|  1|    Northland Region| 2229|
|2018|  0|European|  1|     Auckland Region| 1356|
|2018|  0|European|  1|      Waikato Region|  180|
|2018|  0|European|  1|Bay of Plenty Region|  738|
|2018|  0|European|  1|     Gisborne Region|  630|
|2018|  0|European|  1|  Hawke's Bay Region| 1188|
|2018|  0|European|  1|     Taranaki Region| 2157|
|2018|  0|European|  1|       Tasman Region|  177|
|2018|  0|European|  1|       Nelson Region| 2823|
|2018|  0|European|  1|  Marlborough Region| 1020|
|2018|  0|European|  1|   West Coast Region|  516|
|2018|  0|European|  1|   Canterbury Region|  222|
|2018|  0|European|  1|        Otago Region|  219|
|2018|  0|European|  1|    Southland Region|  234|
|2018|  0|European|  1|    Waimate District|  ..C|
|2018|  0|European|  1|                null|19563|
|2018|  0|European|  1|Kapiti Coast Dist...|19557|
|2018|  0|European|  2|Total - New Zeala...|  768|
+----+---+--------+---+--------------------+-----+
only showing top 20 rows
```

Figure 6.18: Displaying the CSV DataFrame

This command will display the first 20 rows of the DataFrame created in the previous step.

6. Show the JSON file using the following command:

```
df_census_json.show()
```

You should get the following output:

```
+---+--------------------+--------+---+----+-----+
|Age|                Area|  Ethnic|Sex|Year|count|
+---+--------------------+--------+---+----+-----+
|  0|Total - New Zeala...|European|  1|2018|  795|
|  0|Total - New Zeala...|European|  1|2018| 5067|
|  0|    Northland Region|European|  1|2018| 2229|
|  0|     Auckland Region|European|  1|2018| 1356|
|  0|      Waikato Region|European|  1|2018|  180|
|  0|Bay of Plenty Region|European|  1|2018|  738|
|  0|     Gisborne Region|European|  1|2018|  630|
|  0|  Hawke's Bay Region|European|  1|2018| 1188|
|  0|     Taranaki Region|European|  1|2018| 2157|
|  0|       Tasman Region|European|  1|2018|  177|
|  0|       Nelson Region|European|  1|2018| 2823|
|  0|  Marlborough Region|European|  1|2018| 1020|
|  0|   West Coast Region|European|  1|2018|  516|
|  0|   Canterbury Region|European|  1|2018|  222|
|  0|        Otago Region|European|  1|2018|  219|
|  0|     Southland Region|European|  1|2018|  234|
|  0|    Waimate District|European|  1|2018|  ..C|
|  0|                null|European|  1|2018|19563|
|  0|Kapiti Coast Dist...|European|  1|2018|19557|
|  0|Total - New Zeala...|European|  2|2018|  768|
+---+--------------------+--------+---+----+-----+
only showing top 20 rows
```

Figure 6.19: Displaying the JSON DataFrame

This command will display the first 20 rows of the DataFrame created in the previous step.

7. Convert the **Census.csv** file to the Avro format using the following code:

```
df_census_csv.write.format("avro").save(\
        "F:/Chapter06/Data/Output/census_csv.avro")
```

A file is created of the following size:

```
Output file size: 70.8 MB
```

This command will write the CSV DataFrame into the Avro file format. The first argument is the path of the output file.

8. Convert the **Census.json** file to Avro format using the following code:

```
df_census_json.write.format("avro")\
        .save("F:/Chapter06/Data/Output/census_json.avro")
```

A file is created of the following size:

```
Output file size: 5.0 MB
```

This command will write the JSON DataFrame into the Avro file format. The function argument is the path of the output file.

9. Verify the saved file using the following code:

```
var df_census_avro = spark.read.format("avro").load(\
    "F:/Chapter06/Data/Output/census_csv.avro")
```

You should get the following output:

```
df_census_avro: org.apache.spark.sql.DataFrame = [Year: int,
Age: int … 4 more fields]
```

This command will create a DataFrame (**df_census_avro**) and load the Avro file content into the DataFrame.

10. Open the Avro file using the following code:

```
df_census_avro.show()
```

You should get the following output:

```
+----+------+------+---+----+-----+
|Year|   Age|Ethnic|Sex|Area|count|
+----+------+------+---+----+-----+
|2013|999999|  null|  2|null|    9|
|2013|999999|  null|  2|null|  ..C|
|2013|999999|  null|  2|null|  339|
|2013|999999|  null|  2|null|  ..C|
|2013|999999|  null|  2|null| 1029|
|2013|999999|  null|  2|null|  264|
|2013|999999|  null|  2|null|  552|
|2013|999999|  null|  2|null|  ..C|
|2013|999999|  null|  2|null|  945|
|2013|999999|  null|  2|null|  699|
|2013|999999|  null|  2|null|  636|
|2013|999999|  null|  2|null|  585|
|2013|999999|  null|  2|null| 1038|
|2013|999999|  null|  2|null| 1329|
|2013|999999|  null|  2|null|  534|
|2013|999999|  null|  2|null| 1146|
|2013|999999|  null|  2|null|  561|
|2013|999999|  null|  2|null|  756|
|2013|999999|  null|  2|null|  546|
|2013|999999|  null|  2|null|  525|
+----+------+------+---+----+-----+
only showing top 20 rows
```

Figure 6.20: Displaying the Avro DataFrame

This command will display the first 20 rows of the DataFrame created in the previous step.

> **NOTE**
>
> To access the source code for this specific section, please refer to https://packt.live/32boIFh.

By completing this exercise, you are now able to create a DataFrame and load the data from CSV and JSON files. You are also able to convert the created DataFrame into the Avro file format and also load the Avro file in a new DataFrame. Let's learn about the ORC file format in the next section.

ORC

Apache **ORC** (an abbreviation for **Optimized Row Columnar**) is a column-oriented data format available in the Apache Hadoop environment. You can conceive the structure of an ORC file as divided into three sub-parts, which are **Header**, **Body**, and **Footer**, as shown in the following figure:

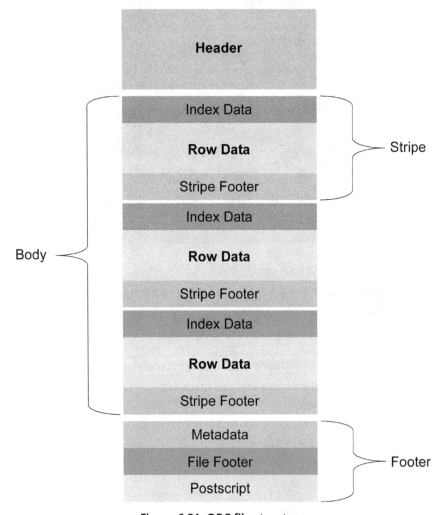

Figure 6.21: ORC file structure

The header is comprised of the keyword **ORC** in case the application requires a determination of the file type while processing.

The body holds the transaction data and indexes, and the transaction data is saved in row chunks known as stripes. The default stripe size is 250 MB, however, increasing the stripe size can increase the read efficiency at the cost of memory.

Each stripe is further apportioned into three sections, that is, the actual data sandwiched between an index section on top and a stripe footer section.

The stripe index holds quick aggregations of the actual data, such as the max and min values, and its row index within that column. The index saves the information required to determine the stripe matching the data needed and the section of the row. The stripe footer includes the encoding information of the columns and the stream directory with its location.

The footer segment is comprised of the layout of the body of the file, statistics about each column, and schema information. It is constructed from three sub-parts, mainly the footer, which is topped by metadata and postscript segments. The metadata segment comprises the statistical information describing columns stored at the stripe level. The file footer has data related to the list of stripes, the number of rows per stripe, and the data type of the column. It has aggregate values for columns, such as **min**, **max**, and **sum**. The postscript segment gives the file information, such as the extent of the file's footer and metadata sections, the file version, and the compression parameters.

Platform Support

When taking a final decision on which file format is most suitable, taking into account the compatibility with the platform is an important consideration. The ORC file format works optimally with Hive since this file format was made for Hive.

Let's now look at a synopsis of the ORC file format:

ORC	
Partitioning	
Schema Evolution	
Compression	
Platform	Hive, Presto

Figure 6.22: ORC attribute synopsis

The synopsis summarizes the performance of the ORC file format with regard to various attributes. The ORC file format is not compatible with schema evolution but works well for applications that require the functionalities of partitioning and compression.

Now we will proceed to the next exercise and create an ORC file from common input file formats.

EXERCISE 6.03: CONVERTING CSV AND JSON FILES INTO THE ORC FORMAT

This exercise uses a dataset in both input formats, that is, CSV and JSON files, and converts it into the ORC format. We will use the same dataset from *Exercise 6.01, Converting CSV and JSON Files into the Parquet Format*. After converting the CSV file, we will read and display the contents of the output Parquet file to validate the conversion.

Perform the following steps to complete the exercise:

1. In the **Chapter06** directory, create the **Exercise06.03** directory to store the files for this exercise.

2. Open your Terminal (macOS or Linux) or Command Prompt window (Windows), move to the installation directory, and open the Spark shell in it using the following command:

```
spark-shell
```

You should get the following output:

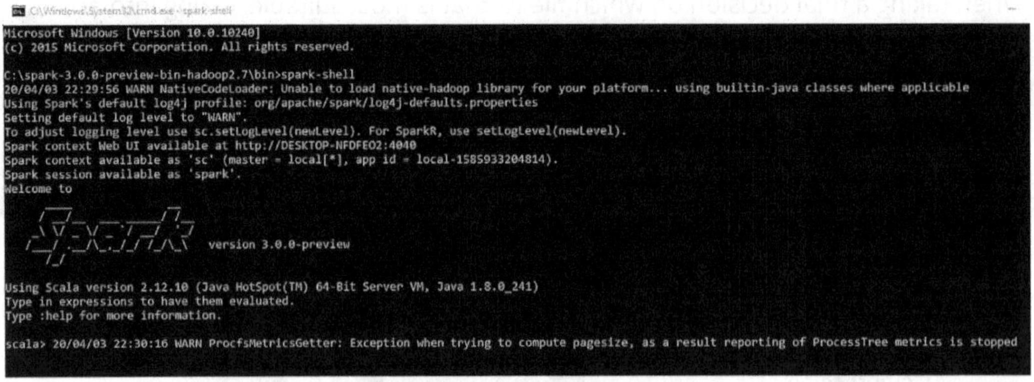

Figure 6.23: Spark shell

3. Read the CSV file using the following code:

```
var df_census_csv = spark.read.options(Map(\
    "inferSchema"->"true","delimiter"->",","header"->"true"))\
    .csv("F:/Chapter06/Data/Census.csv")
```

> **NOTE**
>
> Update the input path of the file according to your local file path throughout the exercise.

You should get the following output:

```
df_census_csv: org.apache.spark.sql.DataFrame = [Year: int,
Age: int … 4 more fields]
```

This command creates a DataFrame variable (**df_census_csv**) and loads the contents of the CSV file into the variable.

4. Read the JSON file using the following code:

```
var df_census_json = spark.read.json(\
    "F:/Chapter06/Data/Census.json")
```

> **NOTE**
>
> Owing to the size of the data, the commands in this exercise can take up to a few minutes to execute.

You should get the following output:

```
df_census_json: org.apache.spark.sql.DataFrame =
[Age: bigint, Area: string … 4 more fields]
```

5. Show the CSV file using the following command:

```
df_census_csv.show()
```

You should get the following output:

```
+----+---+--------+---+--------------------+-----+
|Year|Age|  Ethnic|Sex|                Area|count|
+----+---+--------+---+--------------------+-----+
|2018|  0|European|  1|Total - New Zeala...|  795|
|2018|  0|European|  1|Total - New Zeala...| 5067|
|2018|  0|European|  1|    Northland Region| 2229|
|2018|  0|European|  1|     Auckland Region| 1356|
|2018|  0|European|  1|      Waikato Region|  180|
|2018|  0|European|  1|Bay of Plenty Region|  738|
|2018|  0|European|  1|     Gisborne Region|  630|
|2018|  0|European|  1|   Hawke's Bay Region| 1188|
|2018|  0|European|  1|     Taranaki Region| 2157|
|2018|  0|European|  1|       Tasman Region|  177|
|2018|  0|European|  1|       Nelson Region| 2823|
|2018|  0|European|  1|   Marlborough Region| 1020|
|2018|  0|European|  1|    West Coast Region|  516|
|2018|  0|European|  1|   Canterbury Region|  222|
|2018|  0|European|  1|        Otago Region|  219|
|2018|  0|European|  1|     Southland Region|  234|
|2018|  0|European|  1|     Waimate District|  ..C|
|2018|  0|European|  1|                null|19563|
|2018|  0|European|  1|Kapiti Coast Dist...|19557|
|2018|  0|European|  2|Total - New Zeala...|  768|
+----+---+--------+---+--------------------+-----+
only showing top 20 rows
```

Figure 6.24: Displaying the CSV DataFrame

This command will display the first 20 rows of the DataFrame created in the previous step.

6. Show the JSON file using the following command:

```
df_census_json.show()
```

You should get the following output:

Figure 6.25: Displaying the JSON DataFrame

This command will display the first 20 rows of the DataFrame created in the previous step.

7. Convert the **Census.csv** file to the ORC format using the following code:

```
df_census_csv.write.orc(\
   "F:/Chapter06/Output/census_csv.orc")
```

A file is created of the following size:

```
Output file size: 16.5 MB
```

This command will write the CSV DataFrame into the ORC file format. The first argument is the path of the output file.

8. Convert the **Census.json** file to the ORC format using the following code:

```
df_census_json.write.orc(\
    "F:/Chapter06/Output/census_json.orc")
```

A file is created of the following size:

```
Output file size: 1.1 MB
```

This command will write the JSON DataFrame into the ORC file format. The function argument is the path of the output file.

9. Verify the saved file using the following code:

```
val df_census_orc =
    spark.read.orc("F:/Chapter06/Data/Output/census_csv.orc")
```

You should get the following output:

```
df_census_orc: org.apache.spark.sql.DataFrame = [Year: int,
    Age: int … 4
    more fields]
```

This command will create a DataFrame (**df_census_orc**) and load the ORC file contents into the DataFrame.

10. Open the ORC file using the following code:

```
df_census_orc.show()
```

You should get the following output:

```
+----+------+------+---+----+-----+
|Year|   Age|Ethnic|Sex|Area|count|
+----+------+------+---+----+-----+
|2013|999999|  null|  2|null|    9|
|2013|999999|  null|  2|null|  ..C|
|2013|999999|  null|  2|null|  339|
|2013|999999|  null|  2|null|  ..C|
|2013|999999|  null|  2|null| 1029|
|2013|999999|  null|  2|null|  264|
|2013|999999|  null|  2|null|  552|
|2013|999999|  null|  2|null|  ..C|
|2013|999999|  null|  2|null|  945|
|2013|999999|  null|  2|null|  699|
|2013|999999|  null|  2|null|  636|
|2013|999999|  null|  2|null|  585|
|2013|999999|  null|  2|null| 1038|
|2013|999999|  null|  2|null| 1329|
|2013|999999|  null|  2|null|  534|
|2013|999999|  null|  2|null| 1146|
|2013|999999|  null|  2|null|  561|
|2013|999999|  null|  2|null|  756|
|2013|999999|  null|  2|null|  546|
|2013|999999|  null|  2|null|  525|
+----+------+------+---+----+-----+
only showing top 20 rows
```

Figure 6.26: Displaying the ORC DataFrame

This command will display the first 20 rows of the DataFrame created in the previous step.

NOTE

To access the source code for this specific section, please refer to https://packt.live/309p10I.

By completing the exercise, you will be able to create a DataFrame and load the data from CSV and JSON files. You will also be able to convert the created DataFrame into the ORC file format and load the ORC file in a new DataFrame.

We can observe that the conversions to each file format have resulted in different file sizes owing to the different compression algorithms, schemas, and a few other attributes. Your selection, however, should not just depend upon the size alone but also on the performance of the query execution.

We will now move on to query performance, where we will learn how we can measure the performance of queries on these file formats.

QUERY PERFORMANCE

So far, we have learned how to convert an input file, create DataFrames, and load big data files into these DataFrames. Now, on these DataFrames, we will execute different types of queries to measure the execution performance. To measure the performance, we will create a time function, which will return the time required to execute a query.

The code of the function to do this is as follows:

```
# Function to measure time for query execution

def time[A](f: => A) = {
    val s = System.nanoTime
    val ret = f
    println("Time: "+(System.nanoTime-s)/1e6+" ms")
    ret
}
```

This function, named **Time**, takes an input argument denoted by **A** and then executes the query provided in the argument.

> **NOTE**
>
> The function will result in the output and print the time taken for execution in milliseconds.

Now let's look at the different types of queries that we will execute to measure performance. There are two methods to execute a query over a dataset. The first is to execute queries directly to the DataFrame, and the second is to create a view and run SQL commands over that view. The latter provides the infrastructure to execute complex queries. Let's look into both of these methods.

Method One: DataFrame Operations

In this method, we will use DataFrame functions to query the DataFrame:

```
\\ Define spark dataframe
var dataframe = spark.read.json("F:/Chapter06/Data/Census.json")


\\ This will count the number of records (rows)
\\from the dataframe
dataframe.count()


\\ We can time the query using the time function create above.
time{dataframe.count()}
```

More functions can be found on Apache's website, at https://spark.apache.org/docs/2.2.0/api/scala/index.html#org.apache.spark.sql.functions$.

Method Two: Global Temporary View

1. Create a view from the DataFrame:

   ```
   dataframe.createOrReplaceTempView("view_name")
   ```

   Here, we create a view from the DataFrame by using **createOrReplaceTempView**. It creates or replaces the view name provided to it. The view will be created in Spark's default database and it can be used like a table in Spark SQL. It is session-scoped, which means it will be destroyed after the session is terminated.

2. Formulate the query and use the **spark.sql()** function to execute the query:

   ```
   \\ Defining sql query as a variable

   val sql_query = "Select count(*) from view_name"


   \\ We can time the query using the time
   \\function create above.
   time{spark.sql(sql_query)}
   ```

The SQL function in **SparkSession** facilitates applications to execute SQL queries and returns the result as a DataFrame.

We will move on to the next activity, where we will consider a real-world scenario and apply what we've learned from this chapter.

ACTIVITY 6.01: SELECTING AN APPROPRIATE BIG DATA FILE FORMAT FOR GAME LOGS

This activity focuses on selecting the most suitable big data format from the different file formats available. The selection will be based on size, performance, and current technology to provide the best solution for the given requirements.

You have been recently hired as a data engineer by gaming company XYZ, located in Seattle. They specialize in developing popular and intriguing war games for multiple online users. This results in millions of log records being generated each day, consuming high amounts of storage on the cloud. The trend of the log size and its increasing storage consumption is a concern for the company, and they are determined to resolve this by looking at the various big data file formats available.

The log data comes in JSON format composed of the following columns: **user_id**, **event_date**, **event_time**, **level__id**, and **session_nb**. The company is already utilizing the capabilities of Cloudera and Spark as a part of the big data technology stack.

The analytics team requires the data to be of the smallest size possible that will maintain the operational performance for analytical queries. As a data engineer, the company expects you to analyze and suggest the best file format for this use case, supporting your suggestion with appropriate findings. The decision does not hinge only on compressed file sizes, but on several other factors, so we will store this data in all the file formats that we learned about earlier and perform an operation to find the best-suited format among them for our requirements.

We will be using a sample dataset that was created by the author. The dataset can be found in our GitHub repository at the following location: https://packt.live/2C72sBN.

You need to download the **session_log.rar** file from the GitHub repository. After unzipping it, you will have the **session_log.csv** file.

> **NOTE**
>
> The code for this activity can be found at https://packt.live/2OhlTcH.

Perform the following steps to complete the activity:

1. Load the data from the CSV file.

2. Show the data to verify the load command.

3. Convert the input file into the Parquet, Avro, and ORC file formats and verify the file size of each.

4. Create a function for query performance.

5. Perform and measure the time taken for query execution over each file format for `count` and **GROUP BY** queries.

6. Find the file with the highest compression and quickest query response.

7. Conclude which is the most feasible file format based on the current technology of the company.

> **NOTE**
>
> The solution to this activity can be found on page 616.

SUMMARY

In this chapter, we have learned about common input file formats for big data, such as CSV and JSON. We also learned about popular file formats, namely Parquet, Avro, and ORC, which are useful in the big data environment and looked at essential decision points for making a choice on which to use. We explored the conversion to each of these file formats from the CSV and JSON formats and executed them in a big data environment using Spark and Scala. To strengthen the concept, we executed each format conversion in the respective exercises.

At the end of the chapter, we looked at a real-world business problem and concluded which was the most suitable file format based on the selection criteria learned in this chapter.

In the next chapter, we will extensively cover the vital infrastructure of the big data environment known as Spark. This will lay a strong foundation of the concept and also lead us through the journey of creating our first pipeline in Spark.

7

INTRODUCTION TO ANALYTICS ENGINE (SPARK) FOR BIG DATA

OVERVIEW

This chapter will help you learn the fundamentals of Apache Spark. By combining a sequence of transformations and actions, you will be able to create a pipeline in Spark and run it. We will be using Databricks to launch and use a Spark cluster. By the end of this chapter, you should be comfortable with creating and running a Spark pipeline using a Databricks notebook on a Spark cluster.

INTRODUCTION

What makes Spark one of the most popular analytics engines? How did Spark evolve to become the parallel processing engine of choice? This chapter will help you get answers to these questions and more.

In the previous chapter, we learned about the various big data file formats, including Parquet, AVRO, and ORC, and how to use them. In this chapter, we will solve the challenge of processing large volumes of data that is dynamic, real-time, and grows exponentially in a short period of time. We will learn about systems that can read, write, and process data exponentially faster than sequential processing. This is facilitated by having parallel processing in clusters, which was the origin of **Hadoop** and **MapReduce**. Companies including eBay, Facebook, Twitter, and Google used Hadoop and MapReduce extensively. Later, they found that these systems were not designed for iterative machine learning paradigms. Machine learning models require several iterations to fine-tune the hyperparameters. In the Hadoop and MapReduce framework, data has to be loaded and a separate MapReduce job has to be launched for each iteration, which slows down overall performance.

Researchers at UC Berkeley resolved some of the shortcomings of Hadoop and MapReduce and published their findings in a paper entitled *Spark Cluster Computing with Working Sets*, based on **Resilient Data Sets** (RDDs).

Spark is one of the most popular analytics engines on account of several complementary factors coming together; namely, fast in-memory processing, horizontal scalability, support for the data science community, numerous library packages and support for SQL, streaming, and graph processing. Spark was designed to support parallel processing. With its framework relying on RDDs and lazy evaluation, ensuring that transformations are done only when actions are executed, it became a platform of choice.

The core team of researchers expanded to include an interactive system for running queries, and the addition of libraries, such as MLlib, Spark Streaming, and GraphX. Many of the core team of original contributors to Apache Spark from UC Berkeley got together to start a company named Databricks.

The Databricks business model is to offer a complete analytics engine based on Spark with enterprise-grade support. We will learn how to launch a cluster in Databricks and how to create a Databricks notebook. We will also share some best practices in terms of using Spark effectively as your analytics engine. In the next section, we will introduce Apache Spark and cover Spark's basic terminology, components, and architecture.

APACHE SPARK

Apache Spark is a unified engine that is based on a parallel cluster-based computing model extending the Hadoop MapReduce model. It developed as an open-source project from the research of a Ph.D. student at Berkeley. Apache Spark can handle various parallel and different workloads to process big data.

Spark's foundation layers are low-level granular APIs and structured APIs. The low-level APIs work on RDDs and distributed variables, while the structured APIs work on datasets, DataFrames, and support querying using SQL. Spark's streaming functionalities, unified advanced analytics, and the ecosystem of numerous libraries are built on the foundations of low-level and structured APIs:

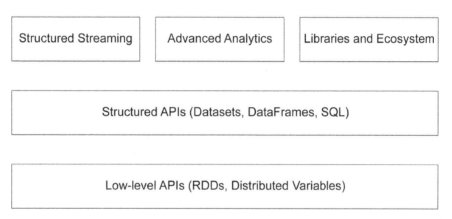

Figure 7.1: Spark architecture

Apache Spark emerged as an attempt to bridge and close some of the gaps in Hadoop/MapReduce. While MapReduce has had its early adoption and traction as a general batch processing engine in companies such as Facebook and Google, the lack of support for iterative multi-pass machine learning modeling and interactive programming, querying, and real-time stream processing was evident.

Matei Zaharia was a Ph.D. student at Berkeley research who got interested in this and developed a programming construct that abstracted several of the data transformations and actions in demand, including streaming, ETL, machine learning, and interactive querying, into a unified platform called Spark.

Spark was initially written in a functional programming language, Scala, and has evolved to support Java and Python. We will use Python as the language of choice for all examples in this chapter, as it is universally popular and used extensively across industries. The ease of setting up Spark and the agile way of developing the platform, by leveraging the open-source community and getting feedback from users all the way through, make Spark a very customer-centric platform.

Databricks, a company that Matei Zaharia co-founded and where he now works as Chief Technologist, provides commercial enterprise distribution and support for Apache Spark, making it one of the most popular and adopted unified analytical platforms. In the next section, we will learn the fundamentals of Spark and cover Spark's terminology.

FUNDAMENTALS AND TERMINOLOGY

An understanding of several Spark concepts and its terminology would be beneficial. These are explained as follows:

- **DataFrame and Datasets**: These are structured tables like collections, with well-defined rows and columns. Every column has the same number of row elements. The difference is that a DataFrame is untyped, while datasets are typed.

- **Lazy Evaluation**: Spark transformations are not executed until there is an action associated with it. So, when we have a transformation call, they are executed immediately. You can force a transformation to be executed by calling an action.

> **NOTE**
>
> Transformations involve taking an input dataset, processing it, and getting an output dataset. Actions are about executing a computation on a dataset and returning a value to the driver program.

- **Spark SQL**: This is Spark's interface for working with structured and unstructured data. Structured data has a schema, a known finite set of columns/ fields for every record/row.

- **MLlib**: This is Apache Spark's machine learning library, with common ML algorithms including classification, regression, clustering, and collaborative filtering. MLlib also supports feature engineering, including feature extraction, transformation, dimensionality reduction, and selection.

- **Structured Streaming**: A scalable and fault-tolerant streaming engine supported by Spark SQL by providing incremental execution and continuous updating.

- **GraphX**: This is a new graph-based abstraction in Spark to help with graph computation, with properties attached to the vertex and edge. It supports several core graph operations.

- **Cluster Manager**: This is a service that acquires and manages resources on a cluster.

- **Spark Context**: Applications in Spark are run as an independent process on a cluster. These are all orchestrated by the Spark context. The Spark context connects to several types of cluster managers. The driver program creates the Spark context and runs the `main()` function.

- **Worker Node**: This is a node that can run application code in a cluster. A worker node interacts with the driver program and the cluster manager.

- **Executor**: This is a process that executes on a worker node, runs tasks, and manages the data.

- **Task**: This is a granular unit of work that can be sent to the executor.

The Spark context in the main driver program orchestrates the various applications that are executed as an independent process. The Spark context connects to various cluster managers. The cluster manager serves as a resource allocation manager across applications. Spark acquires the process known as executors that compute and store application data in the worker node. Spark sends the application code and the tasks to the executors. The cache in the worker node assists in persisting the data in-memory:

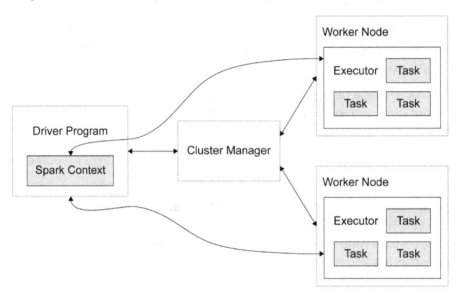

Figure 7.2: Interaction between Spark components

The core construct that makes Spark powerful is originally termed as the concept of RDDs. RDDs have evolved in the terminology of DataFrames and datasets over the years. RDDs are a collection of objects stored in memory or disks across clusters. Data sharing, instead of being on a filesystem in MapReduce, is in-memory in Spark. RDDs are built only via parallel transformations. They restrict the granularity of the transformation operations on data and instead of storing the data in the filesystem, only store the operation and the lineage of the operations. In case of failure, recovery is affected by running the operations on data in the order of lineage sequence in which they have been stored. Spark remembers the sequence of operations used to build the data at each step. So, the user does not have to worry about fault tolerance, as the nodes are automatically rebuilt on failure using the transformation lineage on the data:

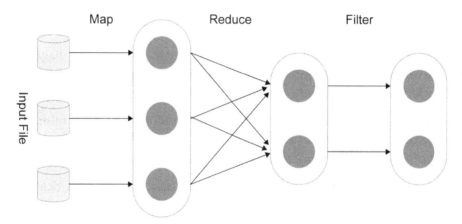

Figure 7.3: RDD framework

The Spark Programming Interface is the framework that allows the user to work with Spark and provides interactive programming using high-level API support for Python, Scala, Java, SQL, and R. It provides the following functionalities:

- Creating and building RDDs

- Performing various computing transformations and operations

- Enabling the user to control the partitioning of the how the RDDs are to be distributed across nodes and how persistence needs to be handled

Also, RDDs support several existing parallel models, including MapReduce, Pregel, and SQL, both individually and as a hybrid combination. In the next section, we will see how Spark works.

HOW DOES SPARK WORK?

Spark has several modular components, as illustrated in the following diagram. Spark Core is the execution engine. It supports in-memory computing. Several applications are supported, and development can be executed in Python, Scala, Java, R, and SQL:

- **Spark SQL**: Data scientists use the Spark SQL module to perform interactive SQL and structured data processing. DataFrames is a programming abstraction that can run as a distributed query engine.

- **Spark Streaming**: Spark Streaming supports interactive and analytical processing on real-time streaming and historical data. It integrates with existing data sources, including HDFC, Flume, Kafka, and Twitter.

- **MLlib**: MLlib is a scalable machine learning library that can execute advanced machine learning algorithms at high speed.

- **GraphX**: Users can build, transform, and process graph-structured data at scale. GraphX supports advanced graphical algorithms at scale:

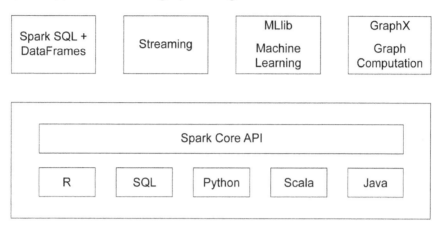

Figure 7.4: Spark components

Spark context is the main entry point for Spark. Spark context connects to various cluster managers and is used to create RDDs and work with variables.

We will use **Apache Spark 2.4** for our examples in this chapter. The data interface is converging toward datasets, which include both DataFrames and RDDs.

In the next section, we will learn about using Databricks and creating a notebook in Databricks.

APACHE SPARK AND DATABRICKS

The most popular integrated platform for learning and using Apache Spark is provided by Databricks. Databricks takes Apache Spark to the next level. It offers five times the performance (compared to Vanilla Apache Spark on the cloud) and integrated Jupyter notebooks in a secure cloud-enabled platform. The core team that developed Apache Spark while at Berkeley is part of Databricks. We will get into the details of the core operations of Spark, namely, transformations and actions. We will use the integrated Jupyter notebooks in Databricks to write the code for this. Databricks enables us to spin Spark clusters on the cloud and connect to it with integrated Jupyter notebooks. So, in the next section, let's set up the Databricks environment and learn to create and use a Jupyter notebook.

EXERCISE 7.01: CREATING YOUR DATABRICKS NOTEBOOK

The best way to learn Spark is by doing exercises and tutorials. You could either set up Spark locally or, even better, use a web-based console. A web-based console will expedite your learning. You can visit the official Spark website https://spark.apache.org/ for more details on how to setup. This exercise aims to create a Jupyter notebook in the Databricks platform, spin up a Spark cluster and connect to it, and be ready to execute the various transformations and actions.

> **NOTE**
>
> Due to the lack of code elements, the code files for this exercise will not be available on GitHub.

Perform the following steps to create your Jupyter notebook in Databricks:

1. Please go to https://databricks.com/try-databricks, enter your details, and click **Sign Up**.

   You should have the following screen:

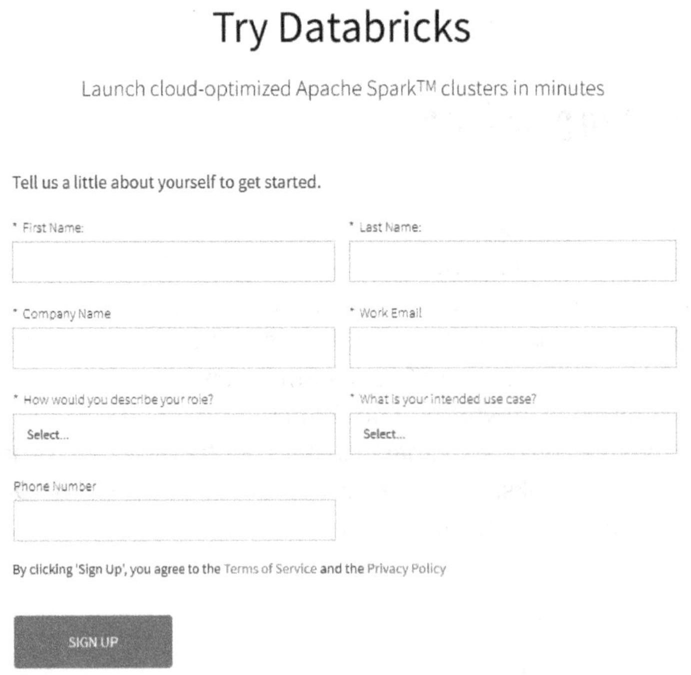

Figure 7.5: Signing up for the Community Edition of Databricks

While signing up, the application will ask for a work email, but it is not mandatory to provide a work email address. You can provide a personal non-work email and it should work as well.

2. Choose **COMMUNITY EDITION** and click `Get Started`, as shown in the following figure:

Try Databricks

Launch cloud-optimized Apache Spark™ clusters in minutes

Select a platform

DATABRICKS PLATFORM - FREE TRIAL

For businesses looking for a zero-management cloud platform built around Apache Spark

- Unlimited clusters that can scale to any size
- Job scheduler to execute jobs for production pipelines
- Fully interactive notebook with collaboration, dashboards, REST APIs
- Advanced security, role-based access controls, and audit logs
- Single Sign On support
- Integration with BI tools such as Tableau, Qlik, and Looker
- 14-day full feature trial (excludes cloud charges)

GET STARTED ON

 OR

COMMUNITY EDITION

For students and educational institutions just getting started with Apache Spark

- Single cluster limited to 6GB and no worker nodes
- Basic notebook without collaboration
- Limited to 3 max users
- Public environment to share your work

GET STARTED

Figure 7.6: Choosing the community version

We selected Community Edition as it is a free version and should suffice for all the learning purposes in this chapter. **`Databricks Platform - Free Trial`** is a commercial version of Apache Spark. You should see a screen like this once you sign up:

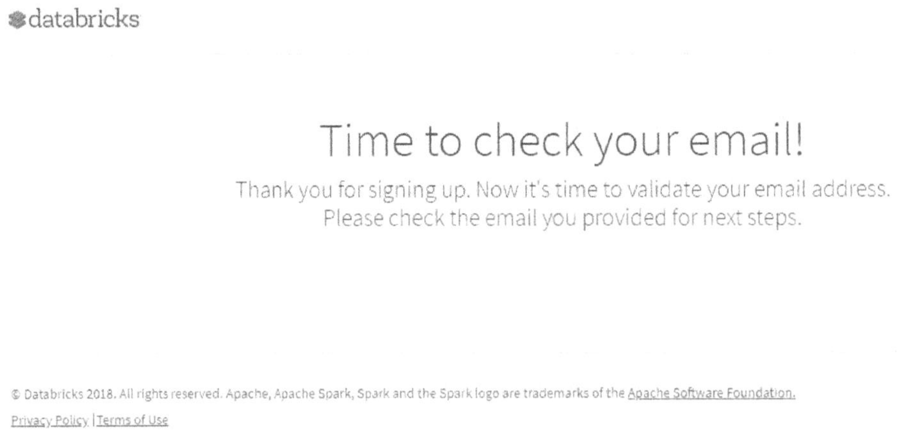
databricks
Time to check your email!
Thank you for signing up. Now it's time to validate your email address.
Please check the email you provided for next steps.
© Databricks 2018. All rights reserved. Apache, Apache Spark, Spark and the Spark logo are trademarks of the Apache Software Foundation.
Privacy Policy | Terms of Use

Figure 7.7: Sharing email details

3. You will receive a verification email (as shown in the following figure) on the email address you signed up with. Click the link to activate your account:

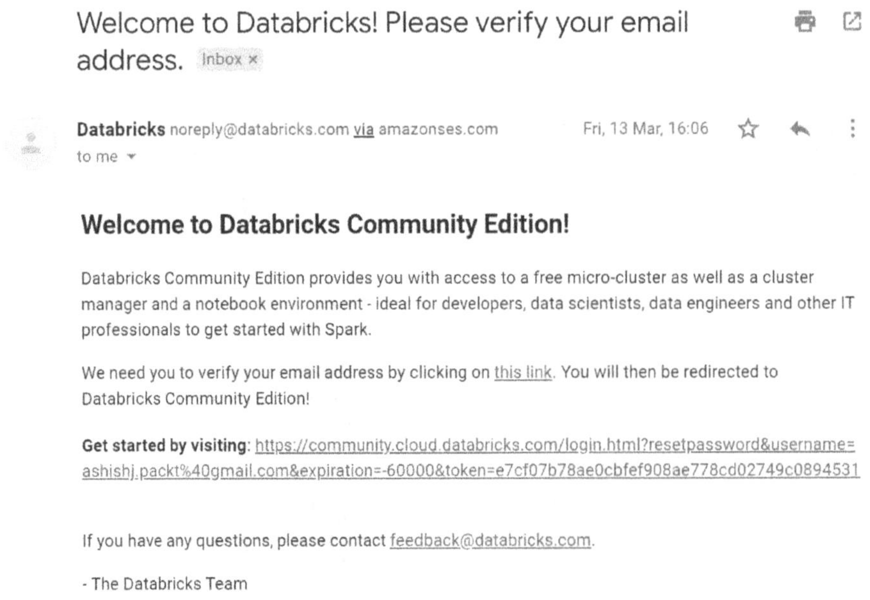
Welcome to Databricks! Please verify your email address. Inbox ×
Databricks noreply@databricks.com via amazonses.com
Fri, 13 Mar, 16:06
to me
Welcome to Databricks Community Edition!
Databricks Community Edition provides you with access to a free micro-cluster as well as a cluster manager and a notebook environment - ideal for developers, data scientists, data engineers and other IT professionals to get started with Spark.
We need you to verify your email address by clicking on this link. You will then be redirected to Databricks Community Edition!
Get started by visiting: https://community.cloud.databricks.com/login.html?resetpassword&username=ashishj.packt%40gmail.com&expiration=-60000&token=e7cf07b78ae0cbfef908ae778cd02749c0894531
If you have any questions, please contact feedback@databricks.com.
- The Databricks Team

Figure 7.8: Email address validation

The link will redirect you to set your password for the account. Please select a password containing a minimum of eight characters, including one uppercase and one special character. You should see a screen like this once you complete the sign-up process:

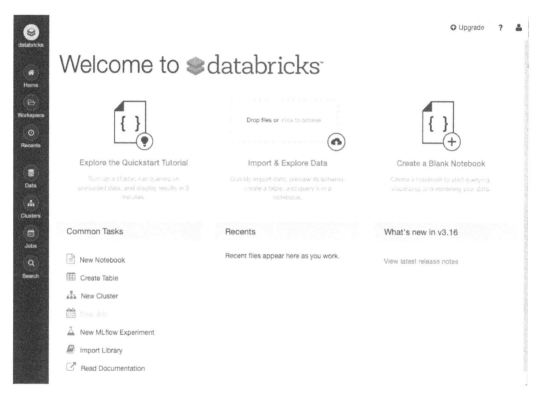

Figure 7.9: Landing page of Databricks

4. Create a cluster by clicking the **Clusters** button on the left pane, and then clicking the **Create Cluster** button, as shown in the following figure:

Figure 7.10: Create a cluster

The cluster needs to be created first and launched so that we can process the data in the cluster. The Jupyter notebook helps you to write the steps associated with processing, but it needs to be connected to a cluster for it to execute.

5. You should have the following screen. Specify the name of the cluster and then click the **Create Cluster** button:

Figure 7.11: Providing the name and other details pertaining to the cluster

6. Upon creation of a cluster, you will be directed to the following screen, where **State** will be **Pending**:

Figure 7.12: Cluster spin in progress

Upon creation, **State** changes to **Running**, as shown in the following figure:

Figure 7.13: The cluster is now active

7. Go to the main home page and click **Create a Blank Notebook**, as shown in the following figure:

Figure 7.14: Creating a Jupyter notebook

8. Create a notebook by specifying the name of the notebook (**SparkNotebook**), and then choose the cluster you would like your notebook to be mapped to, as shown in the following figure:

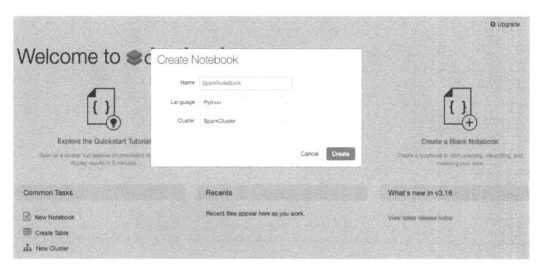

Figure 7.15: Naming the Jupyter notebook

9. Your Jupyter notebook in Databricks is now ready to use:

Figure 7.16: Notebook ready to use

In this exercise, we have learned to create an account in Databricks, spin a Spark cluster, and create a Jupyter notebook in Databricks. In the next section, we will learn about Spark transformations and how to use them.

UNDERSTANDING VARIOUS SPARK TRANSFORMATIONS

Spark supports transformations and actions. Transformations involve taking an input dataset, processing it, and getting an output dataset. Actions are about executing a computation on a dataset and returning a value to the driver program. One such example is the MapReduce construct, where **map** is a transformation, and **reduction** is an action. The transformations are done only when an action is triggered by the driver.

Spark has several core functions. The ETL aspects of the Spark pipeline transformation have already been covered in detail in *Chapter 3, Data Preparation*. The transformation for doing machine learning modeling will be the focus of this section. A typical AI workflow involves identifying the raw data in various formats, including SQL, CSV, and JSON.

Some of the popular transformations are map, filter, sample, union, intersection, and distinct. Applying any of these transformations to a dataset results in another new dataset. Let's now look at transformations in more detail:

- **map(func)**: This transforms an input dataset into an output dataset by applying a func (function) to each element of the input function. A good analogy for understanding this would be multiplying a matrix by a scalar, where each value of the matrix would be multiplied by the scalar value.

- **filter(func)**: This filter selects certain elements based on a criterion, where **func** returns a **true** value. Think of filters like the **SELECT** statement in SQL based on a condition.

- **sample(withReplacement, fraction, seed)**: This picks up a sample of the data with a given random seed generator. With replacement, we replace the first sample we chose before a second sample is chosen. Without replacement, we don't replace the first sample we chose and choose a second sample.

- **union(otherDataset)**: This creates a union between the elements of the input dataset and another dataset that is passed in the argument. For example, the union of {1,2} and {2,3,4} is {1,2,3,4}.

- **intersection(otherDataset)**: This finds the intersection of elements in the source dataset and the dataset that is passed as an argument. For example, the intersection of {1,2} and {2,3,4} is 2.

- **groupByKey([numPartitions])**: **groupByKey** is used on a dataset of key-value pairs. The data can be processed based on the key, over which other actions can be taken. The **groupByKey** with [{a,1}, {a,1}, {a,1}, {b,1},{b,1}] would be {a,3}, {b,2}.

- **cartesian(otherDataset)**: This returns all pairs of combinations (Cartesian multiplication). A Cartesian of {a,b} and {1,2} would be {a,1}, {a,2}, {b,1} and {b,2}.

- **flatMap(function)**: **flatMap** allows for one to many transformations, unlike **Map**, which allows for one-to-one transformation for each of the elements. Splitting lines of a paragraph into various words is a **flatmap** transformation.

We have successfully covered transformations, so let's now explore transformations in more detail through the following exercise.

EXERCISE 7.02: APPLYING SPARK TRANSFORMATIONS TO ANALYZE THE TEMPERATURE IN CALIFORNIA

In this exercise, we will be applying transformations to a dataset that includes temperatures in Celsius for the month of July 2019 in California. We will be converting these temperatures into Fahrenheit and identifying those days when the temperature was higher than 30 degrees Celsius and lower than 25 degrees Celsius.

We will be using a dataset that was created from information at www.accuweather.com/en/us/california/20619/july-weather/2121532?year=2019. The dataset can be found in our GitHub repository at https://packt.live/2C72sBN.

You need to download the **JulyTemp.csv** file from the GitHub repository.

Perform the following steps to complete the exercise:

1. The first step involves logging in to the **COMMUNITY EDITION** of Databricks and accessing the Spark notebook (Jupyter notebook) we created in the previous exercise. You will need to upload the CSV file onto Databricks FileStore first.

2. Input the sample data using the Spark **read** function, as shown in the following code:

```
# temperature of California
#for the month of July 2019
"""https://www.accuweather.com/en/us/california/20619/july-
weather/2121532?year=2019
"""
inTempMonth1inC = spark.read.format('csv')\
                  .options(header='true', \
                  inferSchema='true').load(\
                  '/FileStore/tables/JulyTemp.csv')
display(inTempMonth1inC)
```

You should get the following output:

Day	Temperature
01-Jul-19	27
02-Jul-19	32
03-Jul-19	32
04-Jul-19	31
05-Jul-19	31
06-Jul-19	32
07-Jul-19	28
08-Jul-19	25
09-Jul-19	28
10-Jul-19	29
11-Jul-19	32
12-Jul-19	31
13-Jul-19	30
14-Jul-19	33
15-Jul-19	30
16-Jul-19	33
17-Jul-19	35

Figure 7.17: Input data in Celsius

We have read the CSV file, **JulyTemp**, by using **spark.read.format()** in a **inTempMonth1inC** DataFrame. Setting the **header='true'** parameter makes the function read along with the column headers. Setting **inferSchema='true'** will make Spark infer the data types of the various columns by reading samples of records. **Display()** prints the contents of the DataFrame.

3. Now, transform the sample data from Celsius to Fahrenheit and store the dataset as shown in the following code:

```
#Converting Celsius into Fahrenheit
outTempMonthinF = \
                    inTempMonth1inC.select(\
                    inTempMonth1inC[\
                    'Day'], (inTempMonth1inC[\
                    'Temperature']*9/5)+32)
newcolnames = ['Day','Temperature in F']
for old,new in zip(outTempMonthinF.columns,\
                    newcolnames):
    outTempMonthinF=outTempMonthinF\
                    .withColumnRenamed(old,new)
display(outTempMonthinF)
```

You should get the following output:

Day	Temperature in F
01-Jul-19	80.6
02-Jul-19	89.6
03-Jul-19	89.6
04-Jul-19	87.8
05-Jul-19	87.8
06-Jul-19	89.6
07-Jul-19	82.4
08-Jul-19	77
09-Jul-19	82.4
10-Jul-19	84.2
11-Jul-19	89.6
12-Jul-19	87.8
13-Jul-19	86
14-Jul-19	91.4
15-Jul-19	86
16-Jul-19	91.4
17-Jul-19	95
18-Jul-19	91.4
19-Jul-19	93.2
20-Jul-19	95
21-Jul-19	96.8

Figure 7.18: Data converted into Fahrenheit

We converted the temperature from Celsius into Fahrenheit. The first created a new **outTempMonthinF** DataFrame, from the input DataFrame, by having the temperature column multiplied by a factor of 9/5 and adding 32 to it. We then renamed the column to **Temperature in F** by using the **withColumnRenamed()** function.

4. We will now identify those days where the temperature exceeded **30** degrees Celsius:

```
#Days with temperature greater than or equal
#to 30 degree Celsius
outTempHighinC = \
                inTempMonth1inC.filter(\
                inTempMonth1inC.Temperature >= 30)
display(outTempHighinC)
```

You should get the following output:

Day	Temperature
02-Jul-19	32
03-Jul-19	32
04-Jul-19	31
05-Jul-19	31
06-Jul-19	32
11-Jul-19	32
12-Jul-19	31
13-Jul-19	30
14-Jul-19	33
15-Jul-19	30
16-Jul-19	33
17-Jul-19	35
18-Jul-19	33
19-Jul-19	34
20-Jul-19	35
21-Jul-19	36
22-Jul-19	35
29-Jul-19	33
30-Jul-19	33

Figure 7.19: Higher temperature days

We have used the **filter()** transformation with a condition to filter records with a temperature greater than, or equal to, **30** to show the high temperate days.

5. We will now identify those days where the temperature was lower than **25** degrees Celsius:

```
#finding the days with temperature less than
#or equal to 25 degree Celsius
outTempLowinC = \
                inTempMonth1inC.filter(\
                inTempMonth1inC.Temperature <= 25)
display(outTempLowinC)
```

You should get the following output:

Day	Temperature
08-Jul-19	25
23-Jul-19	25
28-Jul-19	21
31-Jul-19	21

Figure 7.20: Lower temperature days

We have used **filter()** with a condition to filter records where the temperature was lower than, or equal to, **25** to show the low-temperature days.

> **NOTE**
>
> To access the source code for this specific section, please refer to https://packt.live/329p4fP.

In this exercise, we have learned to use transformation with a real-world example of temperature over a period of one month and used **filter** extensively. In the next topic, we will cover Spark actions in detail.

UNDERSTANDING VARIOUS SPARK ACTIONS

Spark actions trigger specified transformations. Transformations create RDDs from another RDD. Actions are the operations that are performed on RDDs to give non-RDD values.

Popular actions include **reduce**, **collect**, **count**, **first**, and **s**. Actions are executed and values of actions are stored back in Spark drivers or external storage systems.

Let's understand transformations in more detail:

- **reduce(func)** : This aggregates the elements of a dataset by executing a function on them. **reduce** works only with commutative and associative functions as it runs in parallel. For example, **reduce** could be taking (a, b) as the two inputs and having a+b as one output. Say if the input data is {1,2,...100}, using the **sum** function on **reduce** would result in {5050}, which is the sum of all the elements of the dataset.

- **collect()** : This returns all the elements in a dataset. This is the equivalent of **select *** in SQL. For example, if the dataset is {1,2,...100}, **count()** would return 100. **collect()** would return the entire dataset {1,2,3,...100}. It retrieves the complete dataset; so you need to use it cautiously on large datasets as it will fetch the entire dataset and store it in-memory. Hence, the entire dataset should fit in the memory.

- **count()** : This counts the number of elements in the dataset. For example, if the dataset is {1,2,3...100}, **count()** would return 100.

- **first()** : This finds the first element in the dataset. For example, if the dataset is {1,2,3...100}, **first()** would return the dataset {1}.

- **take(n)**: This returns the first **n** elements of the dataset. For example, if the dataset is {1,2,3...100}, **take(5)** would return the dataset {1,2...5}.

- **takeSample(withReplacement, num, [seed])** : This returns a sample of size **n**. You may choose to return the sample selected before picking the next sample. For example, if the dataset is {1,2,3...100}, **takeSample(5)** could return a random dataset with five elements, {1,7,11,23, 78}.

- **saveAsTextFile(path)** : This saves the elements of a dataset into a text file in a given path. If the dataset is {1,2,3...100}, **saveAsTextFile(filepath)** would save this to the file path.

- **reduceByKey(func)** : For key-value pair datasets, this applies a function to the ones with the same keys. For example, if we have a dataset {b,b,d,a,b,c,a}, **reduceByKey((x, y) => x+y)** would give {(b,3), (a,2), (d,1), (c,1)}.

- **sortByKey()** : This sorts the input of a key-value pair. For example, if the input is {(b,3), (a,2), (d,1), (c,1)}, the output post sort would be {(a,2), {b,3), (c,1), (d,1)}.

> **NOTE**
>
> The list of all Spark transformations and actions is available at https://spark.apache.org/docs/latest/rdd-programming-guide.html.

SPARK PIPELINE

Spark pipeline is a sequence of transformations and actions typically implemented to achieve a particular objective. By using a combination of transformations and actions, we can create a Spark pipeline. In the following diagram, we have given an example of a Spark pipeline that could be reading a dataset, doing a **map** transformation, filtering it for certain conditions, and then counting the result of the filtered data:

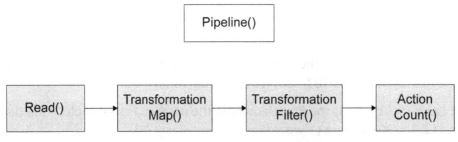

Figure 7.21: Spark pipeline example

For example, you may want to analyze a large blob of text. You could split it into different words using the split **map** function, filter out keywords that are vowels, and perform **countbykey** on the rest to create a dataset for word cloud visualization.

Let's now understand Spark actions better through the next exercise.

EXERCISE 7.03: APPLYING SPARK ACTIONS TO THE GETTYSBURG ADDRESS

In this exercise, we will be applying actions to a text file containing the Gettysburg address. We will upload the text file to the Databricks platform, and then exploring and processing the speech using several Spark functions, including **Filter**, **Count**, **First**, **Take**, and **ReducebyKey**.

We will be using a dataset that was taken from Abraham Lincoln's Gettysburg's address given on November 19, 1863. The dataset for this exercise can be found in our GitHub repository at the following location: https://packt.live/2C72sBN.

You need to download the **GettysburgAddress.txt** file from the GitHub repository.

Perform the following steps to complete the exercise:

1. The first step involves logging in to the **COMMUNITY EDITION** of Databricks.

2. Upload the **GettysburgAddress.text** file to the Databricks platform by clicking the **Drop files or click to browse** button, as shown in the following figure:

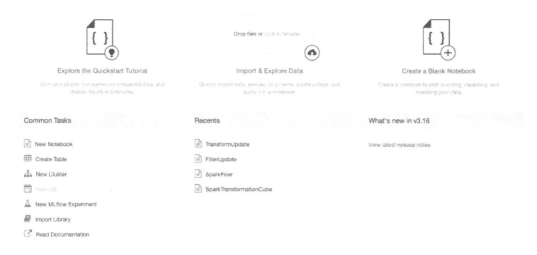

Figure 7.22: Choosing the Drop files or click to browse option in the center

3. Select the file from your local machine to upload the file to Databricks:

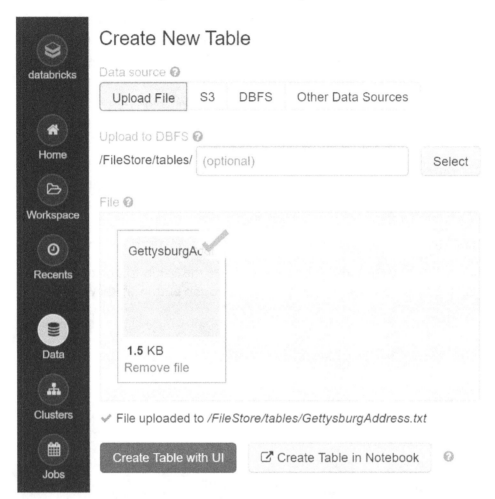

Figure 7.23: Uploading the file to the Databricks platform

Please note the path of the file as **/FileStore/tables/ GettysburgAddress.txt**. The path is shown above the **CreateTable with UI** button.

4. Create a new Jupyter notebook in the Databricks platform and read the **GettysburgAddress.text** file that has been uploaded to Databricks:

```
inputText = \
sc.textFile("/FileStore/tables/GettysburgAddress.txt")
```

5. Find and print the total lines in the dataset using the following code:

```
print("Show the count of lines in input text")
inputText.count()
```

You should get the following output:

```
▸ (1) Spark Jobs
Show the count of lines in input text
Out[316]: 11

Command took 0.30 seconds -- by anand.ns@gmail.com at 01/04/2020, 20:01:02 on SparkCluster
```

Figure 7.24: Number of lines in the input

> **NOTE**
>
> The output included in *Figure 7.24* to *Figure 7.30* will be visible in the Databricks platform. We have uploaded the outputs as text in the GitHub file for your reference.

6. Print the first line of the dataset using the following code:

```
print("Print the first line of the text")
inputText.first()
```

You should get the following output:

```
▸ (1) Spark Jobs
Print the first line of the text
Out[317]: 'Gettysburg Address'

Command took 0.11 seconds -- by anand.ns@gmail.com at 01/04/2020, 20:01:02 on SparkCluster
```

Figure 7.25: First line of the input

7. Let's print the first three lines of the dataset using the following code:

```
print("Print the first 3 lines of the text")
inputText.take(3)
```

You should get the following output:

```
▸ (1) Spark Jobs
Print the first 3 lines of the text
Out[318]: ['Gettysburg Address',
 'Four score and seven years ago our fathers brought forth on this continent a new nation conceived in Liberty and dedica
ted to the proposition that all men are created equal',
 'Now we are engaged in a great civil war testing whether that nation or any nation so conceived and so dedicated can lon
g endure ']
Command took 0.12 seconds -- by anand.ns@gmail.com at 01/04/2020, 20:01:02 on SparkCluster
```

Figure 7.26: First three lines of input

8. Identify those lines with the word **great** in them using the **filter** function, as shown in the following code:

```
#filter the lines with the word great
linesWithGreatWord= inputText.filter(\
                lambda line: "great" in line)


print("Print the count of lines where "\
      "the word Great appears")
linesWithGreatWord.count()
```

You should get the following output:

```
▸ (1) Spark Jobs
Print the count of lines where the word Great appears
Out[319]: 3
Command took 0.23 seconds -- by anand.ns@gmail.com at 01/04/2020, 20:01:02 on SparkCluster
```

Figure 7.27: Count of the lines where the word "great" appears

We identify those lines with the word **great** in them by using the **lambda** expression and then using the **filter** function.

9. Next, print those lines where the word **Great** appears using the following code:

```
print("Print the lines where the word Great appears")
linesWithGreatWord.collect()
```

You should get the following output:

```
 ▸ (1) Spark Jobs
Print the lines where the word Great appears
Out[320]: ['Now we are engaged in a great civil war testing whether that nation or any nation so conceived and so dedicat
ed can long endure ',
 'We are met on a great battlefield of that war We have come to dedicate a portion of that field as a final resting place
for those who here gave their lives that that nation might live ',
 'It is rather for us to be here dedicated to the great task remaining before us that from these honored dead we take inc
reased devotion to that cause for which they gave the last full measure of devotion that we here highly resolve that thes
e dead shall not have died in vain that this nation under God shall have a new birth of freedom and that government of th
e people by the people for the people shall not perish from the earth']

Command took 0.17 seconds -- by ananad.ns@gmail.com at 01/04/2020, 20:01:02 on SparkCluster
```

Figure 7.28: Lines where the word "great" appears

We print the lines where the word "great" appears by using the
`collect()` function.

10. Split the paragraph into various words, and execute a word pair combination as
    shown in the following code:

```
#split the paragraph into various words
wordlist = inputText.flatMap(lambda line: line.split())

#create a word pair combination
wordPairs = wordlist.map(lambda s: (s, 1))
wordPairs.collect()
```

You should get the following output:

```
Out[321]: [('Gettysburg', 1),
 ('Address', 1),
 ('Four', 1),
 ('score', 1),
 ('and', 1),
 ('seven', 1),
 ('years', 1),
 ('ago', 1),
 ('our', 1),
 ('fathers', 1),
 ('brought', 1),
 ('forth', 1),
 ('on', 1),
 ('this', 1),
 ('continent', 1),
 ('a', 1),
 ('new', 1),
 ('nation', 1),
 ('conceived', 1),
 ('in', 1),
 ('Liberty', 1),

Command took 0.15 seconds -- by anand.ns@gmail.com at 01/04/2020, 20:01:02 on SparkCluster
```

Figure 7.29: Output of the word pair combination

First, we create a word list by using the **flatMap** function and then use the **lambda** function with the expression to split the lines. Next, we create a key-value word pair for every word in the word list. The key is the word and the value is **1**.

11. Add up the word pairs, create a word-by-value list, and then sort it in alphabetical order, as shown in the following code:

```
#add up the word pairs and create a word by value list
wordPairsCount = wordPairs.reduceByKey(lambda a, b: a + b)
#sort the list of the word pair key count
#in alphabetical order
print("Printing the word pair key in alphabetical order")
wordPairsCount.sortByKey().collect()
```

You should get the following output:

```
Printing the word pair key in alphabetical order
Out[322]: [('Abraham', 1),
 ('Address', 1),
 ('But', 1),
 ('Four', 1),
 ('Gettysburg', 1),
 ('God', 1),
 ('It', 3),
 ('Liberty', 1),
 ('Lincoln', 1),
 ('Now', 1),
 ('The', 2),
 ('We', 2),
 ('a', 7),
 ('above', 1),
 ('add', 1),
 ('advanced', 1),
 ('ago', 1),
 ('all', 1),
 ('altogether', 1),
 ('and', 6),
```

Figure 7.30: Sorted output of ReduceByKey

We take the word pairs and reduce the key-value pairs by using the **reduceByKey** and the **lambda** expression (**a, b : a+b**). This provides a word count for each word in the input content. Next, we used **sortbykey** to get the values in alphabetical order.

> **NOTE**
>
> To access the source code for this specific section, please refer to https://packt.live/38QFqLz.

In this exercise, we applied various actions, including **Count**, **First**, **Take**, **Collect**, and **ReduceByKey**, to a dataset of strings. In the next activity, we will take a public dataset and solve a real-world problem by applying transformations and actions to it.

ACTIVITY 7.01: EXPLORING AND PROCESSING A MOVIE LOCATIONS DATABASE USING TRANSFORMATIONS AND ACTIONS

Imagine you are a film director based out of San Francisco and have been signed up by Netflix to shoot a web series on how the landscape of San Francisco has changed in the last decade. You are looking for information on crew members, locations, and writers who can help you in this assignment.

The objective is to help you explore locations and crew members from the dataset. You find those movies that were shot after 2015 in San Francisco. From the list of recent movies, you want to identify the popular locations and writers. Lastly, you want to identify the most popular location and most popular writer among the recent moves shot (movies shot after 2015) in San Francisco (SFO).

We will be using a sample dataset that was taken from https://catalog.data.gov/dataset/film-locations-in-san-francisco-b217a. It is intended for public access and use. The dataset can be found in our GitHub repository at the following location: https://packt.live/2C72sBN

You need to download the **Film_Locations_in_San_Francisco.csv** file from the GitHub repository.

> **NOTE**
>
> The code for this activity can be found here: https://packt.live/2PvZSIF.

Perform the following steps to complete the activity:

1. Upload the data from your local repository to the **Databricks filesystem (DBFS)**. All users of DBFS will be able to access a file path.

2. Read this file into a DataFrame.

3. Rename the columns with no spaces between the words.

4. Find the recent movies released in 2015 or later.

5. Find and show the popular locations of recently released movies shot in SFO.

6. Find and show the top three popular locations of recently released movies shot in SFO.

7. Find and show the most popular location of recently released movies shot in SFO.

   You should get the following output:

Locations	count
60 Leavenworth St.	10

Figure 7.31: Most popular location of a recent movie shot in SFO

8. Find and show popular writers of recently released movies shot in SFO.

9. Find and show the top three popular writers of recently released movies shot in SFO.

10. Find and show the most popular writer of recently released movies shot in SFO.

    You should get the following output:

Writer	count
Michael Lannan	168

Figure 7.32: Most popular writer of a recent movie shot in SFO

We have assessed in this activity, our ability to take a public dataset, load it into a Spark cluster, and execute Spark transformations and actions on the dataset. In the next section, we will learn about a number of best practices.

> **NOTE**
>
> The solution to this activity can be found on page 623.

BEST PRACTICES

Now that we have performed exercises to get you started with analytics, let's review some best practices of using Spark. While Spark provides significant performance improvements compared with Hadoop and MapReduce, we need to be aware of some of the best practices to fully derive the value that Spark affords us:

- Use **collect** sparingly; **collect** will try to fetch all the elements in the memory. To validate whether your dataset can be fit into the memory before you used the collect, it is better to use **take** or **take(n)** so that you control the outcome.

- **GroupByKey** is not very efficient, as it involves significant shuffling around; use **ReduceByKey** instead, which aggregates and reduces the amount of shuffling around.

- Use **filter** as a way of pre-processing to clean up the dataset by dropping bad quality data.

- **Map** can be used as a way of pre-processing and imputing values for bad or missing data.

SUMMARY

This chapter provided you with an understanding of Apache Spark. We began with the context of what problems Hadoop and MapReduce resolved, and the gaps that remain. Spark addresses the issue of iterative processing for machine learning algorithms and supports real-time querying and processing of streaming data. We introduced the concept of RDD, which is the core construct of Spark. We also learned how to use the Databricks platform and launch clusters and notebooks in it. We then moved to understanding transformations and actions, which form the key execution steps. Using a combination of transformations and actions, it is possible to create a pipeline. We covered several examples of transformations and actions and how to use them. We learned about transformations, including map, filter, union, and intersection, and also learned how to use actions such as count, collect, reduce, first, and take. We then touched on some of the best practices to keep in mind when using Spark.

In the next chapter, we will cover data system design, and how everything we have learned so far in relation to filesystems, Spark clusters, and others can be integrated to create machine learning and AI pipelines.

8

DATA SYSTEM DESIGN EXAMPLES

OVERVIEW

In this chapter, you will start designing scalable AI systems, putting together information from earlier chapters. We will examine a number of existing system designs and analyze the reasons for specific design choices. We will focus on how to design a mock stock trading system that, in the real world, would rely on AI and data. The system design that we'll apply can be useful for generic systems, too.

By the end of this chapter, you will have seen how important system design is when building systems. You will have internalized core concepts and trade-offs around bottlenecks and scaling. You will also be able to design high availability modular AI systems in a pipeline system that can be generally applied.

INTRODUCTION

In the previous chapter, you learned how to use Apache Spark to process large amounts of data in a pipeline architecture. We'll be looking at a pipeline design again in this chapter and see how we can use pipelines as a powerful system design.

Between 1998 and 2005, the US Navy spent over $1 billion on four separate attempts to implement an **Enterprise Resource Planning (ERP)** system based on SAP AG software. These efforts were regarded as failures, with nearly no value to show for the money that was spent. This shows why having proper designs and plans in place is important, and what can go wrong if the implementation of a system is started before a proper plan is created. Bad design and planning are such a common problem in the software engineering industry that it has led to the much-quoted joke, *"A few weeks of coding can save you hours of planning."*

While it's often tempting to start building systems from the get-go, this approach makes it easy to "paint oneself into the corner" – if you start painting a floor without a plan to get out, you might get stuck. Similarly, if you don't plan properly, you could make mistakes during the implementation phase that require the system to be completely torn down and rebuilt, especially as it scales.

Imagine that you are building an AI system to analyze stock market trades. You might build it and test it out with a few seconds' worth of market data and celebrate when you see that it works as expected. You push it out to a production environment, but it quickly collapses under the real-world load. By the time all the issues that stop it working under real-world loads are discovered, people already rely on the system and are expecting it to work. It's too late to spend time building a proper fix or redesigning the system and the entire project fails.

By taking some time to think and plan at the start of a project, many potential issues can be avoided. For example, in 2014, an air-traffic control system, which had an implementation cost of $2.4 billion, failed because it had insufficient memory to compute a specific calculation about flight paths. With better system design, more memory would have been added, or a safeguard could have been implemented to stop the calculation early if memory use spiked.

In this chapter, we'll walk through designing a scalable system from scratch, focusing on the components that relate closely to AI systems and data storage.

Specifically, we will focus on best practices in system design while looking at the answers to the following questions:

- Why is system design important?

- What are the different components of system design?

- Why is reproducibility important for AI systems, and how can pipelines in system design help achieve this?

- How can we make a system highly available, meaning that it has no downtime and responds quickly to all requests?

Along the way, we will tie together a few different concepts from previous chapters, such as databases, and touch on some concepts that will be covered in more detail in later chapters, such as using cloud resources. We'll understand the need for a good system design in the next section.

THE IMPORTANCE OF SYSTEM DESIGN

An initial system design is important to ensure that it does what is needed. It is hard to estimate how many software projects fail or are late, but it's a large percentage. The **Project Management Institute** (**PMI**) estimates that around 14% of software projects fail completely and almost half are delivered late. Even in small projects, it is common to find out too late that all the stakeholders had vastly different expectations, none of which are met in the final product.

By formally defining what a system will do, what inputs it receives, and what outputs it produces, we can avoid costly mistakes earlier in the process, both in terms of the time needed to fix any mistakes and in terms of lost reputation by making them initially. By clearly defining the different components of a system and how they rely on each other, we can also often identify problems before they happen, and optimize which components have to be built in which order.

A design further allows experts from different fields to collaborate clearly. An architect might look at a system design and raise concerns about scalability, while an information security engineer might look at the same design and raise concerns about how data is secured in a specific component.

Once a system has been built and is running, the design is still important. If any changes have to be made, or if a part of the system breaks, having a clear design can help us figure out if the proposed changes make sense and are safe to implement, or how to fix specific issues.

Overall, a system design helps us in many ways. We can use a system design to do the following:

- Ensure that all the components are compatible

- Discuss trade-offs with different stakeholders

- Make informed decisions before spending time and money on implementation

- Estimate the full cost of the system before things spiral out of control

  Now that we understand why system design is important, let's look at the components that make up system design.

COMPONENTS TO CONSIDER IN SYSTEM DESIGN

No two systems are alike, and therefore no two system designs are alike. However, there are some common components that you will find detailed in system design, such as the features, hardware, data, and architecture of the system, as well as how it handles security, scaling, and so on. We will learn about a few common components in detail. Let's start by understanding the features of a system.

FEATURES

Clear communication regarding the features of the system is integral to any design:

- What does the system do?

- What problems does it solve?

- What inputs can it handle?

- What outputs does it produce?

Everyone must agree on this section. For example, imagine designing Facebook messenger. You might assume that a chat app is very straightforward and doesn't need a complicated design. But if you have several teams working on a chat app, they might all have different ideas about what features to support. Does it handle group chats? Does it display read receipts and show typing indicators? Can images be sent inline? Can chats be exported? What some stakeholders might regard as "obvious" features, others might not even think about. Therefore, it's very important to explicitly describe all of the features and have everyone agree.

HARDWARE

What hardware is used to build the system? We looked at some hardware components in *Chapter 1, Data Storage Fundamentals*, and saw how different hardware was used for different use cases. In a system design document, it is important to detail the hardware requirements of a system and show what hardware will be used.

For example, if you are building an AI data collection system, you must decide early on what chipset and what hard drives you will use. If different teams work on different components, they may not be compatible without prior agreement. Imagine if the machine learning team needs GPUs that consume more electricity than the operations team planned for, or if the platform team gets a $1 million bill for the cloud services that they used, and the finance team was not expecting this. Once again, agreeing on the design prior to implementation is important.

DATA

While we addressed the inputs of the system under the *Features* section, many systems that rely heavily on data need to think carefully about how to store and process this data:

- How does data flow through the system?

- Where is it stored?

- Where might data bottlenecks be?

- How can we keep the data secure?

Imagine that you are building a government system to keep personal records of all the country's citizens. If it's not clearly defined how data is moved between different components and how it is stored, data may likely end up in a place where it was never meant to be. It might be leaked and made public, which could cause all the citizens to have their privacy rights violated, or to be vulnerable to identity theft.

ARCHITECTURE

Often, overlapping with "data" is a more general architecture. What pre-existing patterns are applicable? Some common ones are client-server, peer-to-peer, and message queues, but again, every system is different, and many different patterns can be followed. Similar to building houses, having a bad architecture might not be immediately obvious in software systems. But over time, as changes need to be made, it can become clear that the original design was bad and did not account for future needs. For example, Netscape, one of the most popular original web browsers, realized they had architectural issues and attempted to rewrite the entire code base from scratch for their 6.0 release. This took years, and the original flaws may be one of the reasons why Netscape no longer exists.

SECURITY

While keeping data safe might have been covered in the *Data* section previously, there are many more aspects to a secure system:

- How do we keep our system running, even in the face of malicious actors?

- How do we ensure that it is only used as intended?

- How do we keep audits on how it is used?

A common trend we have seen recently is that of ransomware attacks. An attacker infiltrates a system and encrypts all the files, offering the decryption key in return for a ransom. In 2017, the "WannaCry" attack spread throughout the world, causing, by some estimates, billions of dollars' worth of damage. The **National Health Service (NHS)** in England was badly affected, and the compromised machines affected medical staff and their ability to treat patients. If security isn't baked into a system at all levels, including design, a malicious actor will likely find access and use the system in unintended ways.

SCALING

Finally, we want to make sure that our system design works not only now, but also going forward:

- How can we adapt our system as it grows to ensure that it still works?

- What components can be swapped out as our needs grow?

- Can this be done easily and as necessary?

While good architecture and good design around hardware, data, and features are all part of ensuring that a system can scale, a good system design will address this point explicitly. In 2013, HealthCare.gov, a US website for managing health insurance, was launched without an adequate scaling plan in place. Only around 1% of interested people were able to use it as expected, despite the system costing nearly $2 billion in total. If your system needs to scale to large numbers of people, or to store vast quantities of data, explicitly designing for scale is a must.

Let's look at one way a system can be designed as a pipeline.

EXAMINING A PIPELINE DESIGN FOR AN AI SYSTEM

When your software system is small, such as a small game that you built for your use, it's easy enough to remember all of the pieces mentally and to have a good understanding of how everything works and how it fits together. For complicated systems, such as HealthCare.gov, which we examined previously, it is very normal for no single person to have a complete understanding of how the whole system works. Instead, a few people or a few teams will understand and be responsible for specific components of a system. For example, the responsibility of a global system might be split by region, with one team responsible for America, one for Europe, and so on. Or, responsibility might be split by function, with one team for machine learning, one for data collection, and others for other functions besides.

Let's imagine a hypothetical system that uses information from news articles to inform the end-user of financial trading decisions. It has a list of companies traded on public markets, and it watches for news articles about them. If it finds news articles talking about a new iPhone that Apple released that was well-received, it will buy Apple stock. If it reads about a lawsuit that Uber is facing, it will sell Uber's stock:

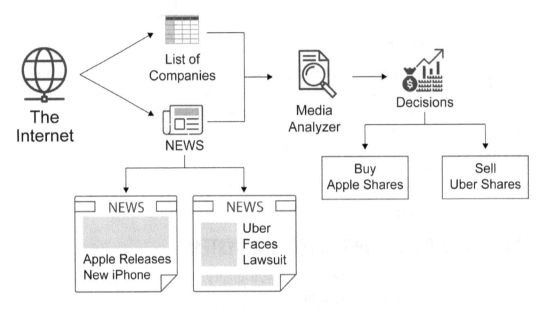

Figure 8.1: A high-level system design diagram

This system would have a lot of functionality and a lot of it would be very complicated. As a very brief summary, it would need to do the following:

- Scrape news articles from a wide range of sources, including Reuters, CNN, Bloomberg, and The Wall Street Journal.

- Clean these news articles by removing adverts, headers, footers, and any other content that isn't central to the specific article.

- Deduplicate these articles, that is, discard copies or near copies of articles when publications use each other's work, resulting in a set of unique articles.

- Analyze the text of each article and perform **Natural Language Processing (NLP)** functions; for example, extracting entities such as company names from the text and performing sentiment analysis to establish whether an article portrays people and companies in good or bad ways.

- Give the processed news information to a decision-making component that decides which stocks to buy, which ones to sell, and what quantities to buy and sell.

- Give that information to a trading component that can execute the trades.

- Have an overall monitoring component that can give insights into each of the previous components and a status report in the case of any issues.

We could try to model all of this as a single system: one huge monolithic code base that executes all of these different functions, and one database that contains all of the data. But this would quickly become hard to maintain, and there's no reason for the news **Scraper** to ever need to be linked directly to the trading component in any case.

One of the cleanest designs a system can have is a pipeline design, where each component takes a specific input and produces a specific output. The next component takes the output of the previous component as input and produces a different output. Each component is independent, apart from the input it receives and the output it produces. If a better component can be created that can handle input and output in the same format, it can easily be switched out:

Figure 8.2: A pipeline architecture

The preceding diagram shows how a pipeline can help with modularity, but pipelines can also help with reproducibility. Let's look at that next.

REPRODUCIBILITY – HOW PIPELINES CAN HELP US KEEP TRACK OF EACH COMPONENT

We discussed how a pipeline design can help with the separation of concerns and interoperability. It is easier to have specific people to be responsible for specific stages of the pipeline, and it is easy to improve the system piece by piece. Another advantage of a pipeline-based design is that it can help us with reproducing results reliably: by using a pipeline design, we can ensure that every input gets processed by each component before being passed on to the next. Because no components can be skipped, we are guaranteed that we will always get the same output, given the same input.

Because data analysis and machine learning often contain many manual steps, including data cleaning, it is easy to lose track of what changes have been made. If a person or a part of the system interferes with another part, it might make changes that could improve the final results. However, on subsequent runs to confirm the validity of the results, it is easy to lose track of what manual changes were applied.

By using a pipeline system, we can ensure that changes happen in a structured and ordered fashion. If each component expects certain inputs and produces certain outputs, it is easier to ensure that, after linking them together, the pipeline itself reliably produces the same overall result from the original input.

While we cannot build this whole system in a single chapter, let's get a feel for system building by implementing a small version of it in the next exercise.

EXERCISE 8.01: DESIGNING AN AUTOMATIC TRADING SYSTEM

In this exercise, we'll build out a small version of the trading system that we described previously. We won't use a real database, and some of the components will be simple mocks (we won't trade real money on real trading platforms), but it will help you get a feel for designing a pipeline-based system. We'll build out the **Scraper**, **Cleaner**, **Deduplicator**, **Analyzer**, and **DecisionMaker** components of the system we described previously.

Perform the following steps to complete this exercise:

1. Create a **Chapter08** directory for all the exercises in this chapter. In the **Chapter08** directory, create the **Exercise08.01** directory to store the files for this exercise.

2. Open your Terminal (macOS or Linux) or Command Prompt (Windows), navigate to the **Chapter08** directory, and type **jupyter notebook**.

3. In the Jupyter notebook, click the **Exercise08.01** directory and create a new notebook file with the Python3 kernel.

4. Define some example URLs of news articles that might be interesting, as shown in the following code:

```
uber_url = "https://www.reuters.com/article/"\
           "us-uber-lawsuit-california/"\
           "uber-is-sued-over-resistance-to"\
           "-california-gig-employment-"\
           "law-idUSKCN1VX1VE"

apple_url = "https://www.reuters.com/article/"\
            "us-apple-macbook/apple-refreshes-"\
            "macbook-pro-laptop-with-16-inch-\"
            "screen-idUSKBN1XN1V8"

apple_url2 = "https://www.reuters.com/article/"\
             "us-apple-macbook/apple-refreshes"\
             "-macbook-pro-laptop-with-16-inch"\
             "-screen-idUSKBN1XN1V8"

article_urls = [uber_url, apple_url, apple_url2]
```

We only define three URLs, all from a single source. Two of them are the same. In the last line, we add all three URLs to an array.

5. Import the **requests** and **string** libraries and the **Counter** module from **collections**, as shown in the following code:

```
import requests
import string
from collections import Counter
```

We imported three libraries that we'll use in various components. The **requests** library is for making HTTP calls to fetch HTML content from the URLs. We'll use the **string** library to clean the texts, and we'll use the **Counter** module from **collections** for some analysis.

6. Define a **Scraper** class that can fetch news articles and extract the entire content, including HTML, as shown in the following code:

```
class Scraper:

    def fetch_news(self, urls):
        article_contents = []
        for url in urls:
            try:
                contents = requests.get(url).text
                article_contents.append(contents)
            except Exception as e:
                print(e)
        return article_contents
```

Here, we defined a **Scraper** class that can fetch news articles. We give it a list of **urls**, and it loops through them. For each **url** element, it attempts to fetch it with the **requests** library and adds the content from that page to an array. Finally, it returns the array. Note that the **Scraper** class takes in an array of URLs (which we have) and outputs an array containing the page contents.

7. Define an **is_clean** function that we'll use in our **Cleaner** module, as shown in the following code:

```
def is_clean(word):
    blacklist = {"var", "img", "e", "void"}
    if not word:
        return False
    if word in blacklist:
        return False
    for i, letter in enumerate(word):
        if i > 0 and letter in string.ascii_uppercase:
            return False
        if letter not in string.ascii_letters:
            return False
    return True
```

This function is outside the main **Cleaner** module. It looks at a word and decides whether it is part of an article. We use a very naive method for this. If the word is in our blacklist, we discard it, as it is probably part of the JavaScript content of the article. If the word is blank, we also discard it.

If both tests pass, we check whether the word has any **uppercase** letters that are not the first letter. If it does, it is probably a function name. Finally, we check whether all the letters in the word are part of the English alphabet. If any other characters are present, we discard the word.

8. Define the full **Cleaner** module, which uses the **is_clean** function, as shown in the following code:

```
class Cleaner:

    def clean_articles(self, articles):
        clean_articles = []

        for article in articles:
            clean_words = []
            try:
                for word in article.split(" "):
                    if is_clean(word):
                        clean_words.append(word)
            except Exception as e:
                print(e)
            clean_articles.append(' '.join(clean_words))
        return clean_articles
```

In this code, we defined a **Cleaner** module with a **clean_articles** function. This function takes the list of articles that the **Scraper** produced and loops through it. For each article, it breaks it into words and keeps only the clean words. It then joins these together again, adds them to a different array, and finally returns the array of cleaned articles.

9. Create the **Deduplicator** module, as shown in the following code:

```
class Deduplicator:

    def deduplicate_articles(self, articles):
        seen_articles = set()
        deduplicated_articles = []
        for article in articles:
            if hash(article) in seen_articles:
                continue
```

```
        else:
            seen_articles.add(hash(article))
            deduplicated_articles.append(article)

    return deduplicated_articles
```

Here, we defined the **Deduplicator** module. It takes in a list of clean articles as an argument. We loop through all of these with a **for** loop and check whether it was seen previously, keeping only new articles. We keep an efficient hash of each article to keep track of what articles we've seen and return a unique set of articles.

10. Create the **Analyzer** module, as shown in the following code:

```
class Analyzer:
    good_words = {"unveiled", "available", \
                  "faster", "stable"}
    bad_words = {"sued", "defiance", "violation"}

    def extract_entities_and_sentiment(self, articles):
        entity_score_pairs = []
        for article in articles:
            score = 0
            entities = []
            for word in article.split(" "):
                if word[0] == word[0].upper():
                    entities.append(word)
                if word.lower() in self.good_words:
                    score += 1
                elif word.lower() in self.bad_words:
                    score -= 1
            main_entities = [i[0] \
            for i in Counter(entities).most_common(2)]
            entity_score_pair = (main_entities, score)
            entity_score_pairs.append(entity_score_pair)
        return entity_score_pairs
```

The **Analyzer** module defines two lists of words: **good_words** and **bad_words** (in a real-world use case, these would be much larger lists). If the article is talking about a new product being unveiled, that is a good sign. If a company is being sued, that is probably bad. It loops through each article with a **for** loop and splits each into a list of words.

For each word, it checks whether the word is an entity (it guesses that a word is an entity if it starts with a capital letter). It then checks whether the word is regarded as a **'good'** or **'bad'** word, based on the lists. If it is good, it adds **1** to the **score** variable. If it is bad, it removes **1**. If the word does not appear in either list, it leaves the **score** variable as-is.

Finally, it finds the two most common entities mentioned in the article and creates a data structure with both entities and the overall score. It returns this as output.

11. Now, create the **DecisionMaker** module using the following code:

```
class DecisionMaker:
    target_companies = set(['Apple', 'Uber', 'Google'])

    def make_decisions(self, entity_score_pairs):
        decisions = []
        for entities, score in entity_score_pairs:
            for entity in entities:
                if entity in self.target_companies:
                    quantity = abs(score)
                    order = "Buy" if score > 0 else "Sell"
                    decision = (order, quantity, entity)
                    decisions.append(decision)
        return decisions
```

This module has a set of target companies. These are the companies whose stock we want to trade. It takes the entity score pairs that we created in the **Analyzer** module as input and turns these into structured trading decisions. If the score is positive for a given entity, it buys that stock. If it is negative, it sells the stock. The more positive or negative the **score** variable is, the more stock it buys or sells. It returns a list of **decisions** array as output.

12. Initialize all components by running the following code:

```
scraper = Scraper()
cleaner = Cleaner()
deduplicator = Deduplicator()
analyzer = Analyzer()
decision_maker = DecisionMaker()
```

We created all five components, and they are now ready to be tested.

13. Fetch the news articles with the **scraper** and print out an excerpt by running the following code:

```
contents = scraper.fetch_news(article_urls)
contents[0][:500]
```

You should get the following output:

```
'<!--[if !IE]> This has been served from cache <![endif]-->\n<!--[if !IE]> Request served from apache server: produs
--i-0a4a08336159d88d2 <![endif]-->\n<!--[if !IE]> Cached on Wed, 26 Feb 2020 23:01:28 GMT and will expire on Wed, 26
Feb 2020 23:16:19 GMT <![endif]-->\n<!--[if !IE]> token: f9fd82a6-e004-4871-85e1-63089475bceb <![endif]-->\n<!--[if !
IE]> App Server /produs--i-0655f4557687834a5/ <![endif]-->\n\n<!doctype html><html lang="en" data-edition="BETAUS">\n
<head>\n\n    <title>\n                    U'
```

Figure 8.3: Output (excerpt) of our raw content

We ran our **Scraper** and output the first **500** characters of the first article. We can see that it fetched content, but that this is messy and full of HTML tags and other information that is not part of the article.

14. Pass these articles to the cleaner for cleaning, as shown in the following code:

```
clean_articles = cleaner.clean_articles(contents)
clean_articles[0][:500]
```

You should get the following output:

```
'This has been served from cache Request served from apache Cached on Feb and will expire on Feb App Server Uber is s
ued over resistance to California employment law Segment snippet included if Page hiding snippet Data Layer Object De
claration New Google Tag Manager new End Google Tag Manager new driver for Uber has sued the company for misclassifyi
ng its drivers as independent hours after California legislators voted to help thousands of those workers and enjoy t
he benefits of produced in Proces'
```

Figure 8.4: Output (excerpt) of our cleaned content

We ran our cleaner and output the first **500** characters of the first article. We can see that a lot of the junk is removed. It isn't perfect as there is still some content in the beginning that is not from the article, but cleaning is a tricky task and at least we can see that the real content appears near the beginning.

15. Check how many articles we have, run the **deduplicator** module, and then check the count of the articles again, as shown in the following code:

```
print(len(clean_articles))
deduplicated = deduplicator.deduplicate_articles(clean_articles)
print(len(deduplicated))
```

You should get the following output:

```
3
2
```

We printed out the length of our **clean_articles** array and noted that we have **3**, one for each of our original URLs. We then ran our **deduplicator** module, which removed the duplicate article, leaving us with the text of **2** articles.

16. Run our **analyzer** module on our clean deduplicated articles, as shown in the following code:

```
entity_score_pairs = analyzer.extract_entities_and_
sentiment(deduplicated)
print(entity_score_pairs)
```

You should get the following output:

```
[(['Uber', 'California'], -18), (['Pro', 'Apple'], 16)]
```

We ran the **analyzer** module on our articles. We can see that it figured out that the first article was mainly about **Uber** and **California** and that it had a negative sentiment. The second article was mainly about **Apple** and **Pro** (the article talks a lot about the new MacBook Pro) and has a positive sentiment.

17. Pass this information to our **DecisionMaker** module to create trade instructions, as shown in the following code:

```
decisions = decision_maker.make_decisions(\
          entity_score_pairs)
print(decisions)
```

You should get the following output:

```
[('Sell', 18, 'Uber'), ('Buy', 16, 'Apple')]
```

We created two decisions from our entity and sentiment pairs. The **DecisionMaker** wants to sell **18** shares of **Uber** and buy **16** shares of **Apple**.

> **NOTE**
>
> To access the source code for this specific section, please refer to https://packt.live/2Ok606j.

All our components are very naive and would not work well in a real-world case. The **Scraper** downloads articles that we give it but can't find them for itself. The cleaner doesn't even attempt to parse HTML and keeps a lot of content that is not relevant. It also discards a lot of "real" words: those with punctuation, brand names with capital letters in the middle of a word, and more.

Our **deduplicator** module only deals with exact duplicates, but in real cases, often, there are small differences between articles that are almost the same. Therefore, hashing is not a good strategy here. Our **analyzer** module uses some hand-picked wordlists that are relevant mainly to the articles that we chose, and it has a very naive entity extractor, relying only on capital letters.

Finally, our **DecisionMaker** does not take information from all the articles into account. If there is 1 very positive article on Apple and 10 slightly negative ones, it might still decide to buy more stock than it sells.

With all of this in mind, we still can see that the system is both very modular and pipeline-based. Each component is responsible only for a single aspect of the entire system, and any of them can be improved without affecting the others, as long as the input and output formats remain the same. The output of each component is fed as input into another component, leaving us with a very neat pipeline.

This is great for maintaining and understanding the system but is also good for reproducing results. Often, in machine learning systems, reproducibility is important, and by having a structured pipeline, you can always feed the same data in to get the same results out.

Another important aspect that a pipeline design can help with is availability, so we'll cover that in the next section.

MAKING A PIPELINE SYSTEM HIGHLY AVAILABLE

While we have seen how a pipeline can help with keeping a design modular and keeping research reproducible, another advantage of a pipeline system is that it can help make a system highly available. This means that the system will remain at least partially operational, even in unexpected scenarios or under high load. For example, popular messaging apps such as Facebook Messenger might occasionally deliver messages more slowly if there are global networking problems, but it is rare for it to not be able to deliver messages at all. Therefore, we say that Facebook Messenger has high availability.

We've already seen how we can easily swap out various components, but in our design, each component relied on the previous component to execute flawlessly. If any step had a problem, the entire system would break down because each component connected directly to the previous and following components.

In the real world, it is common for one component to break independently of the others, or for one specific component to become a bottleneck if it suddenly has to do more work with fewer resources. Instead of a component passing work directly on, it is often good practice to add a queue between components.

A queue is a simple **first-in, first-out** (**FIFO**) data structure that can have work pushed in from one end and taken out the other. Queues can also exist as part of managed services, such as AWS **Simple Queue Service** (**SQS**), which is a fully managed queue service that helps scale large systems.

If each component in a pipeline design puts work onto a queue, then it doesn't matter if the next component breaks down or gets overloaded. The work will simply wait until it is fixed or available again.

Once again, it would be too ambitious to build a real queue-based system in a single chapter, but we can mock out some components and simplify this to gain hands-on experience with designing a system that makes use of queues.

Look at the system from *Figure 8.02* again. In *Exercise 8.01, Designing an Automatic Trading System*, we implemented a version of this system that could scrape data from a limited set of URLs and make trading decisions. We'll build that system again, but because we are focusing on the system design and not the functionality, we will make the requirements even simpler.

We will start with a toy dataset that looks as follows and build a series of components that each modify the dataset in a small way:

```
`['url1-', 'url1-', 'url2-', 'url3-']`.
```

The **Scraper** module will remove the `url` from each string, leaving only the number and the trailing hyphen. The **Cleaner** module will remove the trailing hyphen, leaving only the number. The **Deduplicator** module will delete any duplicated numbers. The **Analyzer** module will turn any even number into a negative number or keep any odd number as-is. The **DecisionMaker** module will choose to "buy" if the number is positive or "sell" if it is negative, and the **trader** module will simply print the result calculated by the **DecisionMaker**.

We'll build the first half of this system in the next exercise.

EXERCISE 8.02: ADDING QUEUES TO A SYSTEM TO MAKE IT HIGHLY AVAILABLE

Imagine that you built a modular system but found that it did not perform well under high load. In this exercise, we'll see how to add a queue data structure between each component of a system to make it highly available. Because the focus now is on adding queues instead of building components, we'll use a slightly simpler implementation of the system. We'll use functions instead of classes for each component, and we'll use static data instead of downloading real news article data. We'll see how queues and threads work in Python and how these can help ensure that a system can cope with higher loads.

Perform the following steps to complete this exercise:

1. Create a directory called **Chapter08** for all the exercises in this chapter. In the **Chapter08** directory, create the **Exercise08.02** directory to store the files for this exercise.

2. Open your Terminal (macOS or Linux) or Command Prompt (Windows), navigate to the **Chapter08** directory, and type **jupyter notebook**.

3. In the Jupyter notebook, click the **Exercise08.02** directory and create a new notebook file with the Python3 kernel.

4. Import the **random** and **time** standard libraries, as well as the **Queue** and **Thread** classes from their respective modules, as shown in the following code:

```
from queue import Queue
from threading import Thread
import random
import time
```

We imported the modules that we will use to design our next mock system. The **queue** and **threading** libraries in Python are designed for parallelization. The **random** library will allow us to simulate random wait times, while the **time** library will be used to simulate heavier work by putting the processes to sleep for short periods.

5. Initialize the mock dataset and put it into a **queue** module, as shown in the following code.

```
urls = ['url1-', 'url1-', 'url2-', 'url3-', \
        'url4-', 'url5-', 'url6-', 'url7-', \
        'url8-', 'url9-', 'url10-']
seen = set()

url_queue = Queue()
for url in urls:
    url_queue.put(url)
```

We created 11 mock URLs and a **seen** set to find duplicates. We then created a **queue** module for our URLs and added each URL to the **queue**.

6. Set up queues for each of the other components, up to and including the **deduplicator**, as shown in the following code:

```
scraped_queue = Queue()
cleaned_queue = Queue()
deduplicated_queue = Queue()
```

We initialized **Queue()** objects for each component to push to when done.

7. Define the **scraper** module as a function, as shown in the following code:

```
def scraper():
    while True:
        time.sleep(random.randrange(0,2))
        url = url_queue.get()
        print("Scraping {}".format(url))
        scraped_queue.put(url[3:])
```

Our **scraper** function is designed to be run in a thread, so we have a while true loop. We use **time.sleep()** to simulate work taking a variable amount of time. The **random.randrange()** call returns an integer in the range [0,2] (so either zero or one). We then pass this random integer to **time.sleep()**, causing an artificial delay of either 0 or 1 second, simulating how some URLs might take longer to download than others. We get an available URL from the first queue for processing and remove the first three characters, leaving just the number and the trailing hyphen.

8. Define similar functions for the **cleaner** and **deduplicator** components, as shown in the following code:

```
def cleaner():
    while True:
        time.sleep(random.randrange(2,4))
        raw = scraped_queue.get()
        print("Cleaning {}".format(raw))
        cleaned_queue.put(raw.replace("-", ""))

def deduplicator():
    while True:
        time.sleep(random.randrange(4,6))
        cleaned = cleaned_queue.get()
        print("Deduplicating {}".format(cleaned))
        if cleaned not in seen:
            deduplicated_queue.put(cleaned)
            seen.add(cleaned)
```

Here, we defined functions for our **cleaner** and **deduplicator** components. They both work very similarly to the **scraper()** function, but the **cleaner** component removes the trailing hyphen and the **deduplicator** component checks whether the cleaned version (only the number) has been seen before.

9. Initialize the threads for each component, as shown in the following code:

```
scraper_worker = Thread(target=scraper)
cleaner_worker = Thread(target=cleaner)
deduplicator_worker = Thread(target=deduplicator)
```

Here, we created three threads, one for each component using the **Thread** class, and passed the respective functions in using the **target** parameter.

10. Add the threads to a list and start each of them, as shown in the following code:

```
threads = [
    scraper_worker, cleaner_worker, deduplicator_worker
]

[t.start() for t in threads]
```

You should get the following output:

```
[None, None, None]

Scraping url1-
Scraping url1-
Cleaning 1-
Scraping url2-
Scraping url3-
Scraping url4-
Scraping url5-
Scraping url6-
Scraping url7-
Scraping url8-
```

Figure 8.5: Excerpt of the output from our queue-based system

We put our three worker threads into an array and called **start()** on each of them. Note that although the scraper works significantly faster than the other two components, the system remains available as the **scraper** puts all its outputs on a **Queue**, waiting to be picked up by the **Cleaner**. If the **Cleaner** were to break completely, the work would continue queuing up, waiting for it to become available again. As the scraper and cleaner complete work from the queue, they output a summary, starting with "Scraping..." or "Cleaning...". The output shows that the scraper is working faster than the cleaner, but that the cleaner doesn't break down under the load. Instead, the work waits in the queue until the cleaner is ready to process it.

> **NOTE**
>
> To access the source code for this specific section, please refer to https://packt.live/38QrnWk.

In this exercise, we saw how a queue-based architecture could be added to a pipeline system architecture to make it highly available. We created a mock system to gain some experience with these concepts.

In the next activity, you will complete the system by building out the remaining components: **analyzer**, **decision_maker**, and **trader**. Based on what we learned in these exercises, we'll implement everything in the following activity.

ACTIVITY 8.01: BUILDING THE COMPLETE SYSTEM WITH PIPELINES AND QUEUES

Imagine that you have been added to a project to build an automated trading system halfway through. The lead engineer has left with only half of the components implemented and it is your job to implement the remaining components to get the system shipped. The previous team implemented the **scraper, cleaner,** and **deduplicator** modules, but it is up to you to implement the **analyzer, decision_maker,** and **trader** modules, and see the system run end to end.

> **NOTE**
>
> The code and the resulting output for this activity have been loaded into a Jupyter notebook that can be found here: https://packt.live/2OISPSA.

Perform the following steps to complete this activity:

1. Import the **random** and **time** standard libraries, as well as the **Queue** and **Thread** classes from their respective modules.

2. Initialize a mock dataset of 10 placeholder URLs, as in the previous exercise, and put each item into a queue.

3. Set up queues for each of the components.

4. Define the **scraper, cleaner,** and **deduplicator** modules, just like we did in *Exercise 8.02, Adding Queues to a System to Make It Highly Available*.

5. Define the **analyzer** module.

6. Define the **decision_maker** module.

7. Define the **trader** module.

8. Initialize the threads for each component.

9. Add all the components to an array and start each thread.

   You should get the following output:

   ```
   [None, None, None, None, None, None]

   Scraping url1-
   Scraping url2-
   Cleaning 1-
   Scraping url3-
   Scraping url4-
   Scraping url5-
   Deduplicating 1Scraping url6-

   Analyzing 1
   Deciding 1
   Trading Buy 1
   Buy 1
   Cleaning 1-
   Scraping url7-
   Scraping url8-
   Cleaning 2-
   Scraping url9-
   Scraping url10-
   ```

   Figure 8.6: Excerpt of the output from the full system

You have now completed the full trading system. You've seen how a queue-based architecture can help with the reproducibility and availability of AI systems.

> **NOTE**
>
> The solution to this activity can be found on page 630.

SUMMARY

Building systems without a design is like trying to build a house without a plan. It might seem like you can just start putting bricks together, but in reality, you need a plan. In this chapter, we looked at the importance of system design. We also showed you how system design can help with many different aspects, including early detection of issues, agreement between key stakeholders on features, and communication between different experts collaborating on one system. We built our modular pipeline-based system to make stock trading decisions based on news about companies. We then saw how adding queues to a pipeline system can increase availability, and we rebuilt another mock of our trading platform using queues and learned how queues could store work for a component that is not yet ready to process it. This pipeline design will be expanded upon in the next chapter.

In the next chapter, you will explore the pipeline concept even further through workflow management platforms and job automation systems, including composing a data pipeline with Airflow DAG.

9

WORKFLOW MANAGEMENT FOR AI

OVERVIEW

In this chapter, we will learn how to create a pipeline by breaking down a job into multiple executable stages. We will implement a simple linear pipeline and then go even further by implementing a multi-stage data pipeline. Then, we will automate the multi-stage pipeline using Bash. Furthermore, to improve efficiency, we will run the pipeline as an asynchronous process using an ETL workflow. Lastly, we will create a **Directed Acyclic Graph** (**DAG**) for the pipeline and implement it using Airflow. By the end of this chapter, you will have created an automated multi-stage pipeline that you can manage with the help of Airflow.

INTRODUCTION

In previous chapters, we introduced different databases for different business-use cases. We also introduced the next-generation compute engine Spark for big data analytics. With these tools, we now have all the necessary building blocks for composing any AI data pipeline:

Figure 9.1: A representative flow chart for a typical data pipeline

A typical data pipeline (not limited to AI) looks like the following:

1. Collect user feedback from a user application.

2. Store all user feedback and data in a data storage system.

3. Extract raw user data from the data storage system.

4. Preprocess raw data into a predefined format so that data science/AI applications can process it.

5. Cook the processed data into a higher-level view so that business people such as product managers can digest it and make data-informed decisions.

Let's imagine you are working in a data-driven company such as Netflix. Data scientists are building data pipelines that generate insights about user behaviors. Based on user behaviors, they also build more pipelines that leverage movie data to build personalized movie recommendation systems. Meanwhile, there are business analysts from different teams, such as the US consumer division, the Netflix original content team, and even the Asia marketing team, who are consuming different analyses from hundreds or even thousands of data pipelines that are being run or constantly updated every day. Now, the question is: how do we manage those data pipelines (at such a scale)?

In a typical business, there could exist hundreds or even thousands of data pipelines that are being run or constantly updated every day. Now the question is: how do we manage those data pipelines (at such a scale)?

Fortunately, this is not a unique problem for one specific company. It's a problem for multiple companies. Thanks to the open-source community, we have a tool called **Airflow** that solves the problem.

In this chapter, you will learn how to build a typical data pipeline and use Airflow to manage the data pipelines you build. After this chapter, you will have the ability to set up and configure a workflow orchestration system to manage data science/AI data pipelines in the industry.

So, let's move on further to understand and create a data pipeline.

CREATING YOUR DATA PIPELINE

If you are interested in data science, you should be somewhat familiar with typical data pipelines. A data pipeline starts with raw data. In this chapter, we will be using data on trending videos in the US for the period from 2017 to 2019. Let's say this raw data is in a flat-file CSV format. There are several columns for each entry. However, not all of the columns of this data are relevant to our data pipeline. We need to only select the columns that are required for our purposes. This step involves cleaning the data from the source files and is called data processing. After data processing, we need to store the clean data in our databases. We select the most appropriate model based on our needs, according to what we found out about when looking at data modeling in *Chapter 5, Data Stores: SQL and NoSQL Databases*. Then, we will perform some queries to deploy this data into production. When the data is up and running, we'll continue monitoring the data to improve its overall performance.

This all often starts with raw data access. You will need to read raw data from a data storage solution, which is often a database. Then you will process the raw data into a format that your machine learning model can understand. You might also want to do some feature engineering with the processed data if you are using traditional machine learning models. After the "laborious" data engineering work, we do the modeling work, which is considered the fun part of the process. After some tuning and tweaking and when we are happy with the model, we then deploy it to production. Once it's in production, we also need to monitor its performance. It's a straightforward process and we can easily illustrate its implementation using the following flow chart:

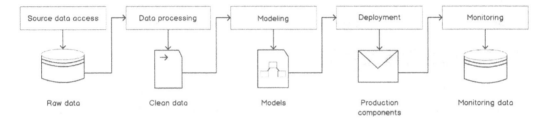

Figure 9.2: A representative flow chart for a typical machine learning modeling process

When we compose a data pipeline, we should ensure that each of the individual steps in the pipeline has the following properties: atomicity, isolation, granularity, and sequential flow.

Looking at the illustration of the data pipeline, notice how we break down a relatively large problem into multiple actionable steps. Each step should be easy to implement. Each represents a single task that should be as atomic as possible. Each step should be relatively encapsulated and be able to stand on its own. When we create a step in a flow, we want to design it in a way such that its complexity is minimized, which means it has the minimum amount of dependencies and parameters. As soon as a task gets complicated, it can be hard to manage and handling failure can be troublesome. We don't need to worry about complexity for now; we will dig into more details later in the chapter.

In the flow chart in *Figure 9.1*, we can easily see that each step in the data pipeline depends on the completion of its previous step. For example, we cannot start the data processing step until we have accessed the source data successfully. The same goes for modeling, deployment, and monitoring. A workflow management system schedules one process after the completion of another. This data pipeline has a linear dependency. Later in this chapter, we will introduce other types of data pipelines that aren't linear, which requires a graph structure to organize the dependencies between tasks.

In short, a complete data pipeline is made up of multiple tasks that are automated by a workflow management system. By the end of this chapter, we will know how to create an end-to-end data pipeline. Before we jump into creating an end-to-end workflow, we should start by creating a single-task job.

EXERCISE 9.01: IMPLEMENTING A LINEAR PIPELINE TO GET THE TOP 10 TRENDING VIDEOS

Imagine you are building a dashboard to show the top 10 daily trending videos in the USA. In this exercise, we will write a Python script to generate a sorted list of the top 10 trending videos.

We will be using a sample dataset that was collected using the YouTube API. The dataset can be found in our GitHub repository at the following location: https://packt.live/2C72sBN.

You need to download the **USvideos.csv.zip** and **US_category_id.json** files from the GitHub repository.

Before proceeding to the exercise, we need to set up a data science development environment. We will be using Anaconda. Please follow the instructions in the *Preface* to install it.

Perform the following steps to complete the exercise:

1. Create a **Chapter09** directory for all the exercises of this chapter. In the **Chapter09** directory, create two directories named **Data** and **Exercise09.01**.

2. Move the downloaded **USvideos.csv.zip** and **US_category_id.json** files to the **Data** directory.

3. Open your Terminal (macOS or Linux) or Command Prompt (Windows), navigate to the **Chapter09** directory, and type **jupyter notebook**. The Jupyter Notebook should look as in the following screenshot:

Figure 9.3: The Chapter09 directory in Jupyter Notebook

4. Select the **Exercise09.01** directory, then click **New** -> **Python 3** to create a new **Python 3** notebook, as shown in the following screenshot:

Figure 9.4: Create a new Python 3 Jupyter Notebook

5. Import the **pandas** module and read the downloaded data using the **read_csv** method, as shown in the following snippet:

```
import pandas as pd
df = pd.read_csv('../Data/Usvideos.csv.zip' ,
  compression='zip')
df.head()
```

You should get the following output:

Out[1]:

	video_id	trending_date	title	channel_title	category_id	publish_time	tags	views	likes	dislikes
0	2kyS6SvSYSE	17.14.11	WE WANT TO TALK ABOUT OUR MARRIAGE	CaseyNeistat	22	2017-11-13T17:13:01.000Z	SHANtell martin	748374	57527	2966
1	1ZAPwfrtAFY	17.14.11	The Trump Presidency: Last Week Tonight with J...	LastWeekTonight	24	2017-11-13T07:30:00.000Z	last week tonight trump presidency|"last week ...	2418783	97185	6146
2	5qpjK5DgCt4	17.14.11	Racist Superman | Rudy Mancuso, King Bach & Le...	Rudy Mancuso	23	2017-11-12T19:05:24.000Z	racist superman|"rudy"|"mancuso"|"king"|"bach"...	3191434	146033	5339
3	puqaWrEC7tY	17.14.11	Nickelback Lyrics: Real or Fake?	Good Mythical Morning	24	2017-11-13T11:00:04.000Z	rhett and link|"gmm"|"good mythical morning"|"...	343168	10172	666
4	d380meD0W0M	17.14.11	I Dare You: GOING BALD!?	nigahiga	24	2017-11-12T18:01:41.000Z	ryan|"higa"|"higatv"|"nigahiga"|"i dare you"|"...	2095731	132235	1989

Figure 9.5: A glimpse of US trending video data

6. Filter the data for the date **17.14.11** by running the following command:

```
df.query('trending_date=="17.14.11" ')
```

You should get the following output:

Out[2]:

	video_id	trending_date	title	channel_title	category_id	publish_time	tags	views
0	2kyS6SvSYSE	17.14.11	WE WANT TO TALK ABOUT OUR MARRIAGE	CaseyNeistat	22	2017-11-13T17:13:01.000Z	SHANtell martin	748374
1	1ZAPwfrtAFY	17.14.11	The Trump Presidency: Last Week Tonight with J...	LastWeekTonight	24	2017-11-13T07:30:00.000Z	last week tonight trump presidency| last week ...	2418783
2	5qpjK5DgCt4	17.14.11	Racist Superman | Rudy Mancuso, King Bach & Le...	Rudy Mancuso	23	2017-11-12T19:05:24.000Z	racist superman|"rudy"|"mancuso"|"king"|"bach"...	3191434 1
3	puqaWrEC7tY	17.14.11	Nickelback Lyrics: Real or Fake?	Good Mythical Morning	24	2017-11-13T11:00:04.000Z	rhett and link|"gmm"|"good mythical morning"|"...	343168

Figure 9.6: US trending videos on the date 2017-11-14

7. Create a **get_trendy_vids.py** Python script under the **Exercise09.01** directory.

> **NOTE**
>
> Besides using Jupyter Notebook, we can also create a Python script in a Terminal with **touch get_trendy_vids.py**. Another recommended option is to use the Visual Studio Code as a text editor to write Python script.

8. Write the following code inside the **get_trendy_vids.py** Python script:

```python
from pathlib import Path
import pandas as pd


def get_topn_viewed(df, date, topn):
    return df.query('trending_date==@date').sort_values('views',
    ascending=False).head(topn)


if __name__ == "__main__":
    # config
    PATH_FILE_IN = Path(__file__).parent.absolute()/'../Data/USvideos.
csv.zip'
    PATH_FILE_OUT = 
        Path(__file__).parent.absolute()/'../Data/top_10_trendy_vids.c
        sv'

    DATE = "17.14.11"
    TOPN = 10

    # read data
    df_data = pd.read_csv(PATH_FILE_IN, compression='zip')
    # get top n trendy
    df_trendy = get_topn_viewed(df_data, DATE, TOPN)
    # save results
    df_trendy.to_csv(PATH_FILE_OUT, index=False)
```

The Python program sorts the **views** column in descending order. It then selects the top 10 trending videos for Nov 14, 2017 (**17.14.11**) and writes to a CSV file named **top_10_trendy_vids.csv**.

> ## NOTE
>
> We have separated the functions from the execution steps. The code under the **if __name__ == "__main__":** statement is executed in sequential order when you run this Python script. The functions defined in the previous step are registered as modules at the global level. This separation makes the code look cleaner and makes it easier for others to understand.
>
> It's best to put all hardcoded variables at the beginning of the program so that they can be modified next time without having to go through all the code to change each value one by one.

9. In your Terminal, navigate to the **Exercise09.01** directory and run the following command to generate the top 10 trending videos:

```
python get_trendy_vids.py
```

The data for the top 10 trending videos are stored in a CSV file named **top_10_trendy_vids.csv** in the **Data** directory.

10. Open the **top_10_trendy_vids.csv** file in Jupyter Notebook using the following command:

```
import pandas as pd
pd.read_csv('../Data/top_10_trendy_vids.csv')
```

You should get the following output:

Out[3]:

	video_id	trending_date	title	channel_title	category_id	publish_time	tags	views
0	2Vv-BfVoq4g	17.14.11	Ed Sheeran - Perfect (Official Music Video)	Ed Sheeran	10	2017-11-09T11:04:14.000Z	edsheeran|"ed sheeran"|"acoustic"|"live"|"cove...	33523622
1	n1WpP7iowLc	17.14.11	Eminem - Walk On Water (Audio) ft. Beyoncé	EminemVEVO	10	2017-11-10T17:00:03.000Z	Eminem|"Walk"|"On"|"Water"|"Aftermath/Shady/In...	17158531
2	9wg3v-01yKQ	17.14.11	Harry Styles - Kiwi	HarryStylesVEVO	10	2017-11-08T13:00:01.000Z	Columbia|"Harry Styles"|"Kiwi"|"Pop"	9632678
3	9t9u_yPEidY	17.14.11	Jennifer Lopez - Amor, Amor, Amor (Official Vi...	JenniferLopezVEVO	10	2017-11-10T15:00:00.000Z	Jennifer Lopez ft. Wisin|"Jennifer Lopez ft. W...	9548677

Figure 9.7: Top 10 trending US videos

We have successfully created a Python job to get the daily top 10 trending US videos.

> **NOTE**
>
> To access the source code for this specific section, please refer to https://packt.live/32elzUm.

> **NOTE**
>
> We use Jupyter Notebook quite often during development. When we work with data, Jupyter Notebook is a great tool to understand and visualize underlying data and its structure. However, as great as Jupyter Notebook is as a development tool, it's best not to use it for production-level applications. It's preferred to use a Python script as the executable for a Python program.

Although the Python job we created in this exercise is very simple, we can still break it down into a multi-stage process. This is illustrated in the following figure:

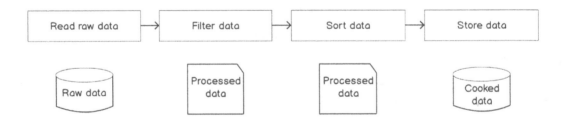

Figure 9.8: A flow chart for fetching the top 10 trending videos

In our Python job, we first use **pandas** to read raw data, then we use the **pandas** DataFrame API to filter and sort data. Finally, we write the output data into a CSV file. These steps constitute a multi-stage process data pipeline.

By completing this exercise, you have successfully implemented your first data pipeline to generate the top 10 trendy US videos in Python using **pandas**. We implemented a three-step data pipeline in a single Python script. In the next exercise, we will expand our data pipeline and create a multi-stage pipeline with multiple Python scripts.

EXERCISE 9.02: CREATING A NONLINEAR PIPELINE TO GET THE DAILY TOP 10 TRENDING VIDEO CATEGORIES

In this exercise, we will increase the complexity of our data pipeline. We will build a pipeline that depends on two data sources: one is the same data from the previous exercise, while the other one is the video category data, which is stored in the **US_category_id.json** file. We will use both data sources to get information on the daily top 10 trending video categories.

During *Exercise 9.01*, *Implementing a Linear Pipeline to Get the Top 10 Trending Videos*, we found that there is a **category_id** column in the dataset, but it's an integer column. We want to find out the literal category for each YouTube video. This means we need to find out the mapping between the actual name of the category and the category ID. Now, your job is to create another Python program that generates the top 10 trending video categories.

Perform the following steps to complete the exercise:

1. Create a new directory, **Exercise09.02**, in the **Chapter09** directory to store the files for this exercise.

2. Open your Terminal (macOS or Linux) or Command Prompt (Windows), navigate to the **Chapter09** directory, and type **jupyter notebook**. The Jupyter Notebook should look as shown in the following screenshot:

Figure 9.9: The Chapter09 directory in Jupyter Notebook

3. In the Jupyter Notebook, click the **Exercise09.02** directory and create a new notebook file with the **Python 3** kernel.

4. We will use a built-in module, **json**, to load the JSON file as shown in the following code:

```
# read cat data
import json

cat = json.load(open('../Data/US_category_id.json', 'r'))
cat
```

You should get the following output:

```
Out[1]:  {'kind': 'youtube#videoCategoryListResponse',
          'etag': '"m2yskBQFythfE4irbTIeOgYYfBU/S730I1t-Fi-emsQJvJAAShlR6hM"',
          'items': [{'kind': 'youtube#videoCategory',
           'etag': '"m2yskBQFythfE4irbTIeOgYYfBU/Xy1mB4_yLrHy_BmKmPBggty2mZQ"',
           'id': '1',
           'snippet': {'channelId': 'UCBR8-60-B28hp2BmDPdntcQ',
            'title': 'Film & Animation',
            'assignable': True}},
          {'kind': 'youtube#videoCategory',
           'etag': '"m2yskBQFythfE4irbTIeOgYYfBU/UZ1oLIIz2dxIhO45ZTFR3a3NyTA"',
           'id': '2',
           'snippet': {'channelId': 'UCBR8-60-B28hp2BmDPdntcQ',
            'title': 'Autos & Vehicles',
            'assignable': True}},
          {'kind': 'youtube#videoCategory',
           'etag': '"m2yskBQFythfE4irbTIeOgYYfBU/nqRIq97-xe5XRZTxbknKFVe5Lmg"',
           'id': '10',
           'snippet': {'channelId': 'UCBR8-60-B28hp2BmDPdntcQ',
            'title': 'Music',
```

Figure 9.10: YouTube video category data

The JSON file gets loaded into the Python environment as a dictionary object. We can see that there are three keys, **['kind', 'etag', 'items']**, at the highest level. Within the **items** key, there is an array of sub-dictionaries. Within one of these sub-dictionaries, we can extract the **title** category from the **snippet** sub-key.

The category information is in the **items** key. The **id** key maps to **category_id** in the previous dataset, **USvideos.csv**.

5. Create a **pandas** DataFrame out of the dictionary object (**cat**) using the following code:

```
import pandas as pd
df_cat = pd.DataFrame(cat)
df_cat
```

You should get the following output:

Out[2]:

	kind	etag	items
0	youtube#videoCategoryListResponse	"m2yskBQFythfE4irbTleOgYYfBU/S730Ilt-Fi-emsQJv...	{'kind': 'youtube#videoCategory', 'etag': '"m2...
1	youtube#videoCategoryListResponse	"m2yskBQFythfE4irbTleOgYYfBU/S730Ilt-Fi-emsQJv...	{'kind': 'youtube#videoCategory', 'etag': '"m2...
2	youtube#videoCategoryListResponse	"m2yskBQFythfE4irbTleOgYYfBU/S730Ilt-Fi-emsQJv...	{'kind': 'youtube#videoCategory', 'etag': '"m2...
3	youtube#videoCategoryListResponse	"m2yskBQFythfE4irbTleOgYYfBU/S730Ilt-Fi-emsQJv...	{'kind': 'youtube#videoCategory', 'etag': '"m2...
4	youtube#videoCategoryListResponse	"m2yskBQFythfE4irbTleOgYYfBU/S730Ilt-Fi-emsQJv...	{'kind': 'youtube#videoCategory', 'etag': '"m2...
5	youtube#videoCategoryListResponse	"m2yskBQFythfE4irbTleOgYYfBU/S730Ilt-Fi-emsQJv...	{'kind': 'youtube#videoCategory', 'etag': '"m2...

Figure 9.11: pandas DataFrame of YouTube video category data

6. Now, extract **title** and **id** from the dictionary object and store them in the new **category** and **id** columns using the **.apply** function, as shown in the following code:

```
df_cat['category'] = df_cat['items'].apply(lambda x:
  x['snippet']['title'])
df_cat['id'] = df_cat['items'].apply(lambda x: int(x['id']))
df_cat
```

You should get the following output:

Out[5]:

	kind	etag	items	category	id
0	youtube#videoCategoryListResponse	"m2yskBQFythfE4irbTleOgYYfBU/S730Ilt-Fi-emsQJv...	{'kind': 'youtube#videoCategory', 'etag': '"m2...	Film & Animation	1
1	youtube#videoCategoryListResponse	"m2yskBQFythfE4irbTleOgYYfBU/S730Ilt-Fi-emsQJv...	{'kind': 'youtube#videoCategory', 'etag': '"m2...	Autos & Vehicles	2
2	youtube#videoCategoryListResponse	"m2yskBQFythfE4irbTleOgYYfBU/S730Ilt-Fi-emsQJv...	{'kind': 'youtube#videoCategory', 'etag': '"m2...	Music	10
3	youtube#videoCategoryListResponse	"m2yskBQFythfE4irbTleOgYYfBU/S730Ilt-Fi-emsQJv...	{'kind': 'youtube#videoCategory', 'etag': '"m2...	Pets & Animals	15
4	youtube#videoCategoryListResponse	"m2yskBQFythfE4irbTleOgYYfBU/S730Ilt-Fi-emsQJv...	{'kind': 'youtube#videoCategory', 'etag': '"m2...	Sports	17
5	youtube#videoCategoryListResponse	"m2yskBQFythfE4irbTleOgYYfBU/S730Ilt-Fi-emsQJv...	{'kind': 'youtube#videoCategory', 'etag': '"m2...	Short Movies	18

Figure 9.12: The new category and id columns

The category's title is stored under the **snippet**, which is stored under the **item** sub-dictionary as mentioned in *Step 4*. Similarly, the **id** category comes from the value of the **id** key. The **pandas.apply** function iterates rows along the **item** column and returns the value that we define in the **lambda** function.

7. Drop the **kind**, **etag**, and **items** columns using the **.drop** function as shown in the following code:

```
df_cat_drop = df_cat.drop(columns=['kind', 'etag', 'items'])
df_cat_drop
```

You should get the following output:

```
Out[6]:
```

	category	id
0	Film & Animation	1
1	Autos & Vehicles	2
2	Music	10
3	Pets & Animals	15
4	Sports	17
5	Short Movies	18

Figure 9.13: Removed unnecessary columns

We have created a **df_cat_drop** mapping table from the JSON file successfully.

8. Read the **USvideos.csv.zip** video data with **pandas**, then filter to **trending_date=="17.14.11"**:

```
# read video data
df_vids = pd.read_csv("../Data/USvideos.csv.zip",
    compression='zip').query('trending_date=="17.14.11"')
```

We created a DataFrame called **df_vids** that contains information about each YouTube video and video category ID. We will join this DataFrame with **df_cat_drop** in the next step.

9. Merge the **df_cat_drop** and **df_vids** DataFrames using the **pandas. merge** method, as shown in the following code:

```
# merge
df_join = df_vids.merge(df_cat_drop, left_on='category_id',
  right_on='id')[
    ['title', 'channel_title', 'category_id', 'category',
      'views']]
df_join
```

You should get the following output:

```
Out[5]:
```

	title	channel_title	category_id	category	views
0	WE WANT TO TALK ABOUT OUR MARRIAGE	CaseyNeistat	22	People & Blogs	748374
1	Me-O Cats Commercial	Nobrand	22	People & Blogs	98966
2	AFFAIRS, EX BOYFRIENDS, $18MILLION NET WORTH -...	Shawn Johnson East	22	People & Blogs	321053
3	BLIND(folded) CAKE DECORATING CONTEST (with Mo...	Grace Helbig	22	People & Blogs	197062
4	Wearing Online Dollar Store Makeup For A Week	Safiya Nygaard	22	People & Blogs	2744430
...	...	...	...	...	...
195	Train Swipes Parked Vehicle	ViralHog	2	Autos & Vehicles	7265
196	Caterham Chris Hoy 60 Second Donut Challenge	Caterham Cars	2	Autos & Vehicles	4850
197	Inside Keanu Reeves' Custom Motorcycle Shop	...	WIRED	2	Autos & Vehicles	704363
198	New Emirates First Class Suite	Boeing 777	...	Emirates	19	Travel & Events	141148
199	L.A. Noire - Nintendo Switch Trailer	Nintendo	20	Gaming	154872

200 rows × 5 columns

Figure 9.14: The joined DataFrame from two datasets

By merging the two DataFrames, we now have category information for each YouTube video. The **pandas .merge** function merges two DataFrames when the **category_id** column from the left DataFrame, **df_cat_drop**, matches the value in the **id** column from the right DataFrame, **df_vid**.

We can use [["col_1", "col_2", ...]] to select specific columns from the DataFrame. Here we selected the `['title', 'channel_title', 'category_id', 'category', 'views']` columns.

Now that we have a category for each YouTube video, we can group each category and sort them based on the total number of views for all of the videos.

10. Group and sort the categories based on the total views using the `.groupby` function, as shown in the following code snippet:

```
df_join.groupby('category')[['views']].sum().sort_values('view
   s', ascending=False).head(10)
```

You should get the following output:

Out[6]:

category	views
Music	97097557
Entertainment	28047357
Comedy	16643490
Howto & Style	15926296
People & Blogs	7753285
Science & Technology	6537832
Film & Animation	6015279
Sports	2755539
News & Politics	1571196
Pets & Animals	1010302

Figure 9.15: Top 10 most viewed YouTube video categories

In **pandas**, we can use **pandas.groupby("column_1")[["column_2"]].sum()** to calculate the sum of the values for each group. Furthermore, we can use **pandas.sort_values("column", ascending=False)** to re-rank the DataFrame.

The **pandas.head(10)** function will return the top 10 rows of the DataFrame.

11. Now, create a **get_trendy_cats.py** Python script in the **Exercise09.02** directory; we will combine all the preceding code snippets in it and save the file:

get_trendy_cats.py

```
1  import json
2  from pathlib import Path
3
4  import pandas as pd
5
6
7  def get_topn_categories(df, date, topn):
8      return df.query('trending_date==@date') \
9          .groupby('category')[['views']].sum() \
10         .sort_values('views', ascending=False).head(topn)
11
12
13 if __name__ == "__main__":
14     # config
15     PATH_FILE_VIDS =
       Path(__file__).parent.absolute()/'../Data/USvideos.csv.zip'
16     PATH_FILE_CAT =
       Path(__file__).parent.absolute()/'../Data/US_category_id.json'
17     PATH_FILE_OUT =
       Path(__file__).parent.absolute()/'../Data/top_10_trendy_cats.csv'
18     DATE = "17.14.11"
19     TOPN = 10
```

The full code is available at https://packt.live/2CpVXtL.

In the Python script, we first import libraries and define the **get_topn_categories** function to calculate the top (Most viewed) YouTube video categories. Under **main**, we use **pandas** to read the CSV file into a DataFrame. We also load the JSON file into a DataFrame and extract the category information. Then, we join them to get the joined DataFrame with category information. Finally, we calculate the top 10 most viewed YouTube video categories and save the file.

12. Run the **get_trendy_cats.py** Python script as shown in the following code:

```
python get_trendy_cats.py
```

A new file, **top_10_trendy_cats.csv**, is created in the **Data** directory.

13. Create a **Python 3** Jupyter Notebook again and use the **pandas.read_csv** function to load the output for the **top_10_trendy_cats.csv** file, as shown in the following code:

```
# check data
pd.read_csv('../Data/top_10_trendy_cats.csv')
```

You should get the following output:

Out[8]:

	category	views
0	Music	97097557
1	Entertainment	28047357
2	Comedy	16643490
3	Howto & Style	15926296
4	People & Blogs	7753285
5	Science & Technology	6537832
6	Film & Animation	6015279
7	Sports	2755539
8	News & Politics	1571196
9	Pets & Animals	1010302

Figure 9.16: Top 10 YouTube video categories

NOTE

To access the source code for this specific section, please refer to https://packt.live/3fsb6t9.

We can represent the job in *Exercise 9.02, Creating a Nonlinear Pipeline to Get the Daily Top 10 Trending Video Categories*, as a multi-stage process. It looks like the following flow chart:

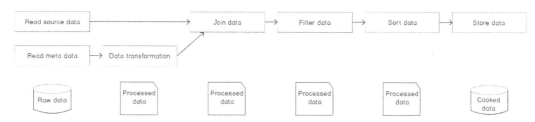

Figure 9.17: A flow chart for fetching the top 10 trending categories

We created a data pipeline that performs read, join, filter, sort, and store operations within a single Python script. Since all of the operations are in one single script, they are closely coupled together, which makes it less flexible and less reusable. In software engineering, a better practice is to write code in a more reusable and modular fashion so that it can be reused or recomposed to accomplish new tasks or larger tasks. In a later exercise, we will be creating a Python script for each operation.

Being able to break down a job into stages allows us to decouple different data operations. There are many benefits to decoupling data tasks. For example, it allows us to reduce code complexity for easier code maintenance. It also increases the overall robustness of the process. In *Figure 9.3*, let's imagine that the **Filter data** step fails. We would have to rerun the entire process from scratch. But if we decoupled each step and cached each step's output data, then we would just need to rerun the steps after the **Join data** step. Therefore, the system would be more manageable and more robust to failure events.

In the world of software engineering, the best practice for managing processes is to use a workflow application, commonly known as a workflow. A workflow is a software application that automates a process or processes to achieve a business-related objective. A process is usually composed of a series of steps to be automated. A workflow comes with many great features, such as automation, scheduling, failure handling, and the flexibility to allow users to introduce new components into an operation. You might wonder why we need a workflow and its awesome features to manage our processes. You will find out more about that in the next section.

CHALLENGES IN MANAGING PROCESSES IN THE REAL WORLD

We have learned about how to create a task and break it into a multi-stage process. Knowing how to do these two things should be enough to create a functioning data pipeline. But when it comes to managing a data pipeline, there's another important thing to know about: job automation. Imagine that someone updated a source CSV file with the most recent data in the workflow illustrated in *Figure 9.01*. Someone would need to jump in to manually rerun the entire workflow and deploy a new version of the model.

AUTOMATION

To ease the burden of managing hundreds of workflows in a company, we want workflows to be fully automated without any extensive human interaction. If any change happens to one step, it should automatically trigger downstream steps to rerun with the new change. In addition to workflow automation, it'd be nice if we could version each run of the workflow so that we could perform retrospective analysis in the future:

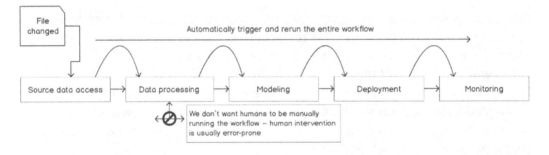

Figure 9.18: A fully automated workflow

FAILURE HANDLING

Another consideration in managing workflows is failure handling. As great as a workflow is when fully automated, it can break. When one of the steps in a workflow breaks, it should not trigger downstream workflow steps. Otherwise, a defective model with potential bugs could be published to production, and we don't want that. For example, in *Exercise 9.02*, *Creating a Nonlinear Pipeline to Get the Daily Top 10 Trending Video Categories*, let's imagine the joining-data step fails and we end up re-ranking video categories based on the wrong values. In the end, we would publish an incorrect report to stakeholders. So, when we manage a workflow, we need to know the status of each step as well as each step's dependencies. A step gets triggered only if its upstream step's status is "successful run." On the other hand, when a step's status is "failure," its downstream step shouldn't run:

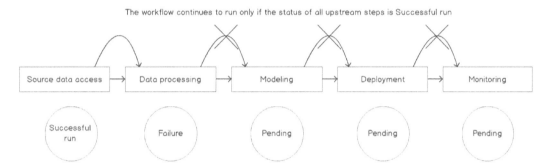

Figure 9.19: A workflow management system should have the notion of status

RETRY MECHANISM

Now, if a step's status is "failure," what should we do about it? A simple solution would be to have a re-try mechanism where the step will start running again after some pre-defined time window (for example, 60 seconds). When a pipeline is executing, it may encounter network issues that trigger pipeline failures. However, we don't want the pipeline to keep re-trying forever if there is a real issue that needs to be addressed; when we go to the ATM to withdraw cash, the card will be locked after three failed PIN attempts.

We can set it to re-try three times. If the step fails three times, then it will escalate to humans, which means we need the workflow management system to send out an email with reports to the engineers who own this workflow:

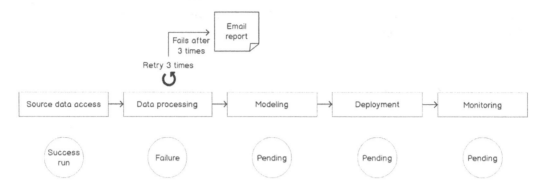

Figure 9.20: A workflow management system should have a re-try mechanism

In short, managing dependencies, scheduling the workflow, handling errors, versioning, and conducting retrospection are common challenges in the world of data science and AI. Many companies face the same pain points and have developed workflow management systems to tackle these problems. If you have experience in the realm of data engineering, these challenges may sound familiar to you. As soon as you start working with data, you will face these pain points and want to solve them.

To better understand why workflow management is needed in the real world, let's create a multi-stage data pipeline. A multi-stage data pipeline is composed of several individual tasks that have dependencies between them. The dependencies between tasks in a pipeline can be complicated and hard to manage. Meanwhile, it leads to an opportunity to optimize a poorly organized pipeline. In the next exercise, we will be building a multi-stage process based on the previous exercise.

EXERCISE 9.03: CREATING A MULTI-STAGE DATA PIPELINE

Recall that in *Exercise 9.02*, *Creating a Nonlinear Pipeline to Get the Daily Top 10 Trending Video Categories*, we created a job that took two files as input and then output the top 10 daily trending video categories. The logic for the entire process sat inside a single Python script, which meant it was a single-step process. To increase the robustness and modularity of the process, we want to break it into multiple steps with files cached at different stages. Now your job is to write multiple Python scripts to represent different steps in the process and achieve the same objective of displaying the top 10 daily trending categories. The single-task process shown in *Figure 9.14* will be converted to a multi-stage process composed of multiple Python scripts, as shown in the following figure:

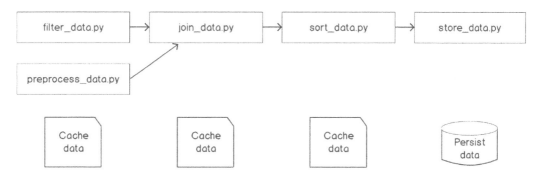

Figure 9.21: A multi-stage process composed of five different Python scripts

Perform the following steps to complete the exercise:

1. Create an **Exercise09.03** directory in the **Chapter09** directory to store the files for this exercise.

2. Open your Terminal (macOS or Linux) or Command Prompt (Windows), navigate to the **Chapter09** directory, and type **jupyter notebook**.

3. Create a **filter_data.py** Python script in the **Exercise09.03** directory and add the following code:

filter_data.py

```
28      # filter
29      df_filtered = filter_by_date(df_data, date)
30      # cache
31      dir_cache = './tmp'
32      try:
33          df_filtered.to_csv(os.path.join(dir_cache,
            'data_vids.csv'), index=False)
34      except FileNotFoundError:
35          os.mkdir(dir_cache)
36          df_filtered.to_csv(os.path.join(dir_cache,
            'data_vids.csv'), index=False)
37      print('[ data pipeline ] finish filter data')
```

The complete code for this step is available at: https://packt.live/38U6v0c.

In **filter_data.py**, we write logic that reads source data from the file path where **USvideos.csv.zip** is located in your filesystem, then filters data to a specific date, such as **17.14.11**. After the data is filtered, it should cache the temporary output file to a working space. In this case, our working space is a temporary directory named **./tmp/**. The **./tmp/** directory will be our workspace for the rest of the data pipeline. Intermediate output files from other steps will be cached in this workspace as well. Other steps will use this workspace for read and write operations.

> **NOTE**
>
> We use the **argparse** module to create this program with a **Command-Line Interface (CLI)**. A CLI enables users to run this program with various configurations without touching the code, which helps encapsulate the program.

4. Create a **preprocess_data.py** Python script in the **Exercise09.03** directory, and add the following code:

preprocess_data.py

```
19      # read data
20      data_cats = json.load(open(filepath, 'r'))
21      # convert json to dataframe
22      df_cat = pd.DataFrame(data_cats)
23      df_cat['category'] = df_cat['items'].apply(lambda x:
        x['snippet']['title'])
24      df_cat['id'] = df_cat['items'].apply(lambda x:
        int(x['id']))
25      df_cat_drop = df_cat.drop(columns=['kind', 'etag',
        'items'])
26      # cache
27      dir_cache = './tmp'
28      try:
29          df_cat_drop.to_csv(os.path.join(dir_cache,
            'data_cats.csv'), index=False)
30      except FileNotFoundError:
31          os.mkdir(dir_cache)
32          df_cat_drop.to_csv(os.path.join(dir_cache,
            'data_cats.csv'), index=False)
33      print('[ data pipeline ] finish preprocess data')
```

The complete code for this step is available at: https://packt.live/306wPR2.

In **preprocess_data.py**, the program reads the video category data as a DataFrame from the file path where **US_category_id.json** is located. It then uses **pandas.apply** functions to extract the video **category**, **title**, and **id** from the **item** column. Finally, the program outputs the mapping table between the **id** and the **title** category to our workspace, **./tmp/**.

5. Create a **join_data.py** Python script in the **Exercise09.03** directory, and add the following code:

```
import pandas as pd

def join_cats(df_vids, df_cats):
    return df_vids.merge(df_cats, left_on='category_id',
        right_on='id')
if __name__ == "__main__":
    import os
    import sys
    from os import path
    # read data from cache
    try:
```

```
        df_vids = pd.read_csv('./tmp/data_vids.csv')
        df_cats = pd.read_csv('./tmp/data_cats.csv')
    except Exception as e:
        print('>>>>>>>>>>>> Error: {}'.format(e))
        sys.exit(1)
    # join data
    df_join = join_cats(df_vids, df_cats)
    # cache joined data
    df_join.to_csv('./tmp/data_joined.csv', index=False)
    print('[ data pipeline ] finish join data')
```

In **join_data.py**, we map the **id** category from the YouTube video data to the **title** category from the **data** category. Notice that we did not use a CLI for this join step because this program depends on the completion of previous steps and assumes that input data has already been cached in the pipeline workspace, **./tmp/**.

> **NOTE**
>
> The **try ... except** block is recommended for use here because the program expects that **data_vids.csv** and **data_cats.csv** are available. When files are not available, which is not expected, we want this program to exit with status code 1 to tell the data pipeline to abort the operation as well as to print the error message.

6. Create a **sort_data.py** Python script in the **Exercise09.03** directory, and add the following code:

sort_data.py

```
1 import sys
2 import pandas as pd
3 import argparse
4
5 def get_topn_cats(df_join, topn):
6     return df_join.groupby('category')[['views']].sum() \
7         .sort_values('views', ascending=False).head(topn)
8
9 def parse_args():
10    parser = argparse.ArgumentParser(
11        prog="exercise 3",
12        description="rank categories")
13    parser.add_argument('-n', '--num', type=int, default=10,
      help='how many?')
14    return parser.parse_args()
```

The complete code for this step is available at: https://packt.live/3iXdtq3.

In **sort_data.py**, the program reads the **data_joined.csv** file. Then, it sums up the video views for each category and ranks the categories in terms of total views in descending order to show the top 10 trending categories. We wrap the logic that performs the group-by sum operations in a function called **get_top_cats**. This will make the code more readable, reusable, and modular.

7. Create a **store_data.py** Python script in the **Exercise09.03** directory, and add the following code:

store_data.py

```
19    # read data from cache
20    try:
21        file_cached = './tmp/data_topn.csv'
22        df_join = pd.read_csv(file_cached)
23    except Exception as e:
24        print('>>>>>>>>>>> Error: {}'.format(e))
25        sys.exit(1)
26
27    # cache joined data
28    df_join.to_csv(filepath, index=False)
29
30    # clean up tmp
31    shutil.rmtree('./tmp')
32
33    print('[ data pipeline ] finish storing data')
```

The complete code for this step is available at: https://packt.live/3frBJhP.

In **store_data.py**, the program will read the output file from the last step and write the file to a new location to persist the final output data from this pipeline. Since the final output data is persisted in a given location, the cached data that was created in the previous steps is no longer needed. At the end of the pipeline, we use the built-in **shutil** library to remove the **./tmp** directory along with all the files inside it. This program's CLI requires an argument for the output file path in which users want to store the output data.

8. Now, run the following commands in sequential order in your Terminal (assuming you are in the **Exercise09.03** directory):

```
python filter_data.py --file ../Data/USvideos.csv.zip --date
    17.14.11
```

You should get the following output:

```
[ data pipeline ] finish filter data
```

Figure 9.22: Output of filter_data.py

```
python preprocess_data.py --file ../Data/US_category_id.json
```

You should get the following output:

```
[ data pipeline ] finish preprocess data
```

Figure 9.23: Output of preprocess_data.py

```
python join_data.py
```

You should get the following output:

```
[ data pipeline ] finish join data
```

Figure 9.24: Output of join_data.py

```
python sort_data.py
```

You should get the following output:

```
[ data pipeline ] finish sort data
```

Figure 9.25: Output of sort_data.py

```
python store_data.py --path ../Data/top_10_trendy_cats.csv
```

You should get the following output:

```
[ data pipeline ] finish storing data
```

Figure 9.26: Output of store_data.py

A new **top_10_trendy_cats.csv** file is created in the **Data** directory.

9. Open a new **Python 3** Jupyter Notebook, import **pandas**, and use the **read_csv** function to read the **top_10_trendy_cats.csv** file, as shown in the following code:

```
import pandas as pd
pd.read_csv('../Data/top_10_trendy_cats.csv')
```

You should get the following output:

```
Out[1]:
```

	category	views
0	Music	97097557
1	Entertainment	28047357
2	Comedy	16643490
3	Howto & Style	15926296
4	People & Blogs	7753285
5	Science & Technology	6537832
6	Film & Animation	6015279
7	Sports	2755539
8	News & Politics	1571196
9	Pets & Animals	1010302

Figure 9.27: Top 10 most viewed video categories

> **NOTE**
>
> To access the source code for this specific section, please refer to https://packt.live/3emvKK0.

By completing the exercise, you have successfully implemented a multi-stage data pipeline that is composed of five Python scripts. Each Python script on its own is a modular and reusable operation. For example, the **filter_data.py** script implements the filter operation. Another script, **preprocess_data.py**, implements the preprocess operation. With more of these modular and reusable scripts for different types of data operations, we can pipe different operations together to compose any data pipeline we want.

AUTOMATING A DATA PIPELINE

You may think that multi-stage jobs are complicated. Users are required to run multiple commands in a specific sequence to complete tasks. One of the principles of workflow management is the minimization of human interaction. Human interaction is usually error-prone. If someone runs commands in the wrong order, there will be different results. We want to remove this manual process, which means we need to automate this job.

Bash is a Unix shell. It's a command language that can be used directly at the command line. Often, people use Bash as glue code to stitch different software systems or tools together, as well as using it for the automation of jobs.

In the next exercise, we will leverage Bash to automate the multi-stage data pipeline of *Exercise 9.03*, *Creating a Multi-Stage Data Pipeline*.

EXERCISE 9.04: AUTOMATING A MULTI-STAGE DATA PIPELINE USING A BASH SCRIPT

In the last exercise, we created four Python scripts, one for each stage of a multi-stage data pipeline. Furthermore, we ran each of them manually to store the data in a CSV file.

This exercise aims to automate the stages of the pipeline using a Bash script.

Perform the following steps to complete the exercise:

1. Create a new **Exercise09.04** directory in the **Chapter09** directory to store the files for this exercise.

2. Create a **run_job.sh** Bash script in the **Exercise09.04** directory and add the following code:

`run_job.sh`

```
1   #!/bin/bash
2
3   set -e
4
5   # set config
6   DATE=17.14.11
7   SOURCE_FILE=../Data/USvideos.csv.zip
8   CAT_FILE=../Data/US_category_id.json
9   OUTPUT_FILE=../Data/top_10_trendy_cats.csv
10  SRC_DIR=../Exercise09.03
11
12  echo "[[ JOB ]] runs on date $DATE with file located in
      $SOURCE_FILE and metadata located in $CAT_FILE"
13  echo "[[ JOB ]] result data will be persisted in $OUTPUT_FILE"
```

The complete code for this step is available at: https://packt.live/32aF3Kw.

In the **run_job.sh** Bash script, the first line (as always) is **#!/bin/bash** to indicate that it's a Bash script. We use a Bash built-in command, **set -e**, to tell the Bash program to exit immediately if any command within the pipeline returns a non-zero status.

We also set a bunch of variables, **DATE, SOURCE_FILE, CAT_FILE, OUTPUT_FILE**, and **SRC_DIR**, at the top of the script.

The next lot of commands look similar to the commands used in *Exercise 9.03, Creating a Multi-Stage Data Pipeline*, for the sequential execution of our data pipeline. The only difference is the additional **echo** command, which writes messages to standard out. During the running of the Bash script, echo messages show us which step of the pipeline is currently running.

At the bottom of the script, the last six lines of code are making sure that the program exits with the appropriate status code. When the program exits with code **1**, this means the program failed with an unexpected error. This increases the robustness of the system in terms of error-handling.

3. Open your Terminal (macOS or Linux) or Command Prompt (Windows), navigate to the **Chapter09/Exercise09.04** directory, and run the following command:

    ```
    sh run_job.sh
    ```

    You should get the following output:

```
[[ JOB ]] runs on date 17.14.11 with file located in ../Data/USvideos.csv.zip and metadata located in ../Data/US_category_id.json
[[ JOB ]] result data will be persisted in ../Data/top_10_trendy_cats.csv
[[ RUNNING JOB ]] step 1: filter source data
[ data pipeline ] finish filter data
[[ RUNNING JOB ]] step 1.1: preprcess metadata
[ data pipeline ] finish preprocess data
[[ RUNNING JOB ]] step 2: join data
[ data pipeline ] finish join data
[[ RUNNING JOB ]] step 3: rank categories
[ data pipeline ] finish sort data
[[ RUNNING JOB ]] step 4: persist result data
[ data pipeline ] finish storing data
[[ JOB ]] END
```

Figure 9.28: A successful pipeline run in the Terminal via Bash

When we see the **[[JOB]] END** message, it means the Bash script ran until the end without returning a non-zero exit code. In other words, the pipeline is running successfully.

Now, we can check the output data that is saved in **Chapter09/Data/top_10_trendy_cats.csv location.**

4. Open a new **Python 3** Jupyter Notebook, import **pandas**, and use the **read_csv** function to read the **top_10_trendy_cats.csv** file, as shown in the following code:

```
import pandas as pd
pd.read_csv('../Data/top_10_trendy_cats.csv')
```

You should get the following output:

```
Out[1]:
```

	category	views
0	Music	97097557
1	Entertainment	28047357
2	Comedy	16643490
3	Howto & Style	15926296
4	People & Blogs	7753285
5	Science & Technology	6537832
6	Film & Animation	6015279
7	Sports	2755539
8	News & Politics	1571196
9	Pets & Animals	1010302

Figure 9.29: Top 10 most viewed video categories

By completing this exercise, we learned to use a Bash script to automate a multi-stage data pipeline. We used Bash script built-in commands and variables to store file path information and other parameters that are used in the pipeline. To log the status for each step in the pipeline, the **echo** command was used.

Bash is used as glue code to compose different operations together to control jobs, or in our case, to automate a job.

> **NOTE**
>
> To access the source code for this specific section, please refer to https://packt.live/3fqHplO.

In *Exercise 9.04*, *Automating a Multi-Stage Data Pipeline Using a Bash Script*, when we ran the **run_job.sh** Bash script in Terminal, the program was executing each line in a sequence. The next line won't be executed until the current line is done. For example, the program won't run **step 1.1 preprocess metadata** until **step 1 filter source data** is finished. This is called **synchronous** execution. Because we ran this Bash script on our local machine with a single-thread process, by default it was a synchronous process. But **step 1** and **step 1.1** don't need to run in a specific sequence; they can run as parallel steps as well. The job will be more efficient if **step 1** and **step 1.1** can run at the same time, rather than running in sequence. This would make it **asynchronous** execution.

Most real-world applications are asynchronous processes, especially in the world of distributed computing. In a single cluster, many compute instances are executing different tasks at the same time to achieve efficient computing. But the next question for us is this: if a program is not executed in a synchronous manner, how can we automate this program? In the next section, we will discuss solutions for managing asynchronous data pipelines.

AUTOMATING ASYNCHRONOUS DATA PIPELINES

A workflow management system should have the ability to automate asynchronous processes without any human interaction. Let's use the most common type of job in data engineering, the **Extract**, **Transform**, and **Load** (**ETL**) workflow, as an example to illustrate how it works and how to automate it:

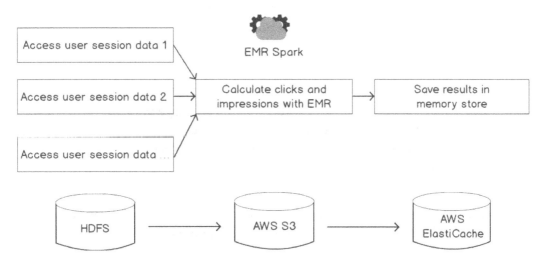

Figure 9.30: A typical ETL workflow

The objective of an ETL pipeline is to output analytics reporting to inform business analysts what is trending right now based on clicks and impression data, which is very similar to the YouTube trending video data pipeline we created earlier. However, an ETL pipeline usually involves performing data operations such as extracting, transforming and loading data at scale.

Let's imagine that our source data, `USvideos.csv.zip`, is a 100+ terabytes dataset, which is very common in the era of big data. We won't be able to work with a flat CSV file anymore. Data of such size will be stored in specialized data systems such as the **Hadoop Distributed File System** (**HDFS**) and other cloud storage systems. The data won't be stored in a single file. It will be partitioned into many files so that the data is manageable as well as fault-tolerant. So, our data pipelines will have to read data from many sources simultaneously in the first step.

In the processing step of our data pipeline, `pandas` is no longer a viable option for processing data at such a scale. We will use a big data processing engine such as Spark, which is offered through **Amazon Elastic MapReduce** (**EMR**), to perform group, aggregation, and sort operations using the map-reduce paradigm.

At the end of the pipeline, the pipeline will persist the final output data to a location. In our example, the location is a cloud memory storage system, Amazon ElastiCache. It's a key-value store for fast access, and it will be talked about in more depth in the next chapter.

The process in *Figure 9.21* is a typical ETL workflow, which usually involves extracting raw data from multiple data sources, transforming raw data into a digestible format, then loading the results into another data store.

Each step runs inside a compute instance. You can think of a compute instance as an independent self-contained environment with resources such as CPU and RAM that aren't shared with others. Notice that the first step in this ETL pipeline is very similar to that of the pipeline in *Exercise 9.04*, *Automating a Multi-Stage Data Pipeline Using a Bash Script*, where tasks in the first step didn't have dependencies with each other and were parallelized. Each task in the first step can be finished at a different time, but the second step won't be triggered until the last task in the first step is finished. So, this pipeline is an asynchronous process.

We need to introduce a new mechanism to understand the dependencies between tasks and be able to trigger the next task when its dependent tasks are completed.

There are multiple approaches to automating asynchronous processes. Let's look at one of them here:

1. We need a mechanism that triggers downstream tasks in a workflow when a certain criterion is met. In this case, this mechanism checks for the user session data becoming available in AWS S3, then it will trigger the next task.

2. We also need to make this mechanism understand the dependencies between tasks so that tasks get triggered in the intended sequence. Practitioners in the industry have figured out that the best way to define workflows with task dependencies is to use **Directed Acyclic Graphs** (**DAGs**). We will study DAGs later in this chapter.

Based on the two requirements, we need to add two more tasks to the workflow:

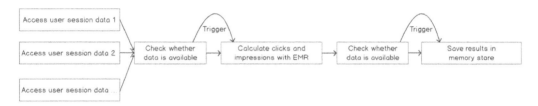

Figure 9.31: Adding two more tasks for workflow automation

Let's zoom in on the workflow again and take a look at the newly added tasks. We can see that the two new added tasks are different than the original tasks in nature. They act as a sensor and trigger as they sense whether a condition is met. While the first and fifth steps are simply moving data from one place to another, the third step is performing some actions on the data.

In workflow management, we use the term **operator** as a basic unit of abstraction to define tasks. A workflow is defined in a DAG and operators are nodes in the DAG that represent tasks. There are three types of operators: **action**, **transfer**, and **sensor**.

We can customize any workflow using these three operators. Most ETL workflows are constructed with these three operators. Here's a general illustration of how they function:

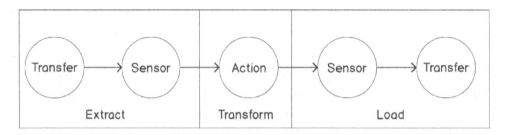

Figure 9.32: An ETL workflow using different types of operators

In the next exercise, we will further improve our data pipeline by introducing the use of asynchronous processes to speed up the runtime. With asynchronous processes, we also need to implement a sensor operation for triggering downstream operations in the pipeline.

EXERCISE 9.05: AUTOMATING AN ASYNCHRONOUS DATA PIPELINE

Let's try to implement the triggering (sensor operator) mechanism in the Bash script and automate the job in *Exercise 9.04, Automating a Multi-Stage Data Pipeline Using a Bash Script*.

Perform the following steps to complete the exercise:

1. Create an **Exercise09.05** directory in the **Chapter09** directory to store the files for this exercise.

2. Create a **run_job.sh** Bash script in the **Exercise09.05** directory and add the following code:

```
#!/bin/bash

set -e

# set config
DATE=17.14.11
SOURCE_FILE=../Data/USvideos.csv.zip
CAT_FILE=../Data/US_category_id.json
```

```
OUTPUT_FILE=../Data/top_10_trendy_cats.csv
SRC_DIR=../Exercise09.03

echo "[[ JOB ]] runs on date $DATE with file located in
  $SOURCE_FILE and metadata located in $CAT_FILE"
echo "[[ JOB ]] result data will be persisted in $OUTPUT_FILE"

# run job
echo "[[ RUNNING JOB ]] step 1: filter source data"
python $SRC_DIR/filter_data.py --file $SOURCE_FILE --date
  $DATE &

echo "[[ RUNNING JOB ]] step 1.1: preprcess metadata"
python $SRC_DIR/preprocess_data.py --file $CAT_FILE &

echo "[[ RUNNING JOB ]] step 2: join data"
python $SRC_DIR/join_data.py

echo "[[ RUNNING JOB ]] step 3: rank categories"
python $SRC_DIR/sort_data.py

echo "[[ RUNNING JOB ]] step 4: persist result data"
python $SRC_DIR/store_data.py --path $OUTPUT_FILE

if [ "$?" = "0" ]; then
    rm -rf ./tmp/
    echo "[[ JOB ]] END"
else
    rm -rf ./tmp/
    echo "[[ JOB FAILS!! ]]" 1>&2
    exit 1
fi
```

At first glance, **run_job.sh** looks the same as the script in *Exercise 9.04, Automating a Multi-Stage Data Pipeline Using a Bash Script*. But it's not the same – you'll notice the difference in the **python $SRC_DIR/filter_data.py --file $SOURCE_FILE --date $DATE** & and **python $SRC_DIR/ preprocess_data.py --file $CAT_FILE** & commands. We added the **&** symbol at the end of the command line, which directs the shell to run the command in the background; that is, it runs in a separate sub-shell, as a job, asynchronously. When the first two commands in the pipeline are running in the background, Bash will immediately return the status code **0** for true. Once the status code **0** is returned, the next command starts to run. This means that the Bash script doesn't wait for the first two steps to finish but continues to execute the remaining commands in the script. You have probably noticed that this isn't a desirable behavior and will cause errors. We will find out its solution in the next step of the exercise.

3. Open your Terminal (macOS or Linux) or Command Prompt (Windows), navigate to the **Chapter09/Exercise09.05** directory, and run the following command:

    ```
    sh run_job.sh
    ```

    You should get the following output:

    ```
    [[ JOB ]] runs on date 17.14.11 with file located in ../Data/USvideos.csv.zip and metadata located in ../Data/US_category_id.json
    [[ JOB ]] result data will be persisted in ../Data/top_10_trendy_cats.csv
    [[ RUNNING JOB ]] step 1: filter source data
    [[ RUNNING JOB ]] step 1.1: preprcess metadata
    [[ RUNNING JOB ]] step 2: join data
    >>>>>>>>>>> Error: File b'./tmp/data_vids.csv' does not exist
    [ data pipeline ] finish preprocess data
    [[ RUNNING JOB ]] step 3: rank categories
    >>>>>>>>>>> Error: File b'./tmp/data_joined.csv' does not exist
    [[ RUNNING JOB ]] step 4: persist result data
    [ data pipeline ] finish filter data
    >>>>>>>>>>> Error: File b'/Users/Kevin/Documents/repos/Data-Storage-for-Artificial-Intelligence/Chapter09/Exercise05/../Exercise03/tmp/data_topn.csv' does not exist
    [[ JOB FAILS!! ]]
    ```

    Figure 9.33: Pipeline error without a sensor operator

    The error occurs in **step 2** while running the **join_data.py** script. It's complaining about the **data_vids.csv** file not existing in the workspace. This is because **step 1 filter_data.py** hasn't finished executing and so the **data_vids.csv** file wasn't written to the **./tmp/** directory.

    To solve this error, we need to implement a sensor operator. It will ensure that until the files are available, **step 2 join_data.py** is not triggered.

4. In the **run_job.sh** Bash script, add the **while** construct (highlighted) before **step 2** to implement the sensor operator as shown in the following code:

run_job.sh

```
22    echo "[[ BLOCKING JOB ]] additional step: check cached
   files and trigger next step"
23 while [ ! -f ./tmp/data_cats.csv ] || [ ! -f
   ./tmp/data_vids.csv ]
24 do
25   sleep 1
26 done
27
28 echo "[[ RUNNING JOB ]] step 2: join data"
29 python $SRC_DIR/join_data.py
30
31 echo "[[ RUNNING JOB ]] step 3: rank categories"
32 python $SRC_DIR/sort_data.py
33
34 echo "[[ RUNNING JOB ]] step 4: persist result data"
35 python $SRC_DIR/store_data.py --path $OUTPUT_FILE
```

The complete code for this step is available at: https://packt.live/32eKzfg.

The **while** loop code block acts as a sensor operator between steps 1 and 2. When the commands of **step 1** are running in the background, it will check whether the **data_cats.csv** and **data_vids.csv** files are available in **tmp/**. If they're not found, it will sleep for 1 second and continue to check the **while** condition until the files are available. When the files are available, the Bash program will exit the **while** loop and execute the commands in **step 2**.

5. Save the updated **run_job.sh** Bash script and run it again using the following command:

   ```
   sh run_job.sh
   ```

   You should get the following output:

Figure 9.34: A successful asynchronous pipeline run with a sensor operator

By completing the exercise, you have successfully implemented an asynchronous multi-stage data pipeline with a Bash script. We first learned how to use the **&** symbol in Bash to direct the shell to run a command in the background asynchronously. Then we implemented a **while** loop that acted as a sensor operator to block the **step 2** commands until the **step 1** commands had finished.

> **NOTE**
>
> To access the source code for this specific section, please refer
> to https://packt.live/3iXdYQX.

Now we can write a Bash program to automate not only synchronous jobs but asynchronous jobs as well. This is great if the job never fails. However, task jobs in real life can fail. Imagine that the tasks of `step 1` kept failing; the program would be stuck at the `[[ BLOCKING JOB ]]` step and the rest of the steps would be blocked forever. We need more sophisticated mechanisms in a production-level workflow management system. A workflow management system should be able to monitor the life cycle and health of each task in each step as well as the overall status of the entire job.

It should be able to handle errors as well as re-try tasks. This is possible using **Airflow**, an open-source workflow management tool. In the next section, we will use the Airflow DAG API, to create a DAG for our data pipeline, and the Airflow Operator API, to implement different data operations for the tasks in our data pipeline. Also, using the Airflow scheduler, we will run our DAG and monitor our DAG's status using its UI.

WORKFLOW MANAGEMENT WITH AIRFLOW

So far, we have learned how to create data pipelines and different types of workflows, including linear and non-linear ones. We define and implement workflows in Python scripts and use Bash to automate workflows. However, that is not enough for us to be able to manage workflows on a large scale. We are going to take workflow management to the next level by solving the following problems:

- Can we find a standardized way to define workflow dependency instead of writing a customized Bash script?

- Can we define data operations with a consistent interface instead of writing a Python program with a customized CLI?

- Can we have a standardized way to log the pipeline's running status?

- Can we monitor a running workflow? Can we schedule workflows?

The answer to all of these problems is Airflow. Airflow is a horizontally scalable, distributed workflow management system that allows us to specify complex workflows using Python code. It is an open-source project developed at Airbnb and it has a growing open source community. It's very easy to use and it has great documentation, so it's a perfect tool for us to manage workflows at scale with.

When we use Airflow to manage jobs, we first need to describe the job with a DAG. A DAG is a collection of tasks organized in a specific order. A DAG describes how you want Airflow to run your workflow. For example, you may want your workflow to run every day until the year 2021, or you may want Airflow to send you an email if the workflow fails, and so on. The DAG is there to make sure your workflow runs at the right time and in the right order, and that it handles errors whenever needed.

Let's consider the following workflows:

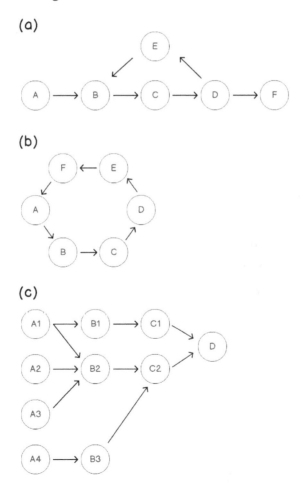

Figure 9.35: DAG and non-DAG workflows

Which workflow is described with a DAG? Only workflow **(c)** is using a DAG. Workflow **(a)** and **(b)** are cyclic workflows. It's very challenging to automate cyclic workflows because circular dependency causes infinite execution loops.

In the preceding charts, the DAG doesn't describe what **A**, **B**, **C**, **D**, **E**, and **F** are or what they are doing. A DAG doesn't concern itself with what the workflow is doing. It doesn't concern itself with each step or task. When we define a workflow, we also need to describe what the job should be doing with **operators**. An operator describes a single task in a workflow. It's usually atomic and can stand on its own. Airflow provides operators for most common tasks, such as **BashOperator**, **PythonOperator**, **Sensor**, and others.

In the next exercise, we will get our hands dirty with Airflow and create a DAG as well as operators for our data pipeline.

EXERCISE 9.06: CREATING A DAG FOR OUR DATA PIPELINE USING AIRFLOW

Now that we have learned about Airflow, a powerful piece of open-source software for managing our workflow, let's start to write a DAG Python file for the workflow we created in previous exercises to generate daily top 10 trending video categories.

Before proceeding to the exercise, we need to set up Airflow in the local-dev environment. Please follow the instructions in the *Preface* to install it.

Perform the following steps to complete the exercise:

1. Create an **Exercise09.06** directory in the **Chapter09** directory to store the files for this exercise.

2. Open your Terminal (macOS or Linux) or Command Prompt (Windows), navigate to the **Chapter09** directory, and type **jupyter notebook**.

3. Create a **top_cat_dag.py** DAG Python script in the **Exercise09.06** directory and add the following code:

> **NOTE**
>
> A DAG Python script should be named with **_dag** as a suffix; it's a convention in Airflow. This will help Airflow recognize it as a DAG file and register it in its metadata store.

```python
import json
import os
import shutil
import sys
from datetime import datetime

import pandas as pd
from airflow import DAG
from airflow.operators.python_operator import PythonOperator


def filter_data(**kwargs):
    # read data
    path_vids = kwargs['dag_run'].conf['path_vids']
    date = str(kwargs['dag_run'].conf['date'])

    print(os.getcwd())
    print(path_vids)

    df_vids = pd.read_csv(path_vids, compression='zip') \
        .query('trending_date==@date')
    # cache
    try:
        df_vids.to_csv('./tmp/data_vids.csv', index=False)
    except FileNotFoundError:
        os.mkdir('./tmp')
        df_vids.to_csv('./tmp/data_vids.csv', index=False)
def preprocess_data(**kwargs):
    # read data
    path_cats = kwargs['dag_run'].conf['path_cats']
    data_cats = json.load(open(path_cats, 'r'))
    # convert json to dataframe
    df_cat = pd.DataFrame(data_cats)
    df_cat['category'] = df_cat['items'].apply(lambda x:
      x['snippet']['title'])
    df_cat['id'] = df_cat['items'].apply(lambda x:
      int(x['id']))
    df_cat_drop = df_cat.drop(columns=['kind', 'etag',
      'items'])
    # cache
```

```
    try:
        df_cat_drop.to_csv('./tmp/data_cats.csv')
    except FileNotFoundError:
        os.mkdir('./tmp')
        df_cat_drop.to_csv('./tmp/data_cats.csv')
def join_data(**kwargs):
    try:
        df_vids = pd.read_csv('./tmp/data_vids.csv')
        df_cats = pd.read_csv('./tmp/data_cats.csv')
    except Exception as e:
        print('>>>>>>>>>>>> Error: {}'.format(e))
        sys.exit(1)
    # join data
    df_join = df_vids.merge(df_cats, left_on='category_id',
        right_on='id')
    # cache joined data
    df_join.to_csv('./tmp/data_joined.csv')
def sort_data(**kwargs):
    topn = kwargs['dag_run'].conf.get('topn', 10)
    try:
        df_join = pd.read_csv('./tmp/data_joined.csv')
    except Exception as e:
        print('>>>>>>>>>>>> Error: {}'.format(e))
        sys.exit(1)
    # sort data
    df_topn = df_join.groupby('category')[['views']].sum() \
        .sort_values('views', ascending=False).head(topn)
    # cache joined data
    df_topn.to_csv('./tmp/data_topn.csv')


def store_data(**kwargs):
    # read data from cache
    path_output = kwargs['dag_run'].conf['path_output']
    try:
        df_join = pd.read_csv('./tmp/data_topn.csv')
    except Exception as e:
        print('>>>>>>>>>>>> Error: {}'.format(e))
        sys.exit(1)
```

```
# cache joined data
df_join.to_csv(path_output)
# clean up tmr
shutil.rmtree('./tmp')
```

Remember that a DAG is composed of data operators. In our case, we have data operations such as **filter data**, **preprocess data**, **join data**, **sort data**, and **store data**. In Airflow, we pass data operations to the DAG using Airflow's **PythonOperator**. When we define **PythonOperator**, we need to pass it a Python callable object, which is simply a Python function. So, the functions in the code will be used to create **PythonOperator** in later steps.

4. Add the following code to the **top_cat_dag.py** script to create **DAG**:

```
# create DAG
args = {
    'owner': 'Airflow',
    'description': 'Get topn daily categories',
    'start_date': datetime(2017, 11, 14),
    'catchup': False,
    'provide_context': True
}

dag = DAG(
    dag_id='top_cat_dag',
    default_args=args,
    schedule_interval=None,
)
```

When we instantiate a **DAG** object, it requires **dag_id**, **schedule_interval**, and **args**. We create an **args** dictionary object first, then we create the **DAG** using **args**.

5. Add the following code to the **top_cat_dag.py** script to create the **PythonOperator** instance to compose a real workflow:

```
op1 = PythonOperator(
    task_id='filter_data',
    python_callable=filter_data,
    dag=dag)


op2 = PythonOperator(
    task_id='preprocess_data',
    python_callable=preprocess_data,
    dag=dag)


op3 = PythonOperator(
    task_id='join_data',
    python_callable=join_data,
    dag=dag)


op4 = PythonOperator(
    task_id='sort_data',
    python_callable=sort_data,
    dag=dag)


op5 = PythonOperator(
    task_id='store_data',
    python_callable=store_data,
    dag=dag)
```

When we instantiate a **PythonOperator** object, we need to give it a **task_id** argument, a unique namespace that distinguishes it from other operations. It also requires a **python_callable** argument, a Python function that defines the data operation. We also need to provide the **DAG** object to attach this **PythonOperator** to the DAG.

6. Add the following code to the **top_cat_dag.py** script to add dependency information for the DAG:

```
[op1, op2] >> op3 >> op4 >> op5
```

We use Python bit shift operators to define a dependency between operators. You can tell that the bit shift operator points to downstream tasks. Operators being in brackets, **[]**, means that they are parallel and not dependent on each other. In our case, operators 1 and 2 can be run in parallel first. When operators 1 and 2 are finished, operator 3 is allowed to run. When operator 3 is finished, then it will run operator 4. The same goes for operator 5.

7. After completing the steps, the **top_cat_dag.py** DAG script should have the following code:

top_cat_dag.py

```
87  # create DAG
88  args = {
89      'owner': 'Airflow',
90      'description': 'Get topn daily categories',
91      'start_date': datetime(2017, 11, 14),
92      'catchup': False,
93      'provide_context': True
94  }
95
96  dag = DAG(
97      dag_id='top_cat_dag',
98      default_args=args,
99      schedule_interval=None,
100 )
101
102 op1 = PythonOperator(
103     task_id='filter_data',
104     python_callable=filter_data,
105     dag=dag)
```

The complete code for this step is available at https://packt.live/38Ypua7.

We first import several Python built-in functions and Airflow functions. In the **DAG** script, after the module import section, we define data operations for the data pipeline. In the second section of the **DAG** script, we define the **DAG** and its **PythonOperators** as well as the operator dependency.

> **NOTE**
>
> In the **DAG** script, we set `'provide_context': True` in the **args** argument to instantiate the **DAG**. This will allow us to pass parameters to the DAG's operators when we trigger the **DAG** to run. You will notice this in the following steps.

8. Run the following commands line by line in your Terminal:

```
# airflow needs a home, ~/airflow is the default,
# but you can lay foundation somewhere else if you prefer
# (optional)
export AIRFLOW_HOME=~/airflow

# install from pypi using pip
pip install apache-airflow

# initialize the database
airflow initdb

# start the webserver, the default port is 8080
airflow webserver -p 8080

# start the scheduler
airflow scheduler

# visit localhost:8080 in the browser and enable the example dag in
the home page
```

If you have already installed Airflow, then you can skip the first two commands, $ export AIRFLOW_HOME=~/airflow and $ pip install apache-airflow. If this is the first time that you are using Airflow, you should run the $ airflow initdb command to initialize the Airflow database. It only needs to be run the first time. The $ airflow webserver -p 8080 command is used to launch the Airflow UI, which provides users with a graphical interface to interact with Airflow systems. The $ airflow scheduler command is used to launch the scheduler, which is responsible for getting your pipelines running on your local dev machine. The scheduler is required if you want to trigger DAGs to run.

> **NOTE**
>
> Please continue to stay in the same working directory, **Exercise09.06**, while you are launching Airflow because Airflow will treat your current working directory as its working directory. Later, our file path locations are based on this working directory. If you are using another working directory, you may get a **file not found** error.
>
> When you launch an Airflow scheduler and it throws an **attempt to write a readonly database** error, it means the user who launched the Airflow scheduler does not have write permission to the Airflow SQLite database file, **airflow.db**. One way to resolve the permission issue is to issue the **sudo chmod 644 ~/airflow/airflow.db** command in your terminal and restart the Airflow scheduler. Run the following command to copy your DAG script to the Airflow home directory:

```
cp . /top_cat_dag.py ~/airflow/dags/
```

For Airflow to register the DAG in its system, we need to put the **top_cat_dag.py** DAG file in the Airflow home directory, **~/airflow/dags.**

> **NOTE**
>
> The Airflow webserver and scheduler should be stopped when you copy a DAG script to **~/airflow**. We can stop them by using **Ctrl + C** in the Command Prompt for Windows/Linux or **Cmd + C** for macOS. After the DAG script has been moved to the Airflow home directory, **~/airflow**, we can launch the Airflow scheduler and webserver again. When the Airflow scheduler is launched again, it will register new DAGs from the DAG scripts in its home directory, **~/airflow**. So, when we issue the **airflow list_dags** command, we can see the new DAGs that were added.

9. Run the following command to list all the DAGs:

```
airflow list_dags
```

You should get the following output:

```
[2020-01-01 14:40:55,656] {__init__.py:51} INFO - Using executor SequentialExecutor
[2020-01-01 14:40:55,657] {dagbag.py:403} INFO - Filling up the DagBag from /Users/Kevin/airflow/dags


-------------------------------------------------------------------
DAGS
-------------------------------------------------------------------
example_bash_operator
example_branch_dop_operator_v3
example_branch_operator
example_http_operator
example_passing_params_via_test_command
example_pig_operator
example_python_operator
example_short_circuit_operator
example_skip_dag
example_subdag_operator
example_subdag_operator.section-1
example_subdag_operator.section-2
example_trigger_controller_dag
example_trigger_target_dag
example_xcom
latest_only
latest_only_with_trigger
test_utils
top_cat_dag
tutorial
```

Figure 9.36: List of DAGs that are registered in your Airflow system

Airflow also supports the use of the CLI. Its CLI syntax is as follows: **airflow action [options]**. The **list_dags** action lists all of the available DAGs that are registered in the system. Notice that our DAG, **top_cat_dag**, is second from the bottom.

10. Run the following command in the Terminal to trigger our DAG:

```
airflow trigger_dag -c '{"path_vids": "../Data/USvideos.csv.zip",
"path_cats": "../Data/US_category_id.json", "date": "17.14.11",
"path_output": "../Data/top_10_trendy_cats.csv"}' 'top_cat_dag'
```

The **trigger_dag** action will trigger the DAG to run immediately. The CLI requires the DAG ID, **top_cat_dag**. However, there is also a **-c** positional argument, which is for passing parameters into each operator in our DAG.

11. To check the status of your DAG, you need to open your browser and go to http:// localhost:8080/ to open the UI of the Airflow system. Look for the **top_cat_dag** DAG and click the **Graph View** icon, as shown in the following figure:

Figure 9.37: Our DAG information is shown in the Airflow UI

12. Now, view the log of any task by clicking on the task and clicking the **View Log** button as shown in the following figure:

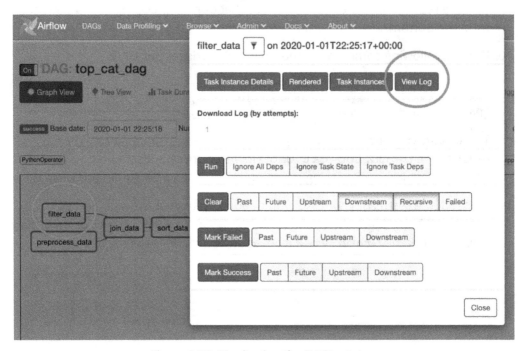

Figure 9.38: Monitoring the DAG's status

By completing this exercise, you have successfully implemented a DAG in Airflow as well as having completed the successful running of a workflow. A DAG is essentially a standardized interface for expressing a multi-stage data pipeline. We first created the DAG object. Then, we implemented our data operations with Airflow using **PythonOperator**. We also learned how to use Python bit shift operators to define dependencies in a workflow. After the DAG script was finished, we launched the Airflow UI and scheduler and got our DAG registered with Airflow. Lastly, we successfully triggered our DAG to run and finish.

Now we can launch Airflow as a workflow management system to automate our workflows as well as to use for monitoring purposes. But we have just started to scratch the surface of what Airflow can do. There are more features that we can explore in Airflow.

> **NOTE**
>
> To access the source code for this specific section, please refer to https://packt.live/2On0D6t.

In the next activity, we will be synthesizing everything we have learned in this chapter and composing a new workflow using Airflow.

ACTIVITY 9.01: CREATING A DAG IN AIRFLOW TO CALCULATE THE RATIO OF LIKES-DISLIKES FOR EACH CATEGORY

Imagine that you are making a YouTube video and you want to optimize the number of likes you receive. You'd like to find out which YouTube video category tends to have more "likes" than "dislikes" So, you decide to find the likes-dislikes ratio based on various categories, such as sports, music, comedy, and so on. Depending on the findings, you will create your video.

In the YouTube video dataset, **USVideos.csv.zip**, there is a **likes** column and a **dislikes** column for each YouTube video. The table is shown here:

	title	channel_title	category_id	likes	dislikes	views	comment_count
0	WE WANT TO TALK ABOUT OUR MARRIAGE	CaseyNeistat	22	57527	2966	748374	15954
1	The Trump Presidency: Last Week Tonight with J...	LastWeekTonight	24	97185	6146	2418783	12703
2	Racist Superman | Rudy Mancuso, King Bach & Le...	Rudy Mancuso	23	146033	5339	3191434	8181
3	Nickelback Lyrics: Real or Fake?	Good Mythical Morning	24	10172	666	343168	2146
4	I Dare You: GOING BALD!?	nigahiga	24	132235	1989	2095731	17518

Figure 9.39: The likes and dislikes columns from the YouTube video data

This activity aims to find out the average ratio of **likes** to **dislikes** for each video category. You will need to first sum up **likes** and **dislikes** for each category. Then, you will calculate the likes-dislikes ratio for each category. Finally, you will write the result to a CSV file.

> **NOTE**
>
> The code and the resulting output for this activity can be found in a Jupyter Notebook here: https://packt.live/32puIdZ.

Perform the following steps to complete the activity:

1. Read the **USvideos.csv** and **US_category_id.json** files using the **pd.read_csv** and **json.load** functions respectively.

2. Extract the required columns using a DataFrame and a dictionary object.

3. Join the DataFrames using the **merge** function and display the file using the **head** function:

	video_id	trending_date	title	channel_title	category_id	publish_time	tags	views	likes
0	2kyS6SvSYSE	17.14.11	WE WANT TO TALK ABOUT OUR MARRIAGE	CaseyNeistat	22	2017-11-13T17:13:01.000Z	SHANtell martin	748374	57527
1	0mINzVSJrT0	17.14.11	Me-O Cats Commercial	Nobrand	22	2017-04-21T06:47:32.000Z	cute|"cats"|"thai"|"eggs"	98966	2486
2	STl2fl7sKMo	17.14.11	AFFAIRS, EX BOYFRIENDS, $18MILLION NET WORTH -...	Shawn Johnson East	22	2017-11-11T15:00:03.000Z	shawn johnson|"andrew east"|"shawn east"|"shaw...	321053	4451
3	KODzih-pYIU	17.14.11	BLIND(folded) CAKE DECORATING CONTEST (with Mo...	Grace Helbig	22	2017-11-11T18:08:04.000Z	itsgrace|"funny"|"comedy"|"vlog"|"grace"|"helb...	197062	7250
4	8mhTWqWlQzU	17.14.11	Wearing Online Dollar Store Makeup For A Week	Safiya Nygaard	22	2017-11-11T01:19:33.000Z	wearing online dollar store makeup for a week|...	2744430	115426

Figure 9.40: YouTube video data with category information

4. Calculate the total **likes** and **dislikes** grouped by category using the **groupby** function.

5. Create a new **ratio_likes_dislikes** column and calculate the ratio by dividing the likes by the dislikes.

6. Create a **ratio_dag.py** DAG script and add the required code in it.

7. Register the DAG by copying the file into the Airflow Home directory and verify it using the **list_dags** command.

8. Launch Airflow and trigger the DAG file using the **trigger_dag** command.

9. Launch the Airflow UI using the **Airflow webserver** command.

10. In the browser, check the status of the **ratio_dag** file using the **Graph View** icon.

11. List the files in the **../Data/** directory and run the **Ratio_Likes_Dislikes.csv** file to get the following output:

	category	likes	dislikes	ratio_likes_dislikes
0	Shows	1082639	24508	44.174922
1	Pets & Animals	19370702	527379	36.730135
2	Education	49257772	1351972	36.434018
3	Comedy	216346746	7230391	29.921860
4	Howto & Style	162880075	5473899	29.755769
5	Music	1416838584	51179008	27.683979
6	Film & Animation	165997476	6075148	27.324022
7	Sports	98621211	5133551	19.211110
8	People & Blogs	186615999	10187901	18.317414
9	Science & Technology	82532638	4548402	18.145414
10	Autos & Vehicles	4245656	243010	17.471116
11	Travel & Events	4836246	340427	14.206411
12	Entertainment	530516491	42987663	12.341134
13	Gaming	69038284	9184466	7.516853
14	Nonprofits & Activism	14815646	3310381	4.475511
15	News & Politics	18151033	4180049	4.342301

Figure 9.41: Ratio of likes to dislikes for each category

NOTE

The solution to this activity can be found on page 634.

SUMMARY

This chapter covered many concepts of workflow management and job control. We started by creating a simple data workflow with a single Python script. We then added more steps into the workflow and broke the workflow down into a multi-stage workflow. Next, we used Bash to compose as well as automate workflows. Lastly, we studied DAGs and implemented them using the open-source tool Airflow.

With the concepts and techniques that you have learned in this chapter, you will be able to tackle more sophisticated problems in the areas of AI and data science. Moreover, you will continue to learn and build experience on top of what you have gained from this chapter.

In the next chapter, you will learn about data solutions from public cloud providers such as Amazon Web Services. The concepts of implementing data operations and creating a data pipeline will be our building blocks for the next chapter. We will continue to build more sophisticated data storage solutions for use in AI and data science.

10

INTRODUCTION TO DATA STORAGE ON CLOUD SERVICES (AWS)

OVERVIEW

In this chapter, you will learn about the pros and cons of various cloud data storage solutions. You will be able to create, access, and manage your Amazon S3 cloud services. You will learn about the AWS **Command-Line Interface (CLI)** and Python **Software Development Kit (SDK)**, which is used to control **Amazon Web Services (AWS)**. Lastly, you will create a simple data pipeline that reads from and writes to your cloud data storage. By the end of this chapter, you will understand the new paradigm of AI engineering and be able to build your data pipeline solutions using cloud storage services.

INTRODUCTION

In the previous chapter, we learned how to create our own data pipelines and use Airflow to automate our jobs. This enables us to create more data pipelines at scale, which is great. However, when we start scaling the number of data pipelines in our local machine, we will quickly run into scaling issues due to the limitations of a single machine. A single machine might give you 16 CPUs and 32 GB of RAM, which allows up to 16 different data pipelines running in parallel providing a memory footprint of less than 32 GB. In reality, AI engineers need to run hundreds of data pipelines every day to train models, predict data, monitor system health, and so on. Therefore, we need many more machines to support operations on such a scale.

Nowadays, software engineers are building their applications on the cloud. There are many benefits to building applications on the cloud. Some of them are as follows:

- The cloud is flexible. We can scale the capacity up or down as we need.

- The cloud is robust and it has a disaster recovery mechanism.

- The cloud is also managed, which means engineers don't need to worry about wasting time maintaining systems and security updates.

Many tech companies have started moving away from their own data centers and are migrating their applications from on-premises to cloud-native. In short, the cloud is quickly becoming the new normal.

There are three big first-tier public cloud providers: **Amazon Web Services** (**AWS**), **Microsoft Azure**, and **Google Cloud Platform** (**GCP**). AWS has the biggest market share in the cloud services market and it's also the earliest one. Each of those three cloud providers has its strengths and weaknesses in different areas. We will focus on data-related services in AWS in this chapter as it's a market leader and currently being used by major companies. Also, you are more likely to work with AWS than the other two providers when you work in the industry.

In this chapter, you will learn about various cloud storage services, as well as database solutions. You will learn how to use a cloud provider's **Command-Line Interface** (**CLI**) or **Software Development Kit** (**SDK**) to manage and control your cloud storage services and database services. Most importantly, you will be able to build your data pipeline in the cloud.

INTERACTING WITH CLOUD STORAGE

Cloud storage allows developers to store data on the internet through cloud computing providers. Instead of storing and managing data locally, cloud providers manage and operate data storage as a service, which removes the needs for buying or managing your own data storage infrastructure. Thus, this allows you to access your data anytime and anywhere globally.

We are in the era of a data-driven world, and many businesses are heavily relying on data, whether it's for decision making or building intelligence systems. For companies whose business models rely on data, there are three fundamental requirements for data storage solutions: **durability**, **availability**, and **security**. It's extremely costly to operate a data storage infrastructure with those requirements but very cost-efficient to directly use a cloud data storage service. The benefits of using cloud data storage are not only getting those requirements met but also having backup, disaster recovery, and scalability at the ready.

When it comes to data storage, there are generally three types of cloud data storage services, which are shown in the following figure:

	Object Storage	File Storage	Block Storage
Interface	API / HTTP	User	OS
Cost	Low	Mid	High
Performance	Low	Mid	High
Data Type	Pdf, images, video	User data, web content	OS, Databases
Scale	High	Mid	Low
Data Profile	Unstructured cold data	Structured hot and cold data	Structured hot and cold data

Figure 10.1: Three types of data storage

The first type is **object** or **blob storage**. It's vastly scalable, flexible, and REST-based for developing or supporting modern applications. Engineers usually store raw data in this type of storage. The second one is **file storage**, which is for those applications that require shared file access and a filesystem. The way we interact with file storage is similar to how we interact with the filesystem on our local machine. The last one is **block storage**. Block or disk storage, unlike the previous two types, cannot be used on its own. It's usually used as a service attached to other applications, such as databases or virtual machines.

Among the three types of data storage, object storage is the most commonly used in modern software and you will most likely be working with object storage when you work in a tech company. In this chapter, we have a lot of opportunities to work with the **Amazon Simple Storage Service (S3)**.

You may be wondering since there are so many cloud providers, why you should choose AWS?

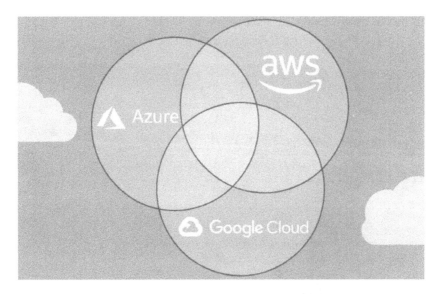

Figure 10.2: Three first-tier public cloud providers

The answer is simple. AWS has a very nice and user-friendly interface for both technical and non-technical people. They also provide free workshops and tutorials about how to leverage their services to build software. AWS is preferable for beginners to learn and use. Knowing how to manage or control services on AWS can help you get on board with the use of other cloud providers faster too.

Now, let's talk about how to access and manage your AWS services. There are mainly two ways to access and control AWS services. The first one is to use the AWS console to interact with its services. The second one, which is also the focus of this chapter, is to programmatically interact with AWS services via the AWS CLI or SDK. As per the following figure, when we log in to the AWS console, we can directly create or terminate services by mouse-clicking and typing in information. However, this approach requires lots of human interaction, which is not a good interface for developing applications. The second approach is for software development purposes. To programmatically access AWS services, you will need to create a role with AWS **Identity and Access Management** (**IAM**) with certain permissions so that you can use AWS CLI or SDK to programmatically control your AWS services:

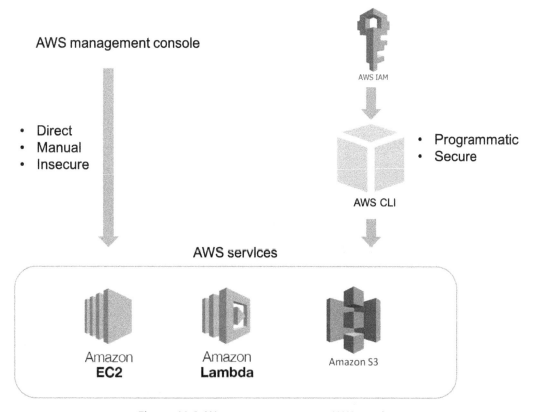

Figure 10.3: Ways to manage your AWS services

Throughout the exercises in this chapter, you will follow the best practices when using cloud services, especially AWS. We will use AWS CLI to interact with AWS services as well as develop workflows using SDK.

In the preceding figure, there are three AWS services. However, we will only be focusing on Amazon S3, which is the object storage service in AWS. To upload your data (photos, videos, documents, and so on) to Amazon S3, you must create an S3 bucket in the AWS region of your choice and give it a globally unique name. AWS recommends that customers choose regions geographically close to them to reduce latency and costs. After an S3 bucket has been created, you can upload any number of objects to the bucket.

In the next exercise, we will create an S3 bucket and upload a file to it.

EXERCISE 10.01: UPLOADING A FILE TO AN AWS S3 BUCKET USING AWS CLI

Imagine that you want to build a data pipeline on a cloud service such as AWS. The first thing you should know is how to upload your files from your local development machine to a cloud storage service such as AWS S3.

We will be using a sample dataset from NYC Open Data. This dataset contains the leading causes of death by sex and ethnicity in New York City (NYC) since 2007. The causes of death are derived from NYC death certificates, which are issued for every death that occurs in NYC. The dataset is provided by the **Department of Health and Mental Hygiene** (**DOHMH**). It was created on November 20, 2013, and the last update was on December 9, 2019.

The dataset can be found in our GitHub repository at the following location:

https://packt.live/2C72sBN.

You need to download the **New_York_City_Leading_Causes_of_Death.csv** file from this book's GitHub repository to your local machine.

Before proceeding with this exercise, we need to set up an AWS account and install AWS CLI, as well as all the necessary configurations. Please follow the instructions on the *Preface* to install it.

Perform the following steps to complete this exercise:

1. Create a directory called **Chapter10** for all the exercises in this chapter. In the **Chapter10** directory, create a **Data** directory.

2. Move the downloaded **New_York_City_Leading_Causes_of_Death.csv** file to the **Data** directory.

3. Open your Terminal (macOS or Linux) or Command Prompt (Windows), navigate to the **Chapter10** directory, and run following command:

```
ls
```

You should get the following output:

```
Data
```

The **ls** command will list the directories and files in the current directory, that is, **Chapter10**. Inside the **Data** directory, there should be a **New_York_City_Leading_Causes_of_Death.csv** file.

4. Similarly, we can check whether there is data in our S3 by running the following command:

```
aws s3 ls
```

If you're using AWS for the first time, then there will be no data in your S3. Therefore, you should see no output from running the preceding command.

Notice that this command is similar to the command in *Step 3*. The AWS CLI follows the pattern of **aws [service name] [action] [object] [flags]**. The command always starts with **aws** and is followed by a specific service. In our case, the service that we are using is **s3**. Next to the service is the action you want to perform on this service. Here, our action is **ls**, which means list files. This **ls** command is similar to the one from Linux bash.

5. Create a bucket by running the following command:

```
aws s3api create-bucket --bucket ${BUCKET_NAME}
```

> **NOTE**
>
> **${BUCKET_NAME}** is a variable in the command, and you should come up with a bucket name for this exercise. In this chapter, let's set **BUCKET_NAME=ch10-data**. You will see the **ch10-data** bucket in the examples throughout the exercises in this chapter.

You should get the following output:

```
{
    "Location": "/ch10-data"
}
```

If you failed to create a bucket, then you will get the following error:

```
An error occurred (BucketAlreadyExists) when calling the CreateBucket operation: The requested bucket name is not a
vailable. The bucket namespace is shared by all users of the system. Please select a different name and try again.
```

Figure 10.4: S3 bucket creation failure

If you receive an error, it's not because you did something wrong. It's because the bucket name **ch10-data** is taken. You just need to give your bucket a more creative name.

The reason why we need to create a new S3 bucket is because an S3 bucket is a store object in S3, which is similar to the concept of a file directory in our operating system. It's a way to organize data. Here, we used **s3api** to create a bucket. The name of the bucket can be anything arbitrary. Here, we called this bucket **ch10-data**. When we create a bucket, it will be created in the **us-east-1** region by default.

6. Verify whether the **ch10-data** bucket has been created by running the following command:

```
aws s3 ls
```

You should get the following output:

```
2020-01-12 16:03:24 ch10-data
```

This means there is a bucket named **ch10-data** in your S3 service. Please ignore the timestamp. The timestamp indicates when the command was run.

7. Upload the **New_York_City_Leading_Causes_of_Death.csv** file to our S3 bucket, **ch10-data**, by running the following command:

```
# upload local file to s3 bucket
aws s3 cp ./Data/New_York_City_Leading_Causes_of_Death.csv
  s3://${BUCKET_NAME}/
```

You should get the following output:

```
upload:Data/New_York_City_Leading_Causes_of_Death.csv to
  s3://ch10-data/New_York_City_Leading_Causes_of_Death.csv
```

We used the **aws s3 cp** command to upload the **New_York_City_Leading_Causes_of_Death.csv** file from our local directory, **./Data**, to the S3 bucket path, that is, **s3://${BUCKET_NAME}/**. The **cp** command is similar to the one from Linux bash. When you upload files, the AWS CLI follows the pattern of **aws s3 [local file path] [s3 bucket path]**.

8. Verify whether the file was uploaded successfully or not by running the following commands:

```
aws s3 ls s3://${BUCKET_NAME}/
```

You should get the following output:

```
2020-01-12 16:15:42 91294
  New_York_City_Leading_Causes_of_Death.csv
```

When you want to list files in an S3 bucket, you can use the AWS CLI by following the pattern of **aws s3 ls [s3 bucket path]**. It will tell you about all the files from that S3 bucket, provided you have the read access privilege to that bucket.

Alternatively, we can use the AWS console to check if the data has been successfully uploaded or not. Let's look into this.

9. Sign in to our AWS console, search for **s3** in the search bar, and click the service, as shown in the following figure:

Figure 10.5: Finding the S3 service on the AWS Management Console

This will redirect us to the S3 console and present all of our S3 buckets.

10. Click on the first bucket, `ch-10-data`, as shown in the following figure:

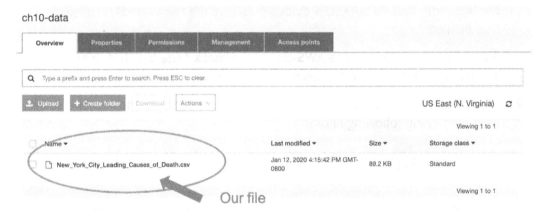

Figure 10.6: List of buckets in our S3

After clicking this bucket, the file should be present inside if it was successfully uploaded, as shown in the following figure:

Figure 10.7: List of files in the ch10-data bucket

We can also use the AWS console to interact with our AWS services. The AWS console provides a human-friendly interface and is very easy to understand and use. If you don't know how to perform certain tasks with the AWS CLI command, you can always fall back to using the AWS console and search the documentation for the specific AWS CLI command you want to perform.

NOTE

To access the source code for this specific section, please refer to https://packt.live/2ZpaKhk.

In this exercise, we learned how to interact with AWS S3 services using the programmatic approach and the AWS console approach. We performed basic tasks on S3 such as creating buckets, listing buckets and files, and uploading files to an S3 bucket.

You will continue to learn about more data tasks in the following exercises. Next, we will learn how to back up our files using S3. We will copy files from one S3 bucket to another.

EXERCISE 10.02: COPYING DATA FROM ONE BUCKET TO ANOTHER BUCKET

Sometimes, we want to back up a piece of data as the source of truth. We can copy the data to another location as a backup so that the original files are free for us to mess with. In this exercise, your task is to create a bucket for backup data and copy the **New_York_City_Leading_Causes_of_Death.csv** file to the backup bucket.

Perform the following steps to complete this exercise:

1. Create a bucket called **backup** using **aws s3api** by running the following command:

```
aws s3api create-bucket --acl private --bucket ${BACKUP_BUCKET}
```

> **NOTE**
>
> **${BACKUP_BUCKET}** is a variable in the command, and you should come up with a bucket name for this backup location in this exercise. In this exercise, let's set **BACKUP_BUCKET=storage-for-ai-data-backup**. You will see the backup bucket named **storage-for-ai-data-backup** throughout this exercise.

If the bucket was created successfully, you should get the following output:

```
{
    "Location": "/storage-for-ai-data-backup"
}
```

If you receive an error again, then it will mostly be because of bucket name collision. This means you just have to give another creative name to this backup bucket.

The preceding AWS CLI command looks different than the one in the previous exercise. We added a new flag, **-acl private**, to limit the access privilege of this bucket to **private**, which means any data inside this bucket cannot be read or written by users who don't have access privilege to the S3 bucket. This is not a mandatory requirement; we added it just to keep it private:

```
Verify if the bucket storage-for-ai-data-backup is created or
not
  by running the following command: aws s3 ls
```

You should get the following output:

```
2020-01-12 16:03:24 ch10-data
2020-01-13 20:44:35 storage-for-ai-data-backup
```

2. Create a copy of the file in the backup bucket by running the following command:

```
aws s3 cp
  s3://${BUCKET_NAME}/New_York_City_Leading_Causes_of_Death.csv
  s3://${BACKUP_BUCKET}/
```

You should get the following output:

```
copy:s3://ch10-data/New_York_City_Leading_Causes_of_Death.csv to
  s3://storage-for-ai-data-
  backup/New_York_City_Leading_Causes_of_Death.csv
```

We used the **aws s3 cp** command once more to copy the files from one location to another. In this case, we copied the **New_York_City_Leading_Causes_of_Death.csv** file from our S3 bucket, **s3://${BUCKET_NAME}/**, to another S3 bucket, **s3://${BACKUP_BUCKET}/**. The **cp** command is similar to the one from Linux bash. When you copy files, the AWS CLI follows the pattern of **aws s3 cp [s3 bucket path for source] [s3 bucket path for target]**.

3. Verify whether the file is in the backup bucket or not by running the following command:

```
aws s3 ls s3://${BACKUP_BUCKET}/
```

If the file was successfully copied to the backup bucket, you should get the following output:

```
2020-01-13 21:20:11 91294
  New_York_City_Leading_Causes_of_Death.csv
```

> **NOTE**
>
> To access the source code for this specific section, please refer to https://packt.live/2ZoZKAl.

In this exercise, we learned how to use the AWS CLI command to perform the copy data task. First, we created a new bucket and then we copied a file from the original S3 bucket to the new S3 bucket.

We will learn how to download files from the S3 bucket using the AWS CLI in the next exercise.

EXERCISE 10.03: DOWNLOADING DATA FROM YOUR S3 BUCKET

Downloading files from an S3 bucket is another very common data task in the day-to-day work of an AI engineer. Whether you are creating a data pipeline or simply performing ad hoc analysis, you will need to download data from cloud storage. In this case, your task is to download data from AWS S3.

Perform the following steps to complete this exercise:

1. Open your Terminal (macOS or Linux) or Command Prompt (Windows), navigate to the **Chapter10/Data** directory, and run the following command:

   ```
   ls
   ```

   You should get the following output:

   ```
   New_York_City_Leading_Causes_of_Death.csv
   ```

   We downloaded this file in *Exercise 10.01, Uploading a File to an AWS S3 Bucket Using AWS CLI*. Now, let's remove it and download this file again from our S3 bucket using the AWS CLI.

2. Remove the file by running the following command:

   ```
   rm New_York_City_Leading_Causes_of_Death.csv
   ```

3. Download the file from our S3 bucket to the **Chapter10/Data** directory by running the following command:

   ```
   aws s3 cp
     s3://${BUCKET_NAME}/New_York_City_Leading_Causes_of_Death.csv
     ./
   ```

> **NOTE**
>
> **${BUCKET_NAME}** is a variable in the command, and you should come up with a bucket name for this exercise. In this chapter, let's set **BUCKET_NAME=ch10-data**. You will see the **ch10-data** bucket in our examples throughout the exercises in this chapter.

You should get the following output:

```
download:s3://ch10-data/New_York_City_Leading_Causes_of_Death.csv
   to ./New_York_City_Leading_Causes_of_Death.csv
```

You may notice that we used the **aws s3 cp** command again to download data. This command copies the data from the source location to the destination location. It doesn't matter if the file location is local or remote in S3. To download the data, we essentially copy data from an S3 location to a local file location. Similarly, when we upload data, we are essentially copying data from a local file location to an S3 location.

4. Verify the downloaded data by running the **ls** command in the current working directory:

```
ls
```

You should get the following output:

```
New_York_City_Leading_Causes_of_Death.csv
```

> **NOTE**
>
> To access the source code for this specific section, please refer to https://packt.live/2WeswC9.

Now that you've completed the previous three exercises, you can see that the AWS CLI provides a very simple way to control your AWS S3 service and files. However, not everybody is a fan of the **bash** command. Sometimes, when we develop a data solution, we try to avoid **bash** because of a lack of testing capability. We want to develop an end-to-end solution in a rather self-contained environment such as developing a data pipeline in Python.

Fortunately, AWS provides many ways for us to access or control its services. Besides AWS CLI, there is an AWS SDK for different programming languages such as Java, Python, Node.js, and so on.

Next, we will be focusing on the Python version of SDK to access and control our S3 services. The Python SDK we are going to use is **Boto3**, which allows Python developers to write software that makes use of services such as Amazon S3 and Amazon EC2.

In the next exercise, you will be scripting a Python program to upload or download data between your local machine and AWS S3.

EXERCISE 10.04: CREATING A PIPELINE USING AWS SDK BOTO3 AND UPLOADING THE RESULT TO S3

In this exercise, we will develop a data pipeline using a Python script. The script will download the dataset from S3, then find the top 10 leading causes of death and upload the resulting data back to S3.

The source data you will work with is located in the **s3://${BUCKET_NAME}/ New_York_City_Leading_Causes_of_Death.csv** S3 bucket.

Before proceeding with this exercise, we need to set up the **Boto3** AWS Python SDK in the local-dev environment. Please follow the instructions on the *Preface* to install it.

Perform the following steps to complete this exercise:

1. Create a directory called **Exercise10.04** in the **Chapter10** directory to store the files for this exercise.

2. Open your Terminal (macOS or Linux) or Command Prompt (Windows), navigate to the **Chapter10** directory, and type in **jupyter notebook**.

3. Create a Python script called **run_pipeline.py** in the **Exercise10.04** directory and add the following code:

```
import os
import shutil
import boto3
import pandas as pd


if __name__ == "__main__":

    # set your bucket name here
```

```
# 'ch10-data' is NOT your bucket. It's just an example here
# you should replace your bucket below
BUCKET_NAME = 'ch10-data'

# create s3 resource
s3_resource = boto3.resource('s3')

# downfile from bucket
try:
    s3_resource.Bucket(BUCKET_NAME).download_file(
        'New_York_City_Leading_Causes_of_Death.csv',
        './tmp/New_York_City_Leading_Causes_of_Death.csv')
except FileNotFoundError:
    os.mkdir('tmp/')
    s3_resource.Bucket(BUCKET_NAME).download_file(
        'New_York_City_Leading_Causes_of_Death.csv',
        './tmp/New_York_City_Leading_Causes_of_Death.csv')
```

In our **run_pipeline.py** script, our first step is to download data from an S3 bucket. We used the Python SDK **Boto3** to connect with AWS. To access S3 services using Boto3, we created an S3 resource object. Then, we accessed a particular bucket from the resource object with the **.Bucket(bucket_name)** method. To download data, we use its **.download_file(source_file, target_file)** method to download the file to our local working directory.

> **NOTE**
>
> You have probably seen the **try-except** block elsewhere. It's generally useful when you expect to catch a known exception and handle the exception gracefully. In this case, we expect that the **tmp/** directory hasn't been created. So, when an exception is thrown, we create the **tmp/** directory first and then perform the download.

4. Add the following code snippet to the **run_pipeline.py** Python script:

```
# read file with pandas
df = pd.read_csv('./tmp/New_York_City_Leading_Causes_of_Death.
csv')

# filter out data with invalid values
```

```
df_filterred = df[df['Deaths'].apply(lambda x:
  str(x).isdigit())]
df_filterred['Deaths'] = df_filterred['Deaths'].apply(lambda
  x: int(x))

# calculate number of deaths for each year
df_agg = df_filterred.groupby('Leading
  Cause')[['Deaths']].sum()

# sort and take top 10
df_top10 = df_agg.sort_values('Deaths',
  ascending=False).head(10)
```

The next step in this script is to perform calculations. After the file has been downloaded to the **tmp/** directory, we use **pandas** to read the file from the local path. Before we calculate the top 10 death causes, we clean up the data because not all the values in the **Deaths** column are integers. After we've filtered out the records that aren't integers, we calculate the number of deaths for each cause. We use the pandas **.groupby** clause to group data into each cause. Then, we apply pandas **.sum()** to sum up the total number of deaths for each cause. Lastly, we use pandas **.sort_values** to sort the data based on the number of deaths in descending order and filter to just the top 10 causes using pandas **.head(n)**.

5. Add the following code snippet to the **run_pipeline.py** Python script to upload the result data:

```
# write new data to new file
df_top10.to_csv('tmp/New_York_City_Top10_Causes.csv')

# upload data to S3
s3_resource.Bucket(BUCKET_NAME).upload_file(
    'tmp/New_York_City_Top10_Causes.csv',
    'New_York_City_Top10_Causes.csv')

# clean up tmp
shutil.rmtree('./tmp')

print('[ run_pipeline.py ] Done uploading result data to S3
  bucket')
```

The final step of this process is to upload the result data to the same S3 bucket. But we need to write the data to a local file first. To do this, we use the **.upload_file** (source file, target file) method of **Boto3** to upload that file. After that, we delete the **tmp** folder and print the message for successful uploading.

6. Now that you've completed the previous steps, the **run_pipeline.py** Python script should contain the following code:

run_pipeline.py

```
13    # create s3 resource
14    s3_resource = boto3.resource('s3')
15
16
17    # downfile from bucket
18    try:
19        s3_resource.Bucket(BUCKET_NAME).download_file(
20            'New_York_City_Leading_Causes_of_Death.csv',
21            './tmp/New_York_City_Leading_Causes_of_Death.csv')
22    except FileNotFoundError:
23        os.mkdir('tmp/')
24        s3_resource.Bucket(BUCKET_NAME).download_file(
25            'New_York_City_Leading_Causes_of_Death.csv',
26            './tmp/New_York_City_Leading_Causes_of_Death.csv')
```

The complete code for this step is available at https://packt.live/38OYU3e.

At the top of the script, we imported all of the necessary modules. Then, we put all of the code snippets inside the **__main__** namespace. The code is executed sequentially under **__main__**.

7. Run the pipeline using the following command:

```
python run_pipeline.py
```

You should get the following output:

```
run_pipeline.py:27: SettingWithCopyWarning:
A value is trying to be set on a copy of a slice from a DataFrame.
Try using .loc[row_indexer,col_indexer] = value instead

See the caveats in the documentation: http://pandas.pydata.org/pandas-docs/stable/indexing.html#indexing-view-versus-copy
    df_filterred['Deaths'] = df_filterred['Deaths'].apply(lambda x: int(x))
[ run_pipeline.py ] Done uploading result data to S3 bucket
```

Figure 10.8: The run_pipeline.py script is successfully running

The script will create a **New_York_City_Leading_Causes_of_Death. csv** file and upload it to the cloud. You may wonder why there seem to be lots of warnings and complaints. These warnings are from **pandas** and they suggest their ways of performing certain operations. We could suppress the log from a specific library such as **pandas**, but this is outside the scope of this book. In our case, we only care about the last line of the output. When we see the last line in the output, it means the script has been executed to the end of the script without any errors occurring.

8. Verify whether the data has been successfully uploaded or not by running the following command:

```
aws s3 ls s3://ch10-data
```

You should get the following output:

```
2020-01-12 16:15:42 91294
  New_York_City_Leading_Causes_of_Death.csv
2020-01-15 23:22:04 513 New_York_City_Top10_Causes.csv
```

> **NOTE**
>
> To access the source code for this specific section, please refer to https://packt.live/38RXftC.

To sum up what we have done so far: we are now able to perform most of the common data tasks such as downloading, uploading, and copying files between our local dev machine and cloud data storage solutions such as S3 using the AWS CLI and the Python SDK, whose library is called **boto3**.

Cloud data storage solutions are more than just S3 on AWS. AWS also provides other data storage solutions such as relational database services and NoSQL database services. In the next section, we will introduce AWS **Relational Database Service** (**RDS**).

GETTING STARTED WITH CLOUD RELATIONAL DATABASES

A relational database is still one of the most used pieces of software in the industry. Many companies use it as a production database to store user information or session logs. Given that **Structured Query Language** (**SQL**) is universally well known, many companies also use it as an analytics engine with which business analysts query business metrics to analyze the health of their business or product.

In the AI industry, we also need a relational database to store metadata information for AI modeling experiments. For example, during the model development period, AI engineers often perform model experiments with different training data, different model hyperparameters, and different configurations.

After releasing a production model, AI engineers also want to keep track of what models have been released and their corresponding prediction performance. So, being able to store and organize your AI-related data in a relational database will be tremendously helpful for reproducing research results or performing any retrospective experiment.

However, setting up and operating your databases involves provisioning compute and storage, installing OSes and databases, as well as updating and maintenance. Luckily, AWS RDS makes it extremely easy for us to use a relational database. With this service, we are just a few clicks away from setting up, operating, and scaling our databases:

Figure 10.9: Features of AWS RDS

If you are looking for a solution to store data in a way so that the data is always available, fault-tolerant, and queryable, AWS RDS meets all of these requirements. There are six familiar database engines to choose from regarding AWS RDS, which are Amazon Aurora, PostgreSQL, MySQL, MariaDB, Oracle Database, and SQL Server. Different engines are optimized for different use cases.

In the next exercise, we will create an AWS RDS service via the AWS console. We will use the MySQL engine for this RDS instance.

EXERCISE 10.05: CREATING AN AWS RDS INSTANCE VIA THE AWS CONSOLE

In this exercise, we will use the AWS console to create an AWS RDS instance, that is, a cloud relational database. Specifically, we will use the RDS MySQL engine. When we create new services in AWS, we usually use the AWS console instead of AWS CLI. This is because creating a new service is a one-time exercise and the AWS console makes creating new services very simple and easy.

This exercise requires an AWS account and the AWS RDS service. This is a free-tier service if your AWS account is less than 12 months old.

Perform the following steps to complete this exercise:

1. Log in to the AWS console and search for and click on **RDS** in the **Find Services** search bar, as shown in the following figure:

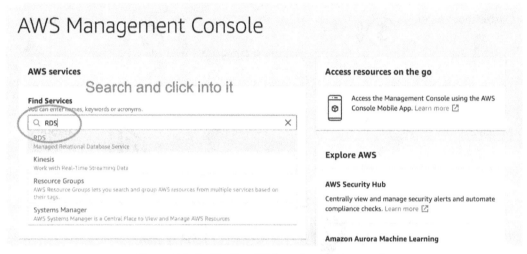

Figure 10.10: Finding RDS on the AWS console

2. In the **Amazon RDS** console, click the **Create database** option, as shown in the following figure:

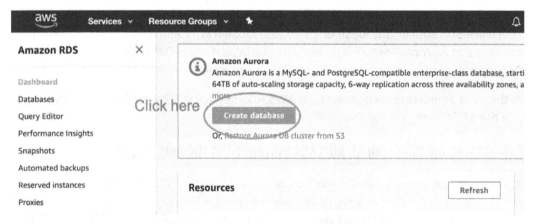

Figure 10.11: Creating a database in the RDS console

3. On the **Create database** page, select the **Standard Create** option, followed by the **MySQL** option, as shown in the following figure:

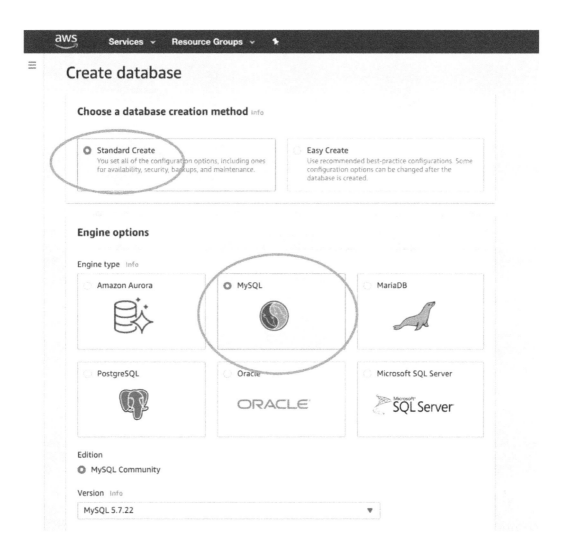

Figure 10.12: Creating a MySQL engine

4. In the **Templates** section, choose **Free tier** so that you won't be charged:

Figure 10.13: Choosing the Free tier

5. In the **Settings** section, set the values for **DB instance identifier** and **Master username** to **my-first-rds** and **my_db_admin**, respectively. Also, clear the **Auto generate a password** checkbox and add a password, as shown in the following screenshot:

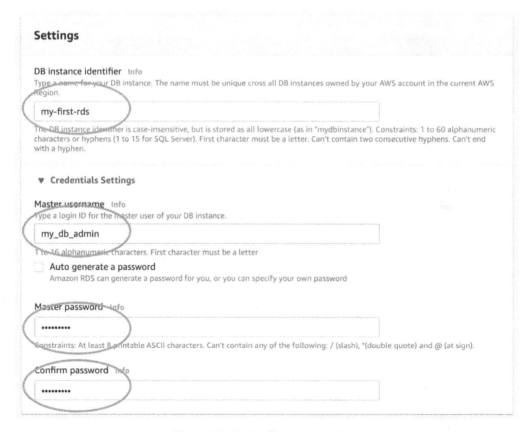

Figure 10.14: Configuring settings

> **NOTE**
>
> Please remember the username and password used in **Credentials Settings**. We will be using these credentials to access the database later.

6. Since we are using **Free tier**, we can leave the **DB instance size**, **Storage**, and **Availability & durability** sections with their default values, as shown in the following figures:

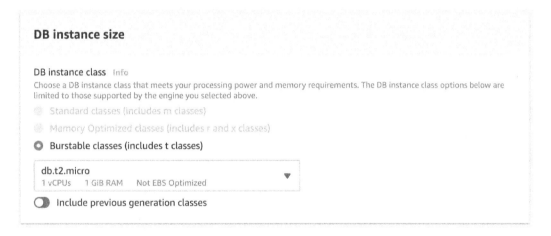

Figure 10.15: Choosing the instance size

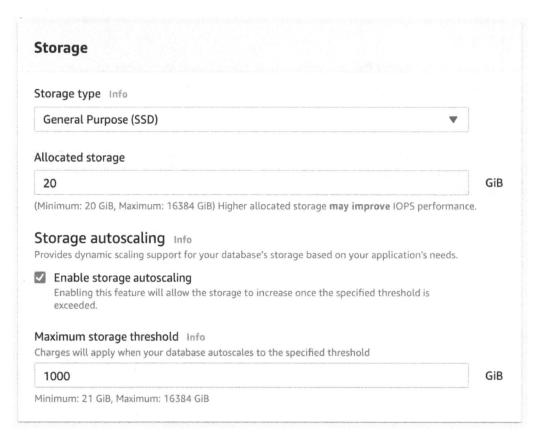

Figure 10.16: Configuring storage

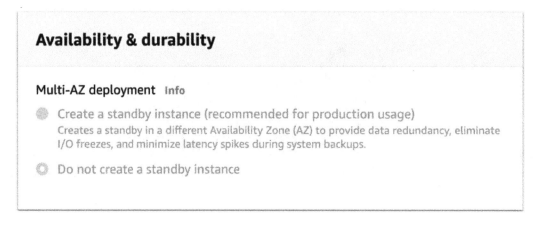

Figure 10.17: Configuring Availability and durability

10. In the **Connectivity** section, expand **Additional connectivity configuration** and change the **Publicly accessible** setting to **Yes**. For the **Availability zone**, we can choose anyone, as shown in the following figure:

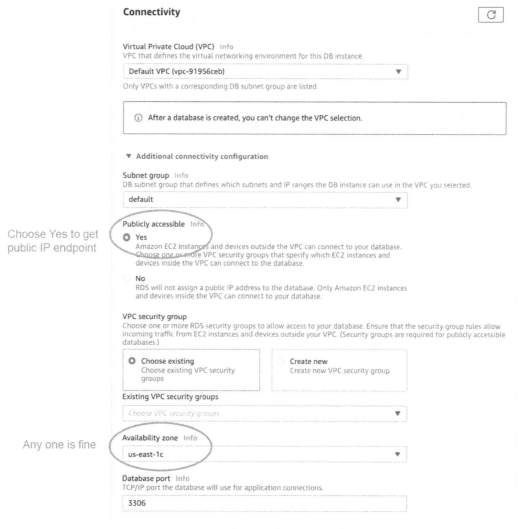

Choose Yes to get public IP endpoint

Any one is fine

Figure 10.18: Configuring connectivity

By choosing **Yes** for **Publicly accessible**, AWS will assign this instance a public IP address, which will allow your local machine to connect to the database. **Availability zone** refers to the different data centers within a region. In this case, we can choose any.

11. Expand the **Additional configuration** section and enter **chapter10** for **Initial database name**. Keep the default settings for the other options as they are:

Figure 10.19: Additional configuration

12. Leave the other sections with their default values and go to the bottom of the console to choose **Create database**, as shown in the following figure:

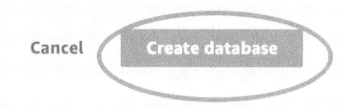

Figure 10.20: Create database

13. Wait for the **Status** of your new database instance to show as **Available**. Then, click the database instance name to show its details:

Figure 10.21: Waiting for the database to be created

When it's finished, you will see the following status:

Figure 10.22: Database created

15. In the **Connectivity & security** section, view the **Endpoint** and **Port** of the database instance:

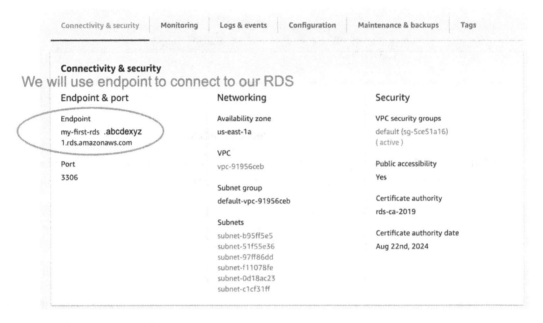

Figure 10.23: Endpoint for RDS connection

> **NOTE**
>
> Remember to delete this RDS service after finishing the exercises in this chapter, as shown in *Step 12*. Otherwise, you will be charged if the instance is running for more than 750 hours per month.

16. (Do this step after you've finished this chapter.) To delete this service, click the **Actions** drop-down and select **Delete**, as shown in the following figure:

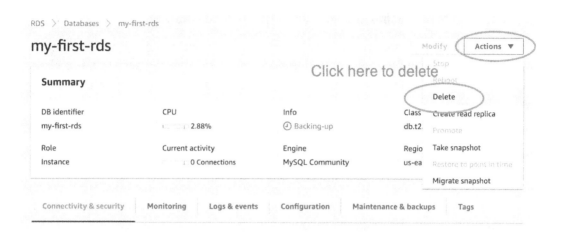

Figure 10.24: Deleting the RDS instance

Now that we have Amazon RDS set up, we can create different types of services with our databases. Traditionally, web developers pair up database services with web services to create websites.

In the industry of AI, engineers usually use relational databases to bookkeep different models' performance and experiment results. Recall from *Chapter 9, Workflow Management for AI*, that we deployed Airflow in our local machine, which is only good for the dev use case. If we want to use Airflow to manage production-level data pipelines, we need to deploy Airflow in the production environment. Instead of using SQLite as our Airflow database, we can use our Amazon RDS as its backend database to manage the metadata for different data pipelines.

But we are not going to deploy Airflow on AWS by ourselves. This type of service deployment or migration usually requires a team of infrastructure engineers. What's more important in this chapter is to be familiar with the AWS CLI so that you can control and manage AWS services.

The way in which programs are connecting to AWS services is through endpoints. To connect to an AWS service, we use an **endpoint**. An endpoint is the URL of the entry point for an AWS web service. Endpoints are usually made up of IP addresses and ports. With endpoints and user login information, we can essentially log in to the RDS instance and talk to the relational database inside.

In the next exercise, you will be using the AWS CLI to manage your Amazon RDS service and access the RDS database via the MySQL client.

EXERCISE 10.06: ACCESSING AND MANAGING THE AWS RDS INSTANCE

One of the common tasks when using cloud services is accessing the service itself. In this exercise, we are going to create the service so that we can access it as well as control it. First, we will use the AWS CLI to talk to the service. Then, we will use the MySQL client to connect to our Amazon RDS instance.

Before proceeding with this exercise, we need to set up a MySQL client. Please follow the instructions on the *Preface* to install it.

Perform the following steps to complete this exercise:

1. Check the Amazon RDS service and find its endpoint for the MySQL client using the AWS CLI, as shown in the following code:

```
# check our RDS instances
aws rds describe-db-instances
```

You should get the following output:

```
{
    "DBInstances": [
        {
            "DBInstanceIdentifier": "my-first-rds",
            "DBInstanceClass": "db.t2.micro",
            "Engine": "mysql",
            "DBInstanceStatus": "available",
            "MasterUsername": "my_db_admin",
            "DBName": "chapter10",
            "Endpoint": {
                "Address": "my-first-rds.c0prwnnsa9ab.us-east-1.rds.amazonaws.com",
                "Port": 3306,
                "HostedZoneId": "Z2R2ITUGPM61AM"
            },
        }
        ...,
        ...,
        ...,
}
```

Figure 10.25: Describing the RDS instance

The preceding output provides all of the information about your RDS services. Here, we only have a single RDS instance, which is named **"my-first-rds"**. The important pieces of information in this giant JSON blob are **"MasterUsername"**, **"Endpoint"**, and its **"Address"** and **"Port"**. We will need these values to log in to our MySQL instance from our local machine.

2. Next, connect our MySQL instance from the local machine using the MySQL client by using the following syntax:

```
mysql --host ${Endpoint.Address} --port ${Endpoint.Port} -u
  ${MasterUsername} -p
```

> **NOTE**
>
> You have to replace **${variable}** with your values in the preceding command. For example, according to the author's credentials, the following command will be used:

```
mysql --host my-first-rds.c0prwnnsa9ab.us-east-
  1.rds.amazonaws.com --port 3306 -u my_db_admin -p
```

It will ask for a password, as shown in the following output:

```
Enter password:
```

You will need to type in the password you created when you created the RDS MySQL instance in *Step 5* of *Exercise 10.05, Creating an AWS RDS Instance via the AWS Console*.

After logging into the MySQL instance, you will observe the following output:

```
Welcome to the MySQL monitor.  Commands end with ; or \g.
Your MySQL connection id is 217
Server version: 5.7.22-log Source distribution

Copyright (c) 2000, 2019, Oracle and/or its affiliates. All rights reserved.

Oracle is a registered trademark of Oracle Corporation and/or its
affiliates. Other names may be trademarks of their respective
owners.

Type 'help;' or '\h' for help. Type '\c' to clear the current input statement.

mysql>
```

Figure 10.26: Successfully logging in to our RDS instance

> **NOTE**
>
> If you are unable to log in to your MySQL instance due to the AWS
> firewall, then you will need to go to the Amazon RDS console to modify the
> `Security group` rules, as shown in *Step 3* to *Step 6*.

3. (Optional) Log in to the AWS console, navigate to the RDS console, and select
   your instance, as shown in the following figure:

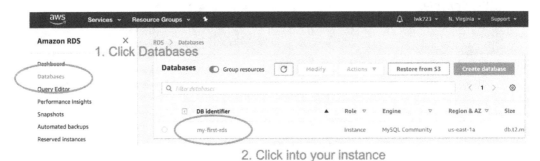

Figure 10.27: RDS console

4. For your instance, you will go to the **Connectivity & security** section and click on the **default (sg-5ce51a16)** group:

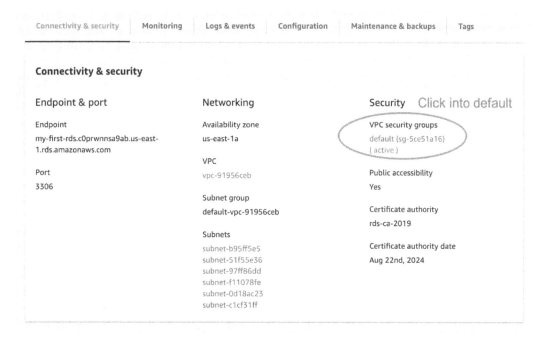

Figure 10.28: Connectivity settings

You will be brought to the **Security Group** console.

In the **Security Group** console, click the **Inbound** tab and then **Edit** to add a new rule, as shown in the following figure:

Figure 10.29: Editing the inbound rule for the security group

Now, we can add a new rule for inbound traffic so that the local machine has access to the Amazon RDS instance.

5. Click the **Add Rule** button, select **My IP** from the **Source** drop-down, and click **Save**, as shown in the following figure:

Figure 10.30: Only allowing My IP to be inbound

After you've saved the new rule, you should be able to connect to your MySQL instance from your local machine. Perform *Step 2* again to connect to your MySQL instance.

We assume that you are now able to log in to your MySQL instance.

6. Query your databases using the following command:

```
show databases;
```

You should get the following output:

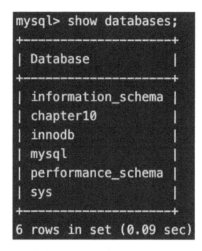

Figure 10.31: Databases in our RDS instance

If we created our instance correctly in *Exercise 10.05, Creating an AWS RDS Instance via the AWS Console*, we should be able to see **chapter10** as one of the databases.

> **NOTE**
>
> You might have deleted this instance after finishing *Exercise 10.05, Creating an AWS RDS Instance via the AWS Console*. If so, you will need to revisit that exercise and recreate the RDS instance for this exercise.

7. Select the **chapter10** database using the following command:

```
use chapter10;
```

You should get the following output:

```
Database changed
```

8. List the tables inside the **chapter10** database by using the following command:

```
show tables;
```

You should get the following output:

```
Empty set (0.09 sec)
```

However, we will get an **Empty set (0.09 sec)** this time because we haven't created any tables in this database. But if we import data into this database, we can query information locally via the MySQL client.

9. Exit the MySQL client using the following command:

```
mysql> exit
```

You should get the following output:

```
Bye
```

To avoid any charges, we will delete our Amazon RDS instance.

10. Navigate to the RDS console within the AWS console, click **Databases** on the left-hand pane, and then click the **my-first-rds** instance, as shown in the following figure:

Figure 10.32: RDS console

11. In the **my-first-rds** instance window, click the **Actions** dropdown and select **Delete** to remove our RDS service, as shown in the following figure:

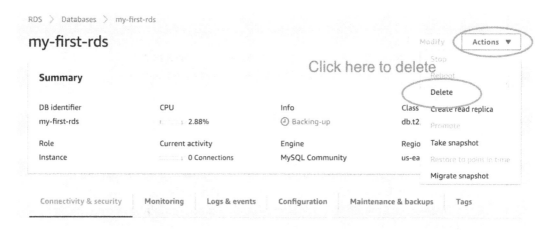

Figure 10.33: Deleting our RDS instance

12. In the confirmation popup, select and clear the respective checkboxes and click **Delete**, as shown in the following figure:

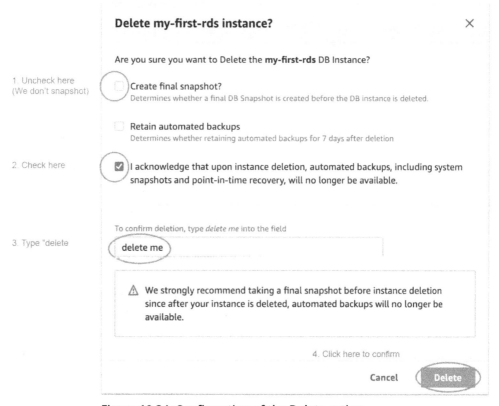

Figure 10.34: Confirmation of the Delete action

> **NOTE**
>
> To access the source code for this specific section, please refer to https://packt.live/3evEbmo.

> **NOTE**
>
> In the real world, you will want to **create a final snapshot** for future recovery. But we don't have any data in this instance, so we don't need to create a snapshot. For more information about database snapshots and how they work, please revisit https://aws.amazon.com/rds/details/backup/.

By completing this exercise, you have successfully implemented the mechanism to access and control the AWS RDS service. First, we used the AWS CLI to get endpoint information. Then, we used the MySQL client to connect with the AWS RDS service to perform database administrative tasks.

Next, we will move on to another type of database, that is, NoSQL databases.

INTRODUCTION TO NOSQL DATA STORES ON THE CLOUD

You may wonder why NoSQL exists when a relational database is powerful enough to store data. As the internet grows, data grows exponentially. And data comes in different shapes and forms. For example, there is more and more long-form document data that needs to be stored somewhere. Not only do we need a database solution to store long-form document data, but it also allows us to query the text data efficiently. Relational databases aren't built for such a use case. Therefore, there are more and more new database technologies emerging, especially in the world of NoSQL databases.

NoSQL databases, also known as non-relational databases, is the other type of database that we can use, as opposed to the relational database. Unlike a relational database, which has a tabular structure and a well-defined entity relationship, NoSQL databases don't require a strict structure or schema. When you are working with a NoSQL database, you generally don't need to think of data in a tabular format with a strict structure. Rather, you will think of data in a loose structure, such as an object in an object-oriented programming language where you can define data in any shape or form you want.

Because of its unstructured nature, NoSQL databases are much more scalable horizontally than traditional relational databases. Therefore, NoSQL databases are especially useful for working with large sets of distributed data:

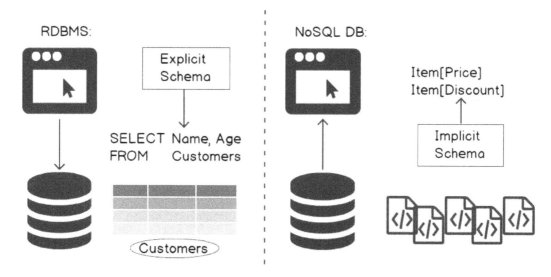

Figure 10.35: Comparison between NoSQL and relational databases

The preceding figure illustrates the difference between NoSQL and relational databases. Relational databases (on the left) work with tabular data with an explicit schema, while NoSQL databases (on the right) work with document data with an implicit schema. "Implicit" schema usually means the schema is more flexible and allows users to define its schema in an ad hoc fashion.

In the world of NoSQL databases, we can summarize different types of NoSQL database into the following four kinds:

- Key-value data store
- Document data store
- Columnar data store
- Graph data store

Different kinds of data stores accommodate different business use cases and data types. They also differ in terms of read or write performance. Let's understand them better in detail.

KEY-VALUE DATA STORES

Key-value data stores are based on the hash map data structure. They implement a data model that pairs a unique key with an associated value. It's extremely simple and performant. Many web applications use a key-value store as their caching system. For example, if engineers want to improve the response time of their website, then they can deploy a caching system alongside the web application to cache user session data or any data that takes time to load.

Some examples of key-value data stores are Redis and Memcached:

Key	Value
Name	Joe Bloggs
Age	42
Occupation	Stunt Double
Height	175cm
Weight	77kg

Figure 10.36: Key-value data stores

In the cloud era, we don't have to stand up our servers to serve Redis or Memcached services. We simply leverage what the cloud providers are offering. For example, AWS offers ElastiCache, which is a fully managed in-memory data store and cache service. You can choose either Redis or Memcached as its in-memory caching engine. Meanwhile, Google Cloud offers Cloud Memorystore, which is equivalent to AWS ElastiCache.

DOCUMENT DATA STORES

The document data store is frequently used when an application needs to store data in a document format. It usually stores data in a nested format. It doesn't require a schema when developers create or update data.

Due to document data growing on the internet and the use of **JavaScript Object Notation (JSON)** becoming prevalent in web development, document data stores have become more and more popular.

Let's take a look at some document examples:

```
record 1: {
    'title': 'where to buy iphone':
    'document':
        {
            'abstract': ...,
            'body': ...
        }
}

record 2: {
    'title': 'sell bitcoin':
    'document':
        {
            'content': ...
        }
}
```

Figure 10.37: JSON document format

Here, we have two document records. You can see that they have different schemas. Both of them can be written into the document store.

Today, some of the most well-known document stores are MongoDB, Riak, and ApacheCouchDB. Cloud providers also provide managed services, such as AWS DocumentDB, Google Cloud Datastore, and Azure Cosmos DB.

COLUMNAR DATA STORE

The columnar data store usually refers to databases that store data by column rather than by row. Traditional databases store data by row because it's easy and intuitive to write data row by row. However, it's less efficient in reading data row by row. When we perform analytical data processing, we usually only select a few columns that are relevant to our analysis. In a row-oriented database, the query engine has to visit every row to fetch the data. However, in the columnar database, the query engine only needs to visit the relevant columns to fetch the data. Therefore, the columnar data store is more efficient in reading data than row-oriented data.

Some well-known examples include Google BigTable, Cassandra, and HBase. Cloud providers offer this type of data store as well. Amazon offers DynamoDB for NoSQL and Redshift for relational.

GRAPH DATA STORE

The graph data store is another alternative NoSQL database. The graph is composed of nodes and edges. Nodes represent records, while edges represent the relationship between the records:

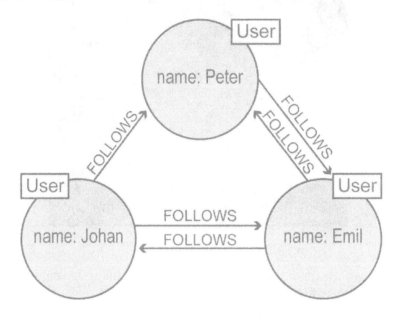

Figure 10.38: Nodes and edges in a graph

In a social network setting, records are usernames and edges indicate whether a user is a follower of another user.

Social networks are a good use case for graph stores. Well-known graph data stores include AllegroGraph, IBM Graph, Neo4j, and Titan. For cloud managed services, Amazon offers Neptune and Google Cloud offers Neo4j.

In short, while relational databases store data in a tabular schema, NoSQL databases are built to store data in document format. We will discuss working with data in the document format in the next section.

DATA IN DOCUMENT FORMAT

For relational databases, data is usually in a tabular form. The file format for tabular data is usually CSV or TSV. However, data in modern web applications is rarely in a tabular form. For example, when we send a tweet on Twitter, we are effectively making a post request to the tweeter's server with the data being a document format. The data in our post request looks as follows:

```
{
  "created_at" :   "Thu Apr 06 15:24:15 +0000 2017" ,
  "id_str" :   "850006245121695744" ,
  "text" :   "1\/ Today we\u2019re sharing our vision for the future of the Twitter API
  "user" :   {
    "id" :   2244994945 ,
    "name" :   "Twitter Dev" ,
    "screen_name" :   "TwitterDev" ,
    "location" :   "Internet" ,
    "url" :   "https:\/\/dev.twitter.com\/" ,
    "description" :   "Your official source for Twitter Platform news, updates & events.
  } ,
  "place" :   {
  } ,
  "entities" :   {
    "hashtags" :   [
    ] ,
    "urls" :   [
      {
        "url" :   "https:\/\/t.co\/XweGngmxlP" ,
        "unwound" :   {
          "url" :   "https:\/\/cards.twitter.com\/cards\/18ce53wgo4h\/3xo1c" ,
          "title" :   "Building the Future of the Twitter API Platform"
```

Figure 10.39: Data in document format

If you are familiar with web programming, you'll notice that this data is in a JSON format. Data in JSON format naturally manifests the structure of the object in object-oriented programming languages such as JavaScript, Python, and Java. Nowadays, the most common data format that web servers are using between their communications is JSON. For web developers, data in JSON format is preferred to work with than data in tabular format, which means NoSQL databases are also preferable over relational databases.

Most NoSQL databases consume data in JSON format. Due to this, we will need to learn how to create a JSON file in the first place. In the next activity, we will create a simple Python job that performs an analysis and returns data in JSON format.

ACTIVITY 10.01: TRANSFORMING A TABLE SCHEMA INTO DOCUMENT FORMAT AND UPLOADING IT TO CLOUD STORAGE

In this activity, we will create a Python job that extracts data from S3, transform the data from table-schema to JSON format, and finally upload the data to S3. Let's imagine that we want to find out the top three death causes for different **Race Ethnicity**.

Let's use the **New_York_City_Leading_Causes_of_Death.csv** file to find out the top three death causes for each **Race Ethnicity**. The data is in tabular format with eight columns, as shown in the following figure:

	Year	Leading Cause	Sex	Race Ethnicity	Deaths	Death Rate	Age Adjusted Death Rate
0	2010	Influenza (Flu) and Pneumonia (J09-J18)	F	Hispanic	228	18.7	23.1
1	2008	Accidents Except Drug Posioning (V01-X39, X43,...	F	Hispanic	68	5.8	6.6
2	2013	Accidents Except Drug Posioning (V01-X39, X43,...	M	White Non-Hispanic	271	20.1	17.9
3	2010	Cerebrovascular Disease (Stroke: I60-I69)	M	Hispanic	140	12.3	21.4
4	2009	Assault (Homicide: Y87.1, X85-Y09)	M	Black Non-Hispanic	255	30	30
5	2012	Mental and Behavioral Disorders due to Acciden...	F	Other Race/ Ethnicity	.	.	.
6	2012	Cerebrovascular Disease (Stroke: I60-I69)	F	Asian and Pacific Islander	102	17.5	20.7

Figure 10.40: View of the New_York_City_Leading_Causes_of_Death.csv file

We are only interested in the **['Leading Cause', 'Race Ethnicity', 'Deaths']** columns. We will use certain values from these columns to find out the top three death causes for each **Race Ethnicity**. In this activity, we will use Jupyter Notebook to explore data and use **pandas** to get the answer. Finally, we will write the result in JSON format. Your final output file should look as follows:

```
'{
    "Hispanic": [cause 1, cause 2, cause 3],
    "Asian and Pacific Islander": [cause 1, cause 2, cause 3],
    ...
}'
```

Figure 10.41: Final output data format

NOTE

The code and the resulting output for this activity has been loaded into a Jupyter Notebook that can be found here: https://packt.live/38TFoCt.

Perform the following steps to complete this activity:

1. Use the **boto3** AWS SDK to download the source data from the **${BUCKET_NAME}** you created in *Exercise 10.01, Uploading a File to an AWS S3 Bucket Using the AWS CLI.*
2. Use the **pandas** library to read the file, perform data exploratory analysis, and examine the quality of the data. This may require some simple data cleaning.
3. Use **pandas**, **groupby**, and aggregation, which we learned about in *Chapter 9, Workflow Management for AI,* to perform the calculation to find out the top three death causes for each **Race Ethnicity**. Remember to use a Python **dictionary** to store the result.
4. Use **json.dump()** to serialize the Python **dictionary** object as you write the data to a JSON file.
5. Use the **boto3** AWS SDK again to upload the output JSON file to the same S3 bucket.
6. Put all of the code snippets together in a single Python script and run it again.
7. Use the AWS CLI to verify if the output JSON file has been created. You should get the following output:

```
2020-01-12 16:15:42      91294 New_York_City_Leading_Causes_of_Death.csv
2020-01-15 23:22:04        513 New_York_City_Top10_Causes.csv
2020-01-19 18:08:17        811 top_causes_per_ethnicity.json
```

Figure 10.42: Final output data has been successfully uploaded to the S3 bucket

NOTE

The solution to this activity can be found on page 644.

SUMMARY

This chapter introduced some of the most well-known cloud storage solutions, specifically AWS S3. It also covered cloud database solutions for both traditional relational databases and NoSQL databases. Some of the common cloud database solutions are AWS RDS, ElastiCache, DocumentDB, GCP Memorystore, Data Store, and BigTable.

We started by using the AWS CLI in a Terminal to perform common data tasks such as creating a bucket, uploading files, and moving files. Later, we move on to the Python environment, where we used the AWS Python SDK to control AWS resources. At the end of this chapter, we leveraged the practical skills we learned in this chapter and composed an end-to-end pipeline that extracts data from S3, transforms data, and uploads data back to S3. The practical concepts you learned about during these exercises will allow you to build data applications or systems.

In the next chapter, you will continue to build on what you learned in the previous chapters. You will learn how to build AI models that can be combined with data storage systems.

> **NOTE**
>
> Please remember to **delete** the RDS service we created in *Exercise 10.05, Creating an AWS RDS Instance via the AWS Console* and *Exercise 10.06, Accessing and Managing the AWS RDS Instance*. Otherwise, you will be billed automatically on your credit card after 750 hours.

11

BUILDING AN ARTIFICIAL INTELLIGENCE ALGORITHM

OVERVIEW

We will start this chapter by introducing the most basic form of machine learning model: the linear regression model. We will use batch gradient descent in NumPy to train a regression model. Then we will get started with the popular deep learning framework PyTorch. Toward the end of the chapter, we will delve into one of the most exciting fields in deep learning research: reinforcement learning, specifically the deep Q-learning algorithm. Lastly, we will learn how to build a deep Q-learning algorithm to solve classic reinforcement learning problems, and we will learn how to improve the algorithm by implementing a double deep Q-learning algorithm in an activity.

By the end of this chapter, you will have a comprehensive understanding of how an **Artificial Intelligence** (**AI**) algorithm is built and trained with different implementations such as NumPy and PyTorch. Both NumPy and PyTorch are Python-based scientific computing packages. AI practitioners use them very frequently in the field of deep learning research. You will learn about the nuts and bolts of AI algorithms along with the best practices in training an AI model.

INTRODUCTION

In previous chapters, we learned about various hardware and software infrastructures for AI practices. We learned about different databases and data solutions as well as their use cases. We learned about big data computing engines such as Spark, which allows engineers to process web-scale data. We also learned about using workflow management systems such as Airflow to manage data pipelines at scale. We also learned a lot about cloud data solutions and how to leverage cloud data storage and perform basic data-related tasks.

This chapter will focus on the science and the mathematical side of artificial intelligence. Without a proper understanding of the theory that underpins AI, we simply cannot build a robust AI application. If we can understand the math and science behind AI, then we will be able to apply different algorithms to solve different real-world problems. With the skills you will gain from this chapter, you will be able to innovate new AI algorithms to solve new problems.

MACHINE LEARNING ALGORITHMS

There are four main types of learning algorithms:

- **Supervised learning algorithm**: This is trained to predict an outcome for a given set of input features. It's well studied and widely used in many areas such as spam classification, fraud detection, and product recommendation.

- **Unsupervised learning algorithm**: This analyzes the underlying patterns or structure of data and groups data into clusters. Examples are outlier detection, fraud detection, and dimensionality reduction.

- **Semi-supervised learning**: This falls between supervised learning and unsupervised learning. It's intended to boost learning accuracy for a supervised learning model by mixing unlabeled data.

- **Reinforcement learning algorithm**: This is trained to play a "game." It learns to take a "smarter" action at each step in a game so that it will eventually win the game. Examples are AlphaGo, robot control, and quantitative trading.

This chapter will focus on supervised learning and reinforcement learning. We will start by learning the fundamental theory and basic techniques in training a supervised learning model. Then we will get hands-on experience in building a reinforcement learning algorithm to play a classic control game in OpenAI Gym.

MODEL TRAINING

There are many different machine learning models for each of those four types of learning algorithms. Machine learning models rely on some forms of mathematical/statistical models. When we train models, it means we use an algorithm to find out the model's unknown parameters. Scientifically speaking, we cannot definitively find out the ground truth for unknown parameters. Instead, we can only estimate the unknown parameters as closely as possible to the ground truth by using mathematical/statistical methods on sample data. Estimating unknown model parameters is equivalent to solving a mathematical equation whose solution comes in one of two forms: closed or non-closed.

CLOSED-FORM SOLUTION

Some algorithms' mathematical models have closed-form solutions. A model with a closed-form solution can be solved by expressing the model parameters analytically in terms of a finite number of certain "well-known" functions. A classic example is a linear regression model. Training a linear regression model is equivalent to solving a quadratic matrix equation in linear algebra. The quadratic matrix equation is derived from a method called **Ordinary Least Squares** (**OLS**). The OLS method essentially tries to minimize the distance between model predictions and ground truth values. Through a series of linear algebra transformations, the minimization solves a quadratic equation, which always has a closed-form solution. To further simplify the problem, let's consider a simple linear regression model, shown in the following figure:

Linear Regression Model	Closed-Form Solution for Weights and Bias
$$y_i = \alpha + \beta x_i + \varepsilon_i.$$	$$\widehat{\beta} = \frac{\sum x_i y_i - \frac{1}{n} \sum x_i \sum y_i}{\sum x_i^2 - \frac{1}{n} \left(\sum x_i \right)^2}$$ $$\widehat{\alpha} = \overline{y} - \widehat{\beta}\,\overline{x},$$

Figure 11.1: Linear regression solved by OLS

In the preceding figure, the left cell is the mathematical form of a simple linear regression that has only two parameters, α, and β. Parameter α is known as the bias and parameter β is known as the weight of input $\mathbf{x}$. The term ε indicates errors that exist in a model. The right cell shows the closed-form solution for OLS.

NON-CLOSED-FORM SOLUTIONS

Note that most other machine learning models do not have closed-form solutions. To train a model with a non-closed-form solution, we need to employ other techniques to estimate model parameters. For example, some of the well-known methods are Newton's method, convex optimization, and stochastic optimization. In the field of machine learning and deep learning, the most common technique is gradient descent from stochastic optimization. If you are not familiar with it, don't worry. In this chapter, we will be spending lots of time on gradient descent. Besides, we will implement gradient descent techniques in our learning algorithms to train models too.

GRADIENT DESCENT

Before we dig into what gradient descent is, let's answer the question of why gradient descent is one of the most important techniques in machine learning. For the sake of simplicity, let's use the linear regression model as an example to illustrate. Let's say we are building a model to predict house prices given the area of the house. The mathematical model is as follows: *predicted house price = bias + weight*area of the house.*

When we initialize the model, we need to initialize the values for the bias and the weight. Usually, we draw random numbers for the initialization of biases and weights. So, now the model becomes the following: *predicted house price = 9 + 5*area of the house.*

Bias = 9 and weight = 5 are not ideal model parameters to predict the house price. Currently, the model could be extremely wrong about its house price prediction. It is not even close to being accurate. At the same time, there might exist an ideal value for the bias and an ideal value for the weight, such that the model predicts house prices accurately. So, now the question is this: how do we train the model to make it become an accurate model?

Training the model means we need to update the values for the bias and the weight in a way such that the model prediction gets closer and closer to being accurate. So, let's measure how wrong the model is by measuring the distance between a model prediction of a house price and the actual price of a house. The smaller the distance, the more accurate the model is. This means we need to update the values for the bias and the weight in such a way that the distance between the model prediction and the actual house price is minimized. Now the model training procedure turns into a minimization problem in the mathematical model, as shown in the following figure:

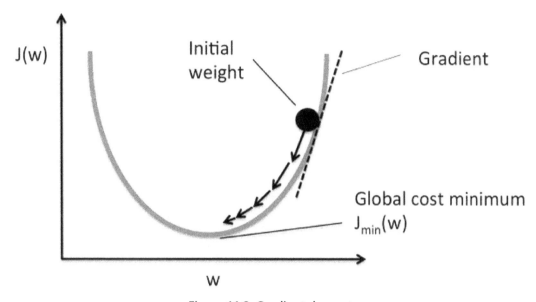

Figure 11.2: Gradient descent

The preceding figure nicely illustrates what gradient is and how gradient descent solves the minimization problem. **w** represents the weights in our house price prediction model, so we can express the distance, $J(w)$, between the prediction and the real house price in terms of **w**. The model with its initial weight is far from being accurate and has a huge distance from the minimum. The fastest way for the model to achieve the minimum is through a process called gradient descent. The gradient here is the steepest direction along the path to the minimum; gradient descent is also known as **steepest descent**. According to *Figure 11.2*, when we train the model, we are updating the values of weights with their gradients iteratively in a way such that $J(w)$ gradually reaches the minimum step by step.

Note that $J(w)$ is usually the loss function to measure the distance between the model prediction and the ground truth. Depending on the model type, there are various loss functions that you can choose from. For example, **Mean Squared Error (MSE)** is usually the go-to loss function for a regression model, and cross-entropy is used for a classifier model. We will be working with regression models, so we will focus on MSE in this chapter. The formula for MSE is as follows:

$$MSE = \frac{1}{n} \sum_{i=1}^{n} \left(Y_i - \widehat{Y}_i \right)^2$$

Figure 11.3: Definition of MSE

In a simple linear regression model, $y=mx+b$, where m is the weight and b is the bias, the gradient of the MSE loss function in the linear regression model is represented by two partial derivatives. One is the partial derivative of MSE with respect to the weight and the other is with respect to the bias. The formula is as follows:

$$\frac{\partial}{\partial m} = \frac{2}{N} \sum_{i=1}^{N} -x_i \left(y_i - \left(mx_i + b \right) \right)$$

$$\frac{\partial}{\partial b} = \frac{2}{N} \sum_{i=1}^{N} - \left(y_i - \left(mx_i + b \right) \right)$$

Figure 11.4: The gradient of MSE with respect to the weight and bias in linear regression

During gradient descent, we will update the weight and bias by subtracting their corresponding partial derivatives. As we update the weight and bias, the loss will continue to reduce until it converges to a small value, just like the global cost minimum $J_{min}(w)$ shown in *Figure 11.2*.

In short, gradient descent is an optimization algorithm used to minimize the distance by iteratively updating a model's weights to the direction of steepest descent as indicated by the gradient.

Let's now get some hands-on experience with gradient descent. We will implement gradient descent in our first exercise.

EXERCISE 11.01: IMPLEMENTING A GRADIENT DESCENT ALGORITHM IN NUMPY

Imagine you are an AI engineer working on a machine learning research project. The first project is a warm-up project and you will create a model and build a gradient descent algorithm to train the model. You may wonder why we would want to build one from scratch when there are already many open source machine learning libraries that have implemented gradient descent. While being practical is crucial to success in engineering, an understanding of the mathematical theory that underpins AI plays a critical role in the success of an AI engineer. You will have a much higher chance of innovating new AI techniques than your peers if your fundamentals are solid.

When we are building a new algorithm from scratch, we want to build it in such a way that it should work for a very simple/naïve model with minimal assumptions. Building something that works for a base case is always a good start. In this case, our base case is a linear regression model. If we build something that doesn't even work for linear regression, then it probably won't work for more complicated models, such as neural networks and deep learning models. At the end of the exercise, our gradient descent algorithm should be able to train the linear regression model and calculate the model's weights to be as close as possible to the ground truth.

To get started with building a gradient descent algorithm, we will use one of the most popular scientific libraries, NumPy. NumPy is a Python library for scientific computing and manipulating multi-dimensional arrays.

Before proceeding to the exercise, we need to set up a data science development environment. We will be using Anaconda. Please follow instructions in the *Preface* to install it.

Perform the following steps to complete the exercise:

1. Create a **Chapter11** directory for all the exercises in this chapter. After creating the **Chapter11** directory, change your working directory from your current working directory to **Chapter11**.

2. In the **Chapter11** directory, launch a Jupyter Notebook in your Terminal (macOS or Linux) or Command Prompt (Windows).

3. After the Jupyter Notebook is launched, let's create a new directory named **Exercise11.01**.

4. Select the **Exercise11.01** directory, then click **New** -> **Python3** to create a new Python 3 notebook.

5. Inside the Python 3 notebook, import all necessary modules as shown in the following code:

```
import numpy as np
import matplotlib.pyplot as plt

plt.style.use('ggplot')
```

We will import the **matplotlib** module for our first exercise. The **matplotlib** module is a very handy tool when it comes to data visualization in the realm of data science. We will be using it to visualize the training data, the model, and the training process in the following steps.

The **plt.style.use('ggplot')** line configures the plotting style for all of the following plots in the notebook. The **ggplot** style will render the plots with a gray background with white grid lines. You will see the plot style in later steps.

6. Next, create some synthetic data for training our linear regression model. Use the **NumPy.random.uniform** method to generate sample data from a uniform distribution and the **NumPy.random.normal** method to generate artificial noise data from a normal distribution as shown in the following code:

```
# set random seed
np.random.seed(9)

# draw 100 random numbers from uniform dist [0, 1]
x = np.random.uniform(0, 1, (100, 1))

# draw random noise from standard normal
z = np.random.normal(0, .1, (100, 1))

# create ground truth for y = 3x - 1
y = 3 * x - 1 + z
```

NumPy supports the drawing of random numbers for various statistical distributions. We used uniform distribution to draw input values for **x**. Uniform distribution is similar to the probability distribution of rolling a die, where every number from 1 through 6 has an equal likelihood of being drawn. In our case, any real number between **0** and **1** is drawn with equal likelihood. We choose to use the standard normal to generate artificial noise data, **z**, to mimic Gaussian noise, which is the most common form of noise in a natural dataset.

Now let's understand the structure of the data. When we are working with an array or matrix, one way to put data into perspective is to examine the array's **shape** property. This tells you the number of elements for each dimension in an array. For example, an array with shape (2, 3) means its first dimension has two elements and its second dimension has three elements, which means the array is a 2 x 3 two-dimensional array.

x is an array with **shape (100, 1)** that we created from a uniform distribution with the parameters **(0, 1)**. **z** is also an array with the shape **(100, 1)** from the normal distribution and it represents noise in the sample data. **y** is also an array with the shape **(100, 1)** created by the model of **3x −1**. Note that **y** is what our model is trying to predict for a given value of **x**. The ground truth model parameters are **weight = 3** and **bias = −1**.

Notice that we also seeded the randomness in the first line with **np.random. seed(9)** so that you can reproduce the same results when you run the same code. In the following steps, we will be pretending that we don't know the values of the model parameters. We will be training our linear regression model to find out the best estimates of the weight and bias.

7. Let's split the syntactic data into two sets as shown in the following code:

```
# split data into train and test
x_train, y_train = x[:80], y[:80]
x_val, y_val = x[80:], y[80:]
```

One set is for model training and the other set is for validating how accurate our model is after training. Let's slice the array to split the data. We are doing an 80/20 split for training and validation, which is the most common model validation strategy. Another model validation strategy is cross-validation. However, it is not included in the scope of this chapter.

8. Let's visualize the training and validation sets using the **matplotlib** data visualization module as shown in the following code:

```
# visualize
fig, axes = plt.subplots(nrows=1, ncols=2, figsize=(12, 4))
axes[0].scatter(x_train, y_train, label='train', color='b')
axes[0].set_title('Train')
axes[0].set_xlabel('x')
axes[0].set_ylabel('y')
axes[0].legend()
axes[1].scatter(x_val, y_val, label='val', color='g')
```

```
axes[1].set_title('Validation')
axes[1].set_xlabel('x')
axes[1].set_ylabel('y')
axes[1].legend()
fig.tight_layout()
```

You should get the following output:

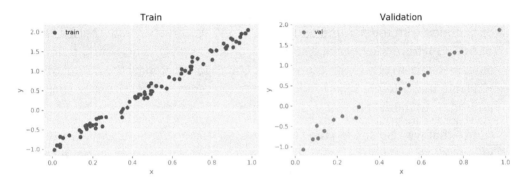

Figure 11.5: Visualization of training and validation data

Let's pretend we don't know that the true model is a linear model with **weight = 3** and **bias = -1** and that we need to create a model to predict the **y** value for any given **x**. According to the data visualization, we can make a strong assumption that the data follow a linear model pattern. So, let's create a linear regression model in the next step.

9. Let's initialize a linear regression model by drawing two random numbers for the model's weight and bias, as shown in the following code:

```
# create trainable parameters for the model
weight = np.random.randn(1)
bias = np.random.randn(1)

print("model with weight = {} and bias = {}"\
      .format(weight, bias))
```

You should get the following output:

```
model with weight = [-1.03600638] and bias = [-0.13392398]
```

10. Let's use this initial linear model to predict **y** values for **x** values in the training data and visualize this model prediction accuracy as shown in the following code:

```
# let's predict y value for the training data using initial model
y_hats = [bias + weight * x for x in x_train]

# let's visualize the initial model
plt.scatter(x_train, y_hats, color='b')
plt.title('Initial model')
plt.xlabel('x')
plt.ylabel('y')
plt.tight_layout()
```

You should get the following output:

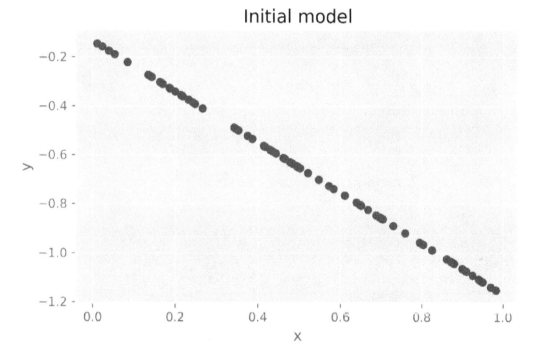

Figure 11.6: Visualization of the newly initialized linear model

Notice that the initialized linear model is almost the opposite of the true model (**y=3x−1**), which is manifested by the training and validation data in *Figure 11.5*. As the initialization of the model is random so the model is opposite to the true model. We will be training the model with the gradient descent technique and updating the weight and bias accordingly. With gradient descent, we adjust the weight and bias step by step so that the model gets closer and closer to the ground truth. You will see how this model improves and gets closer to the ground truth.

11. Next, we will implement the gradient descent algorithm to update the model. The algorithm is implemented in the following code:

```
# set training routine
lr = 1e-1
n_epochs = 500


# keep records
losses = []
val_losses = []
# train model
for epoch in range(n_epochs):
    print("[ epoch ]", epoch)
    # forward pass
    yhat = bias + weight * x_train
    # calculate the error
    error = (y_train - yhat)
    # calcuate loss function
    loss = (error ** 2).mean()
    losses.append(loss)
    print("[ training ] training loss = {}"\
          .format(loss))
    # calcuate the gradients for model's
    #trainable parameters
    bias_grad = -2 * error.mean()
    weight_grad = -2 * (x_train * error).mean()
    # perform gradient descend
    bias = bias - lr * bias_grad
    weight = weight - lr * weight_grad


    # calculate validation loss
    yhat = bias + weight * x_val
```

```
    error = (y_val - yhat)
    val_loss = (error ** 2).mean()
    val_losses.append(val_loss)
    print("[ eval ] validation loss = {}".format(val_loss))

print("linear.weight = {}, linear.bias = {}"\
    .format(weight, bias))
```

You should get the following output:

```
[ epoch ] 494
[ training ] training loss = 0.008616254346014289
[ eval ] validation loss = 0.01318093297273582
[ epoch ] 495
[ training ] training loss = 0.008616169215918002
[ eval ] validation loss = 0.013180202512873695
[ epoch ] 496
[ training ] training loss = 0.008616086227302866
[ eval ] validation loss = 0.013179482350475672
[ epoch ] 497
[ training ] training loss = 0.008616005326299067
[ eval ] validation loss = 0.01317877232874896
[ epoch ] 498
[ training ] training loss = 0.008615926460391929
[ eval ] validation loss = 0.013178072293550766
[ epoch ] 499
[ training ] training loss = 0.008615849578387787
[ eval ] validation loss = 0.013177382093337995
linear.weight = [2.98244477], linear.bias = [-0.97201471]
```

Figure 11.7: Logs of the model training process

NOTE

We will have loads of output data with 500 iteration results. We have only taken the last section of it in the preceding figure.

The output is logging training loss and validation loss for each training epoch so that we can see how the model is improving and the loss is reducing over time. The final weight is **2.98244477** and the bias is **-0.97201471**, which are very close to the ground truth of our model, **y = 3x - 1**.

In the gradient descent algorithm, we create a **for** loop for **500** iterations. This means the weight and bias are being updated 500 times. In deep learning terminology, one iteration of the parameter update with the entire input dataset is also known as one "epoch." In each epoch, we predict **y** values as **yhat** for **x** values for all training data. Then we calculate the **error** and **loss** functions between the ground truth and the model's predictions. We then calculate the loss function's gradient, which is a partial derivative for the **weight** and **bias**. We update the weight and bias by subtracting the *learning rate * gradient*. For each epoch, the gradient always points us downhill and the loss is continuously reduced until it converges on a very small number. After **494** epochs, as shown in *Figure 11.7*, the **training loss** is reduced and converges on a very small number, **0.0086**. When we see that the loss is converged to a small number, it means the model training is finished.

Notice we did not update the weight and bias by subtracting their gradients directly. Instead, we subtract a much smaller amount of the gradient, which is the *learning rate * gradient*; the learning rate is always less than one. The learning rate is a very important hyperparameter of gradient descent. It represents a learning step, and the larger the learning rate, the quicker the loss is minimized. However, the loss may not be able to converge if the learning rate is too large. We set the learning rate to be **0.1** here so that the loss is minimized smoothly.

12. Let's visualize how the MSE loss function is minimized during gradient descent as shown in the following code:

```
# visualize the loss during gradient descent
plt.scatter(range(n_epochs), losses, label='train_loss', color='b')
plt.scatter(range(n_epochs), val_losses, label='val_loss', color='r')
plt.legend()
plt.title('Loss')
plt.xlabel('epochs')
plt.ylabel('mse')
plt.tight_layout()
```

You should get the following output:

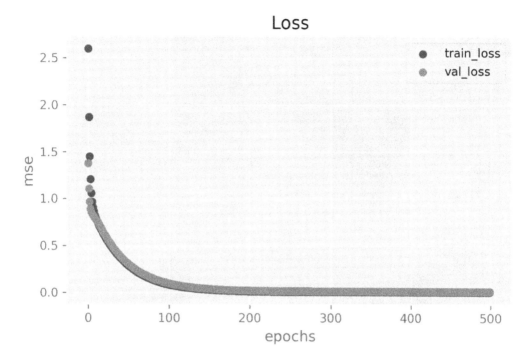

Figure 11.8: Minimizing the loss function using gradient descent

We can see that both the training loss and validation loss drop very quickly from the first epoch to the 100th epoch. After the 100th epoch, the loss starts to converge to a small number and becomes more and more stable. The goal of a training model with a gradient descent technique is to minimize the loss until the point where the loss starts to converge and flatten out. In this case, the model reaches a minimum around the 130th epoch. This means we effectively only need 130 epochs to train this linear regression model.

13. Finally, let's compare the trained model prediction against the ground truth in the validation data by running the following code:

```
# let's predict y value for the validation data using the trained model
y_hats = [bias + weight * x for x in x_val]


# let's visualize the intial model
plt.scatter(x_val, y_hats, label='prediction', \
            color='b')
plt.scatter(x_val, y_val, label='ground truth', \
            color='r')
plt.legend()
plt.title('Trained model')
plt.xlabel('x')
plt.ylabel('y')
plt.tight_layout()
```

You should get the following output:

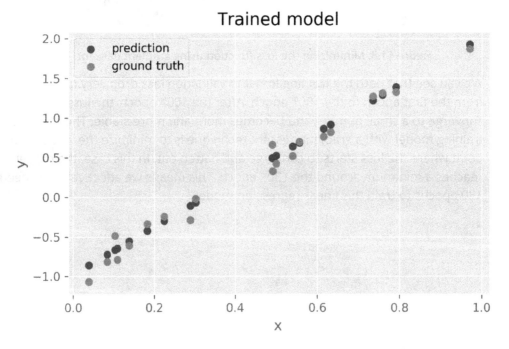

Figure 11.9: Final model prediction versus ground truth

We can see that the model is predicting very closely to the ground truth. The reason why the model isn't able to predict with perfect accuracy is that there is artificial noise in the training data, which was created on purpose. In the real world, there is always noise in the data. The best we can do is to estimate the model weight and bias as closely as possible to the ground truth by minimizing the loss function. That's why gradient descent is such a powerful technique in model training.

> **NOTE**
>
> To access the source code for this specific section, please refer to https://packt.live/38SejQp.

In this exercise, we first created a syntactic dataset and linear regression model using NumPy. Then we implemented a gradient descent algorithm to successfully train the linear regression. At the end, we compared our trained model with the ground truth by using the Matplotlib library to create two scatter plots for comparison.

You need to figure out the gradient function based on the model. We are using the simplest model, linear regression, in this exercise, and it's fairly easy and straightforward to derive gradient by hand in this case. We can analytically calculate the gradient in this case because the OLS technique for training a simple linear regression model has a closed-form solution. But if our model were a deep neural network, then calculating its gradients would be an ordeal. There are thousands or even millions of parameters in a typical neural network model. Calculating the partial derivatives for all its parameters would be the next ordeal that we would have to tackle. This is when deep learning libraries such as TensorFlow and PyTorch come to the rescue. With deep learning libraries, AI engineers can do machine learning/deep learning research at scale with ease. They can innovate new models and algorithms at a much faster iteration speed.

Next, you will be introduced to PyTorch, and we will take you through how PyTorch can help you build an AI algorithm.

GETTING STARTED WITH PYTORCH

PyTorch is one of the most popular open-source deep learning libraries in the world right now. It's known for its fast iteration, model ideation, and prototyping. As a result, many AI researchers or engineers implement their state-of-the-art deep learning models through the PyTorch library or its ecosystem. PyTorch has a large machine learning community and its community continues to grow and mature. Another popular deep learning framework is TensorFlow. TensorFlow gained its popularity a little earlier than PyTorch. Let's compare the differences in their core features:

	☉ PyTorch	🔥 TensorFlow
Adoption	High	High
Graph Definition	Dynamic graph	Static graph
Debugging	Easy	Hard
Development	Easy	Mid
Deployment	Requires integration	TensorFlow serving
Visualization	Requires integration	TensorBoard

Figure 11.10: PyTorch versus TensorFlow

Generally speaking, PyTorch is more development-friendly and TensorFlow is more deployment-friendly. Both are very powerful deep learning frameworks. If you want a better development and research experience, PyTorch is a better fit for you. On the other hand, if you want to deploy your models to production, then TensorFlow is a more mature option. That being said, you can use TensorFlow to achieve everything you can do in PyTorch, but with more effort.

From the user experience perspective, PyTorch is a very intuitive library for building mathematical models, especially for people with a math background. This is one of the reasons why we chose PyTorch for the exercises in this chapter. Using PyTorch helps us to build up the mathematical intuition for building and implementing mathematical algorithms and deep learning models.

NOTE

For more information about PyTorch, please visit https://pytorch.org/.

PyTorch, similar to NumPy, is a Python library for scientific computing. While NumPy serves as a generic scientific computing tool, PyTorch is built for deep learning modeling and neural networks. NumPy makes multi-dimensional array computation extremely efficient on CPUs and PyTorch allows tensors and their computation to run on GPUs. Deep learning models usually consist of tens of layers of tensors, which results in thousands of parameters in a single model. Running computation for thousands of parameters can be very slow sometimes. However, running such computations on a GPU is way more efficient because a GPU has hundreds of small cores to achieve higher parallelization, while a CPU usually has four to eight cores to parallelize.

You may not be familiar with the term "tensor." A tensor is a mapping between two algebraic objects in a vector space. An algebraic object can be a scalar, a vector, or a multi-dimensional array. A neural network model such as an image classifier is made of tensors. A model maps an image's pixels to labels of the objects inside the image. For example, if you feed the pixel data of an image of a car to an image classifier model, then the model will tell you there is a car in the image.

In terms of mathematical representation, a tensor is also an algebraic object in a vector space. Let's revisit our baseline linear regression model to illustrate:

$$y_i = \alpha + \beta x_i$$

Figure 11.11: Linear regression model

In the mathematical expression, **α** is the bias, and **β** is the weight in the linear regression model. In *Exercise 11.01, Implementing a Gradient Descent Algorithm in NumPy*, **α** is a scalar in NumPy and so is **β**. Now, in the world of PyTorch, **α** and **β** are tensors. So is every algebraic object in PyTorch.

Now let's get into the meat of topic about why PyTorch as a deep learning library, compared to NumPy, is more useful for training deep learning models. It's because PyTorch has the **autograd** package. **autograd** is used for automatic differentiation. Recall in *Exercise 11.01, Implementing a Gradient Descent Algorithm in NumPy*, that when we perform gradient descent, we need to calculate the partial derivatives for both the weight and the bias, meaning we need to differentiate the loss function with respect to the weight and the bias. There is no automatic differentiation in NumPy; so we were calculating the partial derivatives by hand and we hardcoded it in the gradient descent algorithm. With PyTorch, the gradient is automatically calculated by its **autograd** library.

To understand how automatic differentiation is achieved in PyTorch's **autograd** package, we need to first understand how a computer interprets a model. A computer interprets a model through a computational graph. Let's use our simple linear regression model, *y=mx+b*, to illustrate the computational graph, as shown in the following figure:

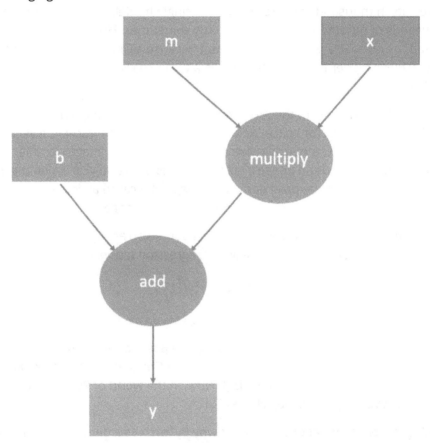

Figure 11.12: Computational graph for a simple linear regression model

In the preceding figure, starting from the top, are the weight (**m**), the bias (**b**), and the input data (**x**). They are also known as the graph leaf nodes in the computational graph. The circles represent the mathematical operators in the graph. Operators connect leaf nodes and produce the final output as **y**, which is the head in the graph.

During model training, we will need to find out the gradients by differentiating the loss with respect to the bias (**b**) and the weight (**m**). When we use PyTorch to build a model, PyTorch's **autograd** package will register nodes and operators into a computational graph so that it keeps track of the gradients and automatically calculates them for you.

Now we know why PyTorch is a good tool for machine learning model training. In the next exercise, let's upgrade our gradient descent algorithm from *Exercise 11.01, Implementing a Gradient Descent Algorithm in NumPy*, and use PyTorch to perform gradient descent instead of NumPy.

EXERCISE 11.02: GRADIENT DESCENT WITH PYTORCH

In this exercise, we will improve our gradient descent algorithm with PyTorch's **autograd** package. Fortunately, we implemented the entire skeleton of the gradient descent algorithm to train the linear regression model in *Exercise 11.01, Implementing a Gradient Descent Algorithm in NumPy*. Now we will only change the code in places that require the use of PyTorch to perform automatic differentiation while keeping most of the skeleton unchanged.

To be more specific about what we need to do in this exercise, we will need to do the following:

- Convert the data and the model from a NumPy object to a PyTorch tensor object.

- Replace the hardcoded gradient calculation part with PyTorch's **autograd** implementation.

- Rerun the model training and verify that the trained model is close to the ground truth.

Before proceeding to the exercise, we need to install the data science environment from Anaconda and the PyTorch package. Please follow the instructions in the *Preface* to install them.

Perform the following steps to complete the exercise:

1. In the **Chapter11** directory, launch a Jupyter Notebook in your Terminal (macOS or Linux) or Command Prompt window (Windows).

2. After the Jupyter Notebook is launched, create a new directory named **Exercise11.02**. Inside the **Exercise11.02** directory, create a Python 3 notebook.

3. Inside the Python 3 notebook, import all necessary modules as shown in the following code:

```
import numpy as np
import torch
import matplotlib.pyplot as plt

plt.style.use('ggplot')
device = 'cuda' if torch.cuda.is_available() else 'cpu'
```

We will be using NumPy and Matplotlib again in this exercise. Besides this, we will use the PyTorch library by importing **torch** into our Jupyter Notebook environment. At the last line, **torch.cuda.is_available()** returns **True** if **torch** detects that there is a GPU device available on your machine; otherwise, it returns **False**. So, the **device** variable is a string with a value of either **cuda** or **cpu**. You will find out how we will use this variable in later steps in the exercise.

4. Generate some syntactic data for the linear regression model as shown in the following code:

```
# set random seed
np.random.seed(9)

# draw 100 random numbers from uniform dist [0, 1]
x = np.random.uniform(0, 1, (100, 1))

# draw random noise from standard normal
z = np.random.normal(0, .1, (100, 1))

# create ground truth for y = 8x - 3
y = 3 * x - 1 + z
```

5. Split the data into two sets, **train** and **test**, as shown in the following code:

```
# split data into train and test
x_train, y_train = x[:80], y[:80]
x_val, y_val = x[80:], y[80:]

print(type(x_train))
```

You should get the following output:

```
<class 'numpy.ndarray'>
```

We use NumPy to create the syntactic data, so the type of data is a **numpy. ndarray** object. In this exercise, we need to work with PyTorch, so we will move the data to a new object type (tensor) that PyTorch can work with in the next step.

6. Next, we will convert the data from a NumPy object to a PyTorch tensor object as shown in following code:

```
# move data from numpy to torch
x_train_tensor = torch.from_numpy(x_train).float().to(device)
y_train_tensor = torch.from_numpy(y_train).float().to(device)
x_val_tensor = torch.from_numpy(x_val).float().to(device)
y_val_tensor = torch.from_numpy(y_val).float().to(device)
print(type(x_train_tensor))
```

You should get the following output:

```
<class 'torch.Tensor'>
```

Now both the training data and the validation data are converted to **torch. Tensor** objects. The **torch.from_numpy()** function call is a very handy method for converting a NumPy array object to a PyTorch tensor object. **.float()** means we want the converted data type to be of the float type. **.to(device)** will move the tensor from its current device to the desired **device**. For example, if a GPU is available on your machine and you want to move your tensor from the CPU to the GPU, then you will use the **.to('cuda')** method. This explains why we set the **device** variable to either **cpu** or **cuda** in *step 3* at the last line.

7. Now we will initialize our linear regression model parameters, **weight**, and **bias**, in PyTorch as shown in the following code:

```
# create trainable parameters for the model
weight = torch.randn(1, requires_grad=True, \
                     dtype=torch.float, device=device)
bias = torch.randn(1, requires_grad=True, \
                   dtype=torch.float, device=device)

print(weight, bias)
```

You should get the following output:

```
tensor([-1.0688], requires_grad=True) tensor([-0.6226],
requires_grad=True)
```

Notice that we set **requires_grad=True** when we initialize the tensors for weight and bias. In PyTorch, tensors with the **requires_grad=True** attribute will be registered into the PyTorch computational graph where any graph leaf tensor is automatically differentiable. This will allow us to automatically compute the gradients for the weight and the bias, which will make gradient descent much easier.

8. After the data and the model are created, we now can train the linear regression model by performing gradient descent. The following code implements the gradient descent algorithm:

ex2-notebook.ipynb

```
# set training routine
lr = 1e-1
n_epochs = 500

# train model
losses = []
val_losses = []
for epoch in range(n_epochs):
    print("[ epoch ]", epoch)
    yhat = bias + weight * x_train_tensor
    error = y_train_tensor - yhat
    loss = (error ** 2).mean()
    losses.append(loss.item())
    print("[ training ] training loss = {}".format(loss))
    # calculate gradients
    loss.backward()
    # update weight and bias
    with torch.no_grad():
        bias -= lr * bias.grad
        weight -= lr * weight.grad
    # zero out grads
    bias.grad.zero_()
    weight.grad.zero_()
```

The full code is available at https://packt.live/30bU0t2.

You should get the following output:

```
[ epoch ] 495
[ training ] training loss = 0.008615845814347267
[ eval ] validation loss = 0.013177354820072651
[ epoch ] 496
[ training ] training loss = 0.008615771308541298
[ eval ] validation loss = 0.013176674023270607
[ epoch ] 497
[ training ] training loss = 0.008615700528025627
[ eval ] validation loss = 0.013176006264984608
[ epoch ] 498
[ training ] training loss = 0.008615627884864807
[ eval ] validation loss = 0.013175344094634056
[ epoch ] 499
[ training ] training loss = 0.008615557104349136
[ eval ] validation loss = 0.0131746931001544
linear.weight = tensor([2.9827], requires_grad=True), linear.bias = tensor([-0.9722], requires_grad=True)
```

Figure 11.13: Logs of the model training process

> **NOTE**
>
> We will have loads of output data with 500 iteration results. We have only shown the last section of it in the preceding figure.

The log output is similar to the one in *Exercise 11.01, Implementing a Gradient Descent Algorithm in NumPy*, where both training loss and validation loss are minimized over time. The final weight is **2.9827** and the bias is **-0.9722**, which are again very close to the ground truth of our model, $y = 3x - 1$.

If we compared the code snippet to *step 11* in *Exercise 11.01, Implementing a Gradient Descent Algorithm in NumPy*, you will notice that the skeleton of the training routine is almost the same and the only difference is the part that updates the gradients, as shown in the following code:

```
# calculate gradients
loss.backward()
# update weight and bias
with torch.no_grad():
    bias -= lr * bias.grad
    weight -= lr * weight.grad
# zero out grads
bias.grad.zero_()
weight.grad.zero_()
```

We calculated the **loss** by performing a series of algebraic operations on tensors with the **requires_grad=True** attribute so the **loss** is also registered in the PyTorch computational graph. The **loss.backward()** line performs backpropagation, which means it's computing gradients for all of the graph leaf tensors in the computational graph. Then the gradient of a graph leaf tensor can be obtained by accessing its **.grad** attribute. For example, **bias.grad** will give you the gradient of the bias.

Note that before we update the values for the bias and weight, the **with torch.no_grad()** line will turn off the computational graph for all operations inside the **with** statement. In this case, the **bias -= lr* bias.grad** line and the **weight -= lr * weight.grad** line only modifies the values of the bias and the weight. The computational graph for their gradients won't interfere.

If we don't execute **with torch.no_grad()** line, the operations for updating the weight and bias will by default be registered into the computational graph, which will mess up the original gradient calculation in the computational graph of our linear regression model.

So, please remember that if we just want to perform simple mathematical calculations and we do not wish to change the model structure as well as its gradient, we should always turn off the computational graph with **torch.no_grad()**.

At the end of the gradient descent, we zero out gradients for both the bias and the weight. This is because the PyTorch computational graph will cache and accumulate gradients for tensors in the graph if you don't zero their gradients. So, we need to zero out gradients after performing **.backward()**.

9. Lastly, let's visualize how the MSE loss function is minimized during the gradient descent as shown in the following code:

```
plt.scatter(range(n_epochs), losses, label='train_loss', color='b')
plt.scatter(range(n_epochs), val_losses, label='val_loss', color='r')
plt.legend()
plt.title('Loss')
plt.xlabel('epochs')
plt.ylabel('mse')
plt.tight_layout()
```

You should get the following output:

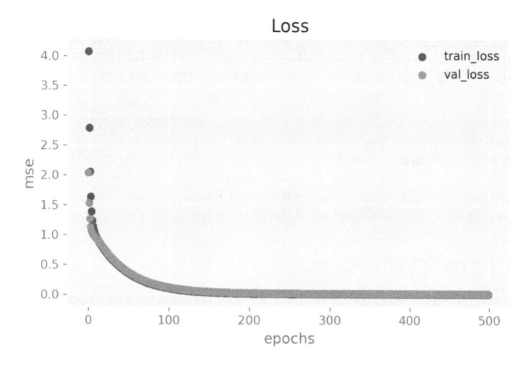

Figure 11.14: Minimizing the loss function by gradient descent

Similar to the process of minimizing loss in *Exercise 11.01, Implementing a Gradient Descent Algorithm in NumPy*, we see that both the training loss and validation loss are again dropping very quickly by the 100th epoch. After the 100th epoch, the loss starts to converge on a small number and becomes more and more stable.

This plot of training loss versus validation loss is a very telling visualization when training deep learning models. When we see both loss functions reduced to the minimum, this usually means 'the model is trained enough and has reached the best fit. If you continue to perform gradient descent after both loss functions reach the minimum, then you will overfit the model and you'll see that the validation loss starts to increase. This is when you need to stop training the model.

NOTE

To access the source code for this specific section, please refer to https://packt.live/30bU0t2.

In this exercise, we created a linear regression model using a PyTorch tensor with the **requires_grad=True** attribute. We then implemented the gradient descent algorithm using PyTorch's **autograd** package. Lastly, we visualized how loss functions behave during the training process and verified that our gradient descent algorithm had successfully trained the linear regression model.

So far, our gradient descent algorithm is using the whole training dataset at each epoch to calculate the gradient for all input data. This type of gradient descent is called **batch gradient descent**. "Batch" refers to the data for gradient computation as a whole batch. In practice, batch gradient descent isn't widely used in the real world. Instead, most industry practitioners use **Stochastic Gradient Descent** (**SGD**) or **mini-batch Stochastic Gradient Descent** (**mini-batch SGD**). We will discuss mini-batch SGD in-depth in the next topic.

MINI-BATCH SGD WITH PYTORCH

Let's recap what we have learned so far. We started by implementing a gradient descent algorithm in NumPy. Then we were introduced to PyTorch, a modern deep learning library. We implemented an improved version of the gradient descent algorithm in PyTorch in the last exercise. Now let's dig into more details about gradient descent.

There are three types of gradient descent algorithms:

- Batch gradient descent

- Stochastic gradient descent

- Mini-batch stochastic gradient descent

While batch gradient descent computes model parameter' gradients using the entire dataset, stochastic gradient descent computes model parameter' gradients using a single sample in the dataset. But using a single sample to compute gradients is very unreliable and the estimated gradients are extremely noisy. So, most applications of stochastic gradient descent use more than one sample, or a mini-batch of a handful of samples, to compute gradients; thus, this strategy is called mini-batch stochastic gradient descent.

You may wonder why most people use mini-batch SGD while batch gradient descent is rarely used. There are primarily two reasons why mini-batch SGD dominates over batch gradient descent in real-world practice:

1. Batch gradient descent is just simply too computationally expensive and the gain isn't justified by the huge computational cost.

2. A real-world model's loss function usually has many local minima. To escape local minima and find the global minima, the randomness in computing the gradient from a mini-batch can help the algorithm escape the local minima without getting trapped.

Let's dig into the deeper details of these points. The reason why batch gradient descent is too computationally expensive is that real-world datasets in the industry are usually web-scale and you won't be able to fit all of that data into a single machine. Imagine a company such as Google, which has billions of user searches per day – its data is easily at the scale of petabytes. Performing gradient calculations on data at such a scale is almost impossible.

Rather than computing the gradient using the whole dataset, mini-batch SGD randomly selects several samples from the dataset and calculates the gradient using these randomly selected samples. Because the computation cost is much reduced, mini-batch SGD is a lot faster than batch gradient descent:

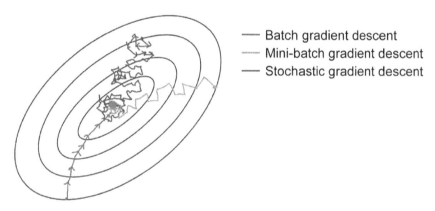

Figure 11.15: Comparison of different gradient descent algorithms

As the preceding figure suggests, batch gradient descent might get you the smoothest gradient descent path given all the data is being used, and mini-batch SGD is less optimal if significantly less data is used. SGD is expected to be the noisiest because It uses only one sample at a time. However, all of them can get you to the final destination – the global minima. Although the gradient estimated by mini-batch SGD is a little bit noisier than batch gradient descent and it takes a longer path to reach the minima, mini-batch SGD is still the preferred algorithm because of its computational advantage.

Having briefly discussed the problem of gradient descent, let's move onto another problem that is encountered in the loss functions of real-world datasets – multiple local minima. The loss function in a real-world deep learning model isn't always convex, nor does it always have a smooth path to the global minima. This results in the batch gradient descent algorithm getting trapped in the local minimum as illustrated in the following figure:

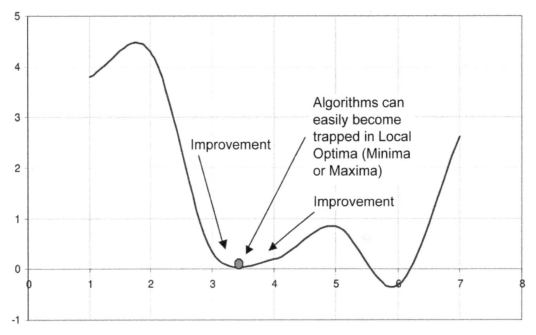

Figure 11.16: The algorithm gets trapped in the local minimum

One way to get the algorithm to escape the local minimum is to introduce some randomness in the gradient descent path so that it can be jerked out of the local minimum and land in a new region that leads to the global minimum:

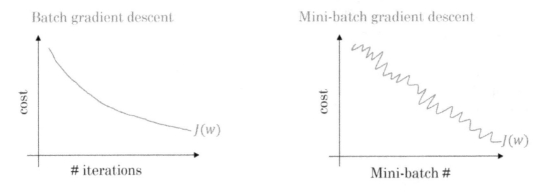

Figure 11.17: Noise causes oscillations in the gradient descent path

As the preceding figure illustrates, batch gradient descent is more desirable in an ideal world where models have perfectly smooth convex loss functions. However, the oscillation behavior exhibited in mini-batch gradient descent is more desirable in reality, where models aren't perfectly convex, as they have multiple local minima.

Now let's implement a mini-batch SGD algorithm. Luckily, PyTorch has the **utils** module, which will help us tremendously in mini-batch SGD implementation. Besides this, we will also improve our model-building skills with best practices by using PyTorch's **optim** and **nn** modules.

Let's briefly introduce these submodules in PyTorch one by one:

- In the **utils** module, there is a **data** module, which contains many very handy helper functions that make passing data to the model very easy.

- The **optim** module implements various state-of-the-art optimization algorithms, including SGD. Instead of manually writing a gradient descent algorithm as we did in previous exercises, we will use the SGD algorithm implemented by the **optim** module in the next exercise.

- The **nn** module is the core module in PyTorch. It implements many high-level model abstractions over the concept of **tensor**. This means the classes in this module are built on top of **tensor** and extended with more methods and attributes.

We discussed that a deep learning model is made up of layers of tensors, which means we can implement different neural network models by putting **tensor** objects together in different ways, just like playing with Lego. The most Pythonic way of creating different models is through writing different classes that inherit from the same base class. In PyTorch, **torch.nn.module** is the base class for all neural network models. Essentially, a model in PyTorch is represented as a user-defined class that inherits from **nn.module**. If this sounds strange to you, we will show you some concrete examples in the next exercise.

> **NOTE**
>
> For more information about PyTorch, please visit its documentation at
> https://pytorch.org/docs/stable/index.html.

In the next exercise, we will implement mini-batch SGD in PyTorch with high-level modules such as the **utils.data**, **optim**, and **nn** modules.

EXERCISE 11.03: IMPLEMENTING MINI-BATCH SGD WITH PYTORCH

In this exercise, we will improve the gradient descent algorithm by implementing mini-batch SGD with high-level PyTorch modules. We will still use a simple linear regression model as our baseline to implement the mini-batch SGD. At the highest level, most of the training procedure is still very similar to *Exercise 11.01, Implementing a Gradient Descent Algorithm in NumPy* and *Exercise 11.02, Gradient Descent with PyTorch*. But this exercise will require lots of code changes because we will be using high-level PyTorch modules for implementation instead of low-level **tensor** objects, as in *Exercise 11.02, Gradient Descent with PyTorch*.

To be more specific about what we need to do in this exercise, we will need to do the following:

1. We will use the **utils.data** module to create data loader objects for generating mini-batches in the training procedure.

2. We will implement our simple linear regression model by defining a model class that inherits from **nn.module**.

3. We will implement mini-batch SGD using the **optim** module.

Perform the following steps to complete the exercise:

1. In the **Chapter11** directory, launch a Jupyter Notebook in your Terminal (macOS or Linux) or Command Prompt window (Windows).

2. After the Jupyter Notebook is launched, create a new directory named **Exercise11.03**. Inside the **Exercise11.03** directory, let's create a Python 3 notebook.

3. Inside the Python3 notebook, import all the necessary modules as shown in the following snippet:

```
# import modules
import numpy as np
import torch
import torch.optim as optim
import torch.nn as nn
from torch.utils.data import DataLoader, \
    TensorDataset, random_split
```

```
import matplotlib.pyplot as plt

plt.style.use('ggplot')
device = 'cuda' if torch.cuda.is_available() else 'cpu'
```

In addition to **numpy** and **matplotlib**, we import **torch.optim** for SGD, **torch.nn** for creating the linear regression model, and **torch.utils.data** for feeding batches into the model.

4. Generate some syntactic data using **numpy** for the linear regression model as shown in the following code:

```
# set random seed
np.random.seed(9)

# draw 100 random numbers from uniform dist [0, 1]
x = np.random.uniform(0, 1, (100, 1))

# draw random noise from standard normal
z = np.random.normal(0, .1, (100, 1))

# create ground truth for y = 8x - 3
y = 3 * x - 1 + z
```

5. After the data is generated, we will convert the data from the form of **numpy** to the form of **TensorDataset** as shown in the following code:

```
# move data from numpy to torch
x_tensor = torch.from_numpy(x).float().to(device)
y_tensor = torch.from_numpy(y).float().to(device)
# create tensor dataset from tensor
dataset = TensorDataset(x_tensor, y_tensor)
print(dataset[0])
```

You should get the following output:

```
(tensor([0.0104]), tensor([-1.0155]))
```

You will notice that **TensorDataset()** is very similar to the Python built-in function **zip()**. It takes as input *tensors (1 tensor or more) and retrieves the nth sample from *tensors to bundle them into a tuple so that the nth sample from *tensors can be retrieved by indexing. In the output, we see that **tensor([0.0104])** is the first sample **x_tensor** and **tensor([-1.0155])** is the first sample **y_tensor**.

The reason why we convert our data to **TensorDataset()** is because **torch.utils.data** helper functions are expecting input data in the form of **TensorDataset** and not any other form. In the next step, you will see how we can create **DataLoader** out of **TensorDataset**.

6. Now let's create **DataLoader** for the training and validation datasets, as shown in the following code:

```
# split data into train and eval
train_dataset, val_dataset = random_split(dataset, [80, 20])


# create data loader
train_loader = DataLoader(dataset=train_dataset, batch_size=8)
val_loader = DataLoader(dataset=val_dataset, batch_size=10)
```

Notice that we use the **random_split** helper function from the **torch. utils.data** module to split the data with a ratio of **80** to **20** for the training and validation datasets. Both **train_dataset** and **val_dataset** are still in the form of **TensorDataset**.

Then we create **DataLoader** for both **train_dataset** and **val_dataset** with batch sizes of **8** and **10** respectively. The batch size of **8** for the training data loader means it will generate **8** randomly selected samples from the training dataset every time it gets invoked.

The batch size for the validation data loader doesn't have to be the same as the one for the training data loader. The batch size for the training data loader could impact our mini-batch SGD's performance. For the validation data loader, the batch size can be an arbitrary number. It does **not** impact the mini-batch SGD. If the whole validation dataset can fit into RAM, then you can set its batch size as the size of the whole validation dataset.

7. Next, we will define our linear regression model by creating a Python class that inherits from **nn.Module** as shown in the following code:

```
# define our linear regression model in pytorch
class LinearRegression(nn.Module):
    def __init__(self):
        super().__init__()
        self.linear = nn.Linear(1, 1)

    def forward(self, x):
        return self.linear(x)
```

```
# initialize our model
model = LinearRegression().to(device)

# state_dict() contains the parameters of the model
print(model.state_dict())
```

To create a model class in PyTorch, you need to implement an **__init__ (self)** constructor and the **forward(self, x)** method. In the constructor, you will initialize attributes for the model class. Notice that the only attribute is **nn.Linear(1,1)**, which is also a model class that inherits from **nn.Module**. The **nn.Linear(1,1)** object is another **nn.Module**. It's a generalized form of a linear model that applies a linear transformation to its input data. In this case, we specify **(1,1)**. This means the linear model will map a piece of one-dimensional input data to another piece of one-dimensional output data. It is equivalent to a simple linear regression, **y = bias + weight * x**.

The other important method you have to implement is **forward(self, x)**. This defines the forward pass for the model. In our case, the model applies a linear transformation for an input, **x**, and returns the transformed value, which is **y_hat = bias + weight * x**.

8. Next, we will create a loss function from the **torch.nn** module and the SGD optimization algorithm from the **torch.optim** module as shown in the following code:

```
# set training routine
lr = 1e-1
n_epochs = 50

# create loss function and optimizer
loss_fn = nn.MSELoss(reduction='mean')
optimizer = optim.SGD(model.parameters(), lr=lr)
```

We set the learning rate to **0.1** and number of epochs to **50**. Our loss function is still MSE. PyTorch provides a variety of loss functions in the **torch.nn** module, so we don't have to manually derive the loss function from the input data.

Our optimizer is **optim.SGD**, which is a standard SGD algorithm. To create one, you need to pass in the model's parameters (**weight** and **bias**), which can be accessed via the **.parameters()** method, and the learning rate. In the next step, you will see how we create a gradient descent algorithm using the loss function and an optimizer.

9. We will define a function that performs the mini-batch SGD as shown in the following code:

```
def train_one_batch(model, loss_fn, optimizer, x_batch, y_batch):
    model.train()
    # forward pass
    yhat = model(x_batch)
    # calculate training loss
    loss = loss_fn(y_batch, yhat)
    # gradient descend
    loss.backward()
    optimizer.step()
    optimizer.zero_grad()
    return loss.item()
```

Let's go through the preceding code line by line. The first line, **model.train()**, is a bit misleading the first time you see it. This method call does **not** train the model. Instead, it just sets the model in training mode. The next line performs the forward pass and outputs the model predictions for the small batch of input data. Then we use the loss function we defined in the previous step to calculate the MSE loss. The next three lines take care of gradient descent. The **loss. backward()** function performs backpropagation to calculate the gradients for the model parameters. The **optimizer.step()** function performs a parameter update, which subtracts the parameter value from the *learning rate * gradient*. The last line, **optimizer.zero_grad()**, zeroes out all of the cached gradients in the computational graph.

10. Next, we will actually start the model training procedure as shown in following code:

```
# train model
losses = []
val_losses = []
for epoch in range(n_epochs):
    print("[ epoch ]", epoch)
    tmp_losses = []
    for idx, (x_batch, y_batch) in enumerate(train_loader, 1):
        x_batch = x_batch.to(device)
        y_batch = y_batch.to(device)
        loss = train_one_batch(model, loss_fn, optimizer, \
                               x_batch, y_batch)
        tmp_losses.append(loss)
```

```
print("[ training ] training loss = {}"\
        .format(sum(tmp_losses)/idx))
losses.append(sum(tmp_losses)/idx)
# eval
with torch.no_grad():
    # set eval mode
    model.eval()
    tmp_val_losses = []
    for idx, (x_batch, y_batch) \
            in enumerate(val_loader, 1):
        x_batch = x_batch.to(device)
        y_batch = y_batch.to(device)
        yhat = model(x_batch)
        tmp_val_losses.append(loss_fn(y_batch, yhat)\
                                .item())
    print("[ eval ] validation loss = {}"\
            .format(sum(tmp_val_losses)/idx))
    val_losses.append(sum(tmp_val_losses)/idx)


print(model.state_dict())
```

You should get the following output:

```
[ epoch ] 45
[ training ] training loss = 0.009553260239772498
[ eval ] validation loss = 0.010500878561288118
[ epoch ] 46
[ training ] training loss = 0.009549114271067083
[ eval ] validation loss = 0.010552239138633013
[ epoch ] 47
[ training ] training loss = 0.009545810241252183
[ eval ] validation loss = 0.010597990360110998
[ epoch ] 48
[ training ] training loss = 0.009543169708922505
[ eval ] validation loss = 0.010638704989105463
[ epoch ] 49
[ training ] training loss = 0.009541058051399886
[ eval ] validation loss = 0.010674883145838976
OrderedDict([('linear.weight', tensor([[3.0278]])), ('linear.bias', tensor([[-1.0106]]))])
```

Figure 11.18: Logs of the model training process

> **NOTE**
>
> We will have loads of output data with 500 iteration results. We have only shown the last section of it in the preceding figure.

This step is somewhat similar to *Step 8* in *Exercise 11.02, Gradient Descent with PyTorch*. The biggest difference is that we are implementing mini-batch SGD in this exercise. For each epoch, the algorithm generates **(x_batch, y_batch)** from **train_loader**. Recall how in *Step 6*, we set the batch size for the training data loader to be **8**, so it will generate 8 samples for each iteration. The heavy-lifting work for model training is done by the **train_one_batch** function, which was defined in the previous step. The remaining logic in this snippet mostly records training losses and validation losses for later visualization.

11. Finally, let's visualize how the MSE loss function is minimized during the gradient descent as shown in the following code:

```
plt.scatter(range(n_epochs), losses, label='train_loss', \
            color='b')
plt.scatter(range(n_epochs), val_losses, label='val_loss', \
            color='r')
plt.legend()
plt.title('Loss')
plt.xlabel('epochs')
plt.ylabel('mse')
plt.tight_layout()
```

You should get the following output:

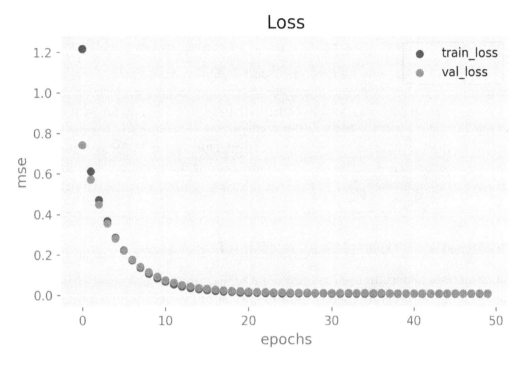

Figure 11.19: Minimizing the loss function with mini-batch SGD

You may notice that it only took 28 epochs to converge in mini-batch SGD, compared to over 100 epochs in batch gradient descent. This is because there are more iterations in each epoch in mini-batch SGD compared to the one iteration in each epoch in batch gradient descent. The number of iterations in each epoch equals the number of samples in the training dataset divided by the number of samples in the mini-batch.

> **NOTE**
>
> To access the source code for this specific section, please refer to https://packt.live/2C7VCMw.

In this exercise, we learned that mini-batch SGD is a better algorithm in the real world than batch gradient descent. To implement mini-batch SGD, we leveraged PyTorch's high-level abstractions, such as `torch.nn` for creating the model class, `torch.utils.data` for creating the data loader, and `torch.optim` for SGD optimization.

So far, we've covered both the theoretical part and the practical part of developing and training machine learning models. For the theoretical part, we learned the core concepts in neural networks such as constructing the loss function for your model and minimizing the loss function via the gradient descent approach. For the practical part, we learned to build and train models in both NumPy and PyTorch. Now that we have the theory and practical skills in machine learning modeling, we are ready to explore a more interesting field in machine learning: reinforcement learning.

We will discuss and implement a reinforcement learning algorithm in the next section.

BUILDING A REINFORCEMENT LEARNING ALGORITHM TO PLAY A GAME

One of the very important breakthroughs in the AI community happened in 2016 when a computer program called **AlphaGo** outplayed the best human professional Go player at the board game Go. This breakthrough shook the world and triggered tremendous hype around AI not only in academia but also in various industries.

The AlphaGo algorithm is an example of a reinforcement learning algorithm. Reinforcement learning is very different from supervised learning. The simple linear regression model from our previous exercises is an example of supervised learning. In previous exercises, we provided the training algorithm with training data input, x, and an output, y. After several epochs of parameter updates, the training algorithm could find estimates for the weight and bias that were extremely close to the ground truth. As you can see, the objective of supervised learning is to minimize the distance between the model's estimate and the ground truth.

Unlike supervised learning, there is no ground truth in reinforcement learning. Imagine you are playing the game of Go. You and your opponent take turns to place a stone on the board at each step. You and your opponent are trying to control the territory by making a boundary to capture one another's stones. Go is a strategy game. There are many ways to win and each way of winning also depends on the moves your opponent has made. When you are playing a strategy game, you are dealing with uncertainty at the present moment, but you know there is a path to winning. Supervised learning requires certainty and ground truth, which is not available in this case, so we will need a new approach to solve this kind of problem. For representational purposes, you can get an idea of the game Go from the following figure:

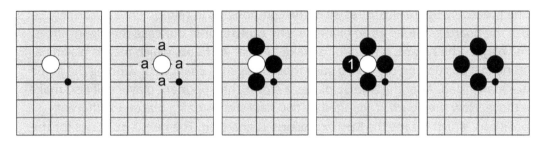

Figure 11.20: The game Go

Reinforcement learning trains an agent that tries to maximize some notion of reward by taking a series of "best" actions given a series of corresponding environment states. When playing Go, every time your opponent makes a move, the state of the board with black and white stones in different positions is a new environment state. Based on the given environment, the agent needs to find an optimal action that will eventually lead to a maximum reward:

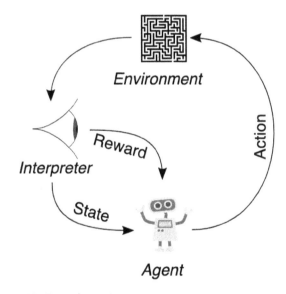

Figure 11.21: Overview of the reinforcement learning algorithm

The preceding figure illustrates all of the components in reinforcement learning. Let's use the board game Go as an example. The **Agent**, or in our case the algorithm, is the player. The **Agent** is trained to make a "best" **Action**, which means placing a stone in a position on the board such that the agent's chances of winning the game are maximized. After the **Agent** places a stone, the opponent player will also make a move. Now the board is changed by two new stones being placed in different positions, which means the **Environment** is changed. The next **Action** is for the **Agent** to learn from the past. However, the **Agent** is a computer algorithm. We need to feed the algorithm with data that the computer can understand and digest. So, we need an **Interpreter**. The **Interpreter** needs to accomplish two things. One is to interpret **Reward** in terms of a score. The other one is translating an **Environment** into a **State**, which should be an array of real numbers and can be digested by a computer algorithm. Now the **Agent** can learn to improve from the past by looking at the past states and rewards. Finally, the **Agent** will make the "best" action again for the given state and the cycle starts over again.

Now let's dig deeper into how the agent learns to improve its game. As the board game Go is being played, the agent records the state and the reward for every iteration so that the agent can learn as they play the game. During its learning time, the agent retrieves a batch of states and rewards from its memory and tries to come up with a policy that can tell it the reward of different actions so that it can pick the action that maximizes the reward. However, the policy is far from being accurate at the beginning of the game. As the game goes on, the policy will improve and make a more accurate prediction for the reward. This is where exploration versus exploitation comes into place. The longer the agent explores, the more the policy can learn. However, the returns of excessive learning will diminish eventually. So, the agent needs to start to exploit the policy after the agent starts to trust the policy. The strategy for trading off between exploration and exploitation is called an *epsilon-greedy strategy*:

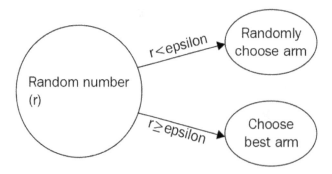

Figure 11.22: Epsilon-greedy policy

Basically, for every iteration, the agent can either take a random action (or "arm") or listen to the policy and make the "best" action that maximizes the reward. To determine which way to go, we will draw a random number from a uniform distribution and compare the random number to the value of epsilon. The value of epsilon and the probability of choosing the random action will decay exponentially as the game goes on. This means the agent gradually leans toward the policy and converges on the policy eventually.

You may wonder how the policy can predict the reward for a given state and choose an action. Let's start with a simple game such as Tic-Tac-Toe. The reward for this game is either a win (+1), a loss (-1), or a draw (0). Depending on which player starts first, you can write a simple computer program to perfectly play the game to not lose. This is because the game's state-space complexity is very small and has a total of 765 different positions. Mathematically speaking, all 765 different states can be represented in the form of a **Markov decision process (MDP)**, which is like a tree where the next state depends on the development of previous states:

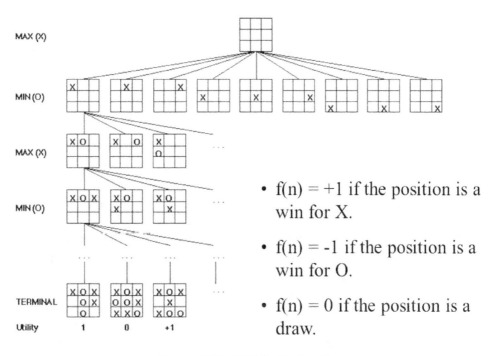

Partial Game Tree for Tic-Tac-Toe

- f(n) = +1 if the position is a win for X.
- f(n) = -1 if the position is a win for O.
- f(n) = 0 if the position is a draw.

Figure 11.23: MDP for Tic-Tac-Toe

As the game tree in the preceding figure illustrates, the space of the states in the game is finite. This means there is always a path of a given state and a chosen action that leads to the ultimate reward. This is our policy – the agent will look at the policy and find out which action leads to a win and it will take that action.

> **NOTE**
>
> You can refer to the following link to find out more about MDPs:
>
> http://www.pitt.edu/~schaefer/papers/MDPTutorial.pdf

For a game like Go, the state space is infinite and unbounded. Its MDP is too large to be represented by an exact mathematical model. This is where neural networks come in handy. We can train a neural network to approximate the game's MDP. The neural network will predict the reward given a state and an action. Note that the reward is a cumulative reward, which is represented by R_{t_0} in the following equation:

$$R_{t_O} = \Sigma_{t=t_0}^{\infty} \gamma^{t-t_o} r_i$$

Figure 11.24: Formula of the cumulative reward

The cumulative reward is the sum of future discounted rewards for each timestamp, **t**. The symbol **γ** is a discounted factor with a value between 0 and 1. In the game of Tic-Tac-Toe, the cumulative reward at the end of the game is +1 if we win, -1 if we lose, or 0 if it's a draw. The reward for each step could be different for different games. The definition of a reward for each step is usually up to us to define. In the game of Tic-Tac-Toe, it would be a fraction of 1. To make the math work, the reward decays exponentially into the future so that the sum will converge on a finite value that is predictable by the neural network. Also, the discounted factor makes the neural network biased toward recent rewards. The action that the agent takes at the current moment has far more impact on the cumulative reward than the actions deep in the future. Just like playing a game of Tic-Tac-Toe, the location you choose to draw an **O** at the current moment will directly impact where your opponent will draw an **X** because your opponent tries to read what you are doing and stop you from winning the game.

This cumulative reward is also known as the **Q value** in the reinforcement learning paradigm. The neural network that approximates the MDP is the **Q-function**, also known as a **deep Q neural network**. In summary, the process in which the neural network is trained to better approximate the Q value is called **deep Q-learning**:

$$\pi^*(state) = \underset{action(a)}{argmax} \; Q^*(state, action)$$

Figure 11.25: Take the best action that maximizes the Q value

The Q function predicts the Q value for a given state and action. The agent will choose the action that maximizes the Q value and the **π*** symbol represents the **best** action for the given state. Imagine that when playing the game of Tic-Tac-Toe, drawing an O at the center at the start of the game will maximize the chance of winning, thus maximizing the Q value.

So far, we still don't know what exactly the Q function is, in just the same way as the agent doesn't know how to play the game to win. However, the Q function always obeys the **Bellman equation**, shown in the following formula:

$$Q^\pi(s,a) = r + \gamma \underset{a}{max} \; Q^\pi(s', \pi(s'))$$

Figure 11.26: Bellman equation

This formula says the approximated Q value at the current step with state=**s** and action=**a** is equal to the current reward plus a discount factor **γ** weighted Q value at the next step (state=**s'**). We can use this Bellman equation to construct the loss function for the Q function as shown in the following formula. In the example of Tic-Tac-Toe, it means we can use the current reward plus the predicted cumulative reward at the next step to approximate the cumulative reward for the current step:

$$Loss = Q^\pi(s,a) - (r + \gamma \underset{a}{max} \; Q^\pi(s', \pi(s')))$$

Figure 11.27: Loss function of the Q-learning algorithm

After the loss function is defined, we can use what we learned in previous exercises to perform gradient descent to find out the best fit for the Q function and use the Q function as our policy to play the game to win.

Now that we've learned the basic theory behind reinforcement learning, it's time to apply it to a problem. In the next exercise, we will implement a reinforcement learning algorithm to play the game of CartPole-v1 from OpenAI Gym.

EXERCISE 11.04: IMPLEMENTING A DEEP Q-LEARNING ALGORITHM IN PYTORCH TO SOLVE THE CLASSIC CART POLE PROBLEM

In this exercise, we will play the game of CartPole-v1. It's a classic control theory problem where the goal is to balance a pole attached to a cart on a frictionless track. A representation of the game is shown in the following figure:

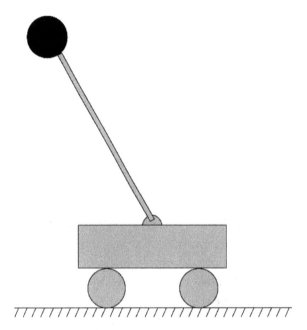

Figure 11.28: Balancing the cart pole from OpenAI Gym

While playing the game, the agent can apply a force of +1 or -1 to the cart to help the pole remain balanced. A reward of +1 is provided for each timestep that the pole remains balanced for. For more details about this game, please visit https://gym.openai. com/envs/CartPole-v1/.

Before proceeding to the exercise, please make sure you have installed the **Gym**. If you haven't installed it, please follow the installation instructions in the *Preface*.

Perform the following steps to complete the exercise.

1. In the **Chapter11** directory, launch a Jupyter Notebook in your Terminal (macOS or Linux) or Command Prompt window (Windows).

2. After the Jupyter Notebook is launched, create a new directory named **Exercise11.04**. Inside the **Exercise11.04** directory, create a Python 3 notebook.

3. Inside the Python 3 notebook, import all the necessary modules and seed the environment as shown in the following code block:

```
# import module
import random
import numpy as np
from itertools import count
from collections import deque

import torch
import torch.nn as nn
import torch.nn.functional as F
import torch.optim as optim

import gym

# make game
env = gym.make('CartPole-v1')

# seed the experiment
env.seed(9)
np.random.seed(9)
random.seed(9)
torch.manual_seed(9)
```

The **env = gym.make('CartPole-v1')** line creates the cart pole game for this exercise. In the last four lines in the snippet, we seed the randomness for those four libraries that we will use in later steps.

4. Let's define our **deep Q network (DQN)** as shown in the following code:

```
# define our policy
class DQN(nn.Module):
    def __init__(self, observation_space, action_space):
        super(DQN, self).__init__()
        self.observation_space = observation_space
        self.action_space = action_space
        self.fc1 = nn.Linear(self.observation_space, 32)
        self.fc2 = nn.Linear(32, 16)
        self.fc3 = nn.Linear(16, self.action_space)

    def forward(self, x):
```

```
        x = F.relu(self.fc1(x))
        x = F.relu(self.fc2(x))
        x = self.fc3(x)
        return x
```

We define a class called **DQN**, which is a simple three-layer feedforward neural network model with **rectified linear unit (ReLU)** activation functions in between each layer. ReLU activation is commonly found in modern neural network architectures.

The **DQN** takes an input state with four features (for example, **tensor([1,0,1,0])**) and output the Q values (for example, **tensor([1.2, 0.4])**) for the action of left force (-1) and the action of right force (+1). The input state with four features describes the position and angle of the cart pole in the game. For the moment, we don't need to understand what each of the numbers means. The algorithm will figure out a way to learn the better action (either a left force or a right force) for a given input state such that the cart pole remains in a balanced state. This is the magic part of machine learning.

5. Next, we will create the agent to play the game as shown in the following code:

ex04-notebook.ipynb

```
# define our agent
class Agent:
    def __init__(self, policy_net):
        MEMORY_SIZE = 10000
        GAMMA = 0.6
        BATCH_SIZE = 128
        EXPLORATION_MAX = 0.9
        EXPLORATION_MIN = 0.05
        EXPLORATION_DECAY = 0.95

        self.policy_net = policy_net
        self.optimizer = optim.RMSprop(\
                        policy_net.parameters(), lr=1e-3)
        self.memory = deque(maxlen=MEMORY_SIZE)
        self.gamma = GAMMA
        self.batch_size = BATCH_SIZE
        self.exploration_rate = EXPLORATION_MAX
        self.exploration_min = EXPLORATION_MIN
        self.exploration_decay = EXPLORATION_DECAY

    def select_action(self, state):
        if np.random.rand() < self.exploration_rate:
            return torch.tensor([[random.randrange(\
                            self.policy_net.action_space)]])
        else:
            with torch.no_grad():
                q_values = self.policy_net(state)
            return q_values.max(1)[1].view(1,1)
```

The full code is available at https://packt.live/2Wh0hTo.

Notice we implemented three methods for the **Agent** class: **select_action()**, **remember()**, and **experience_replay()**. These three methods describe the responsibility of the agent. The agent needs to take action – remember that the game is played in light of the outcome – and learn to play the game better.

At each timestamp, the agent is responsible for choosing the "best" action (left force or right force) to keep the pole balanced. Recall the concept of the epsilon-greedy strategy for tackling the challenge of exploration versus exploitation. **self.exploration_rate** is the epsilon, which starts its value at **EXPLORATION_MAX=0.9** and will decay over time with a factor of **EXPLORATION_DECAY=0.95** until it reaches its minimum value at **EXPLORATION_MIN = 0.05**.

While playing the game, the agent also needs to record past actions and states for training the DQN model. The **remember()** method is called to record the state, action, reward, and the next state for every timestep during the game.

The **experience_replay()** method is used to train our **DQN** policy for every timestamp. It first draws samples from previous timestamps during the game. It then makes two different forward passes to get Q values and expected Q values. Recall the following loss function:

$$Loss = Q^\pi(s, a) - \left(r + \gamma \max_a Q^\pi(s', \pi(s'))\right)$$

Figure 11.29: Loss function of the Q-learning algorithm

The first term in the equation, $Q^\pi(s, a)$, is calculating the actual Q values from states and their corresponding actions. The second term in the equation, $(r + \gamma \max_a Q^\pi(s', \pi(s')))$, is calculating the expected Q values using the optimal actions and their corresponding next states.

Lastly, we use **F.smooth_l1_loss** to calculate the loss function. This loss function is the Huber loss function, which is also very commonly used in training DQNs.

After gradient descent for this learning round is completed, we update the epsilon **self.exploration_rate** to make sure it decays exponentially to the minimum value.

6. Finally, we will implement the deep Q-learning training loop as shown in the following code:

ex04-notebook.ipynb

```
# create policy
observation_space = env.observation_space.shape[0]
action_space = env.action_space.n
policy_net = DQN(observation_space, action_space)

# create agent
agent = Agent(policy_net)

# play game
game_durations = []
for i_episode in count(1):
    state = env.reset()
    state = torch.tensor([state]).float()
    print("[ episode {} ] state={}"\
            .format(i_episode, state))
    for t in range(1, 10000):
        action = agent.select_action(state)
        state_next, reward, done, _ = env.step(action.item())
        if done:
            state_next = None
        else:
            state_next = torch.tensor([state_next]).float()
        agent.remember(state, action, \
        torch.tensor([[reward]]).float(), state_next)
        print("[ episode {} ][ timestamp {} ] state={}, \
            action={}, reward={}, next_state={}"\
            .format(i_episode, t, state, action, \
                reward, state_next))
```

The full code is available at https://packt.live/2Wh0hTo.

Before the training loop starts, we initialize the policy and agent. In the main training loop, the game environment is reset for each episode. An episode ends if the pole loses balance and falls over. For each timestep in an episode, the agent will select an action based on the epsilon-greedy strategy and execute it. After the action is taken, the game returns the reward and new states. The agent will put the information into its memory, as well as executing an experience replay to optimize the policy.

For the lines at the bottom of the previous snippet, we limit the main training loop to 500 episodes. If there's confusion between the concepts of an **episode** and an **epoch**, an **epoch** is a concept for gradient descent whereas an **episode** is a concept in playing games. An **episode** means playing a game through once. 500 episodes means playing a game 500 times.

If the agent requires more than 500 episodes to converge, then this is not a good deep Q-learning algorithm and it is failing to converge. If the agent can last long enough with the number of timesteps greater than **env.spec.reward_ threshold**, which is set to 500 timesteps, in an episode, then the agent wins the game.

After you have run the training loop, you should get the following output:

```
[ episode 184 ][ timestamp 496 ] state=tensor([[1.9317, 0.0974, 0.0831, 1.3612]]), action=tensor([[1]]), reward=1.0,
next_state=tensor([[1.9336, 0.2914, 0.1104, 1.0957]])
[ Experience replay ] starts
[ episode 184 ][ timestamp 497 ] state=tensor([[1.9336, 0.2914, 0.1104, 1.0957]]), action=tensor([[1]]), reward=1.0,
next_state=tensor([[1.9394, 0.4849, 0.1323, 0.8396]])
[ Experience replay ] starts
[ episode 184 ][ timestamp 498 ] state=tensor([[1.9394, 0.4849, 0.1323, 0.8396]]), action=tensor([[1]]), reward=1.0,
next_state=tensor([[1.9491, 0.6780, 0.1491, 0.5912]])
[ Experience replay ] starts
[ episode 184 ][ timestamp 499 ] state=tensor([[1.9491, 0.6780, 0.1491, 0.5912]]), action=tensor([[1]]), reward=1.0,
next_state=tensor([[1.9627, 0.8708, 0.1609, 0.3490]])
[ Experience replay ] starts
[ episode 184 ][ timestamp 500 ] state=tensor([[1.9627, 0.8708, 0.1609, 0.3490]]), action=tensor([[1]]), reward=1.0,
next_state=None
[ Experience replay ] starts
[ Ended! ] Episode 184: Exploration_rate=0.05. Score=500.
[ Solved! ] Score is now 500
```

Figure 11.30: Output logs of the main training loop

> **NOTE**
>
> We will have loads of output data with 500 iteration results. We have only shown the last section of it in the preceding figure.

The output tells us that the agent can keep the pole balanced for **500** timesteps at **episode 184**. According to the game's reward spec, this is a winning solution.

7. Lastly, let's visualize how our deep Q-learning algorithm improves over time by running the following code:

```python
import matplotlib.pyplot as plt
plt.style.use('ggplot')

plt.scatter(range(i_episode), game_durations)
plt.title('Game Duration Over Time')
plt.xlabel('game episode')
plt.ylabel('game duration')
plt.tight_layout()
```

You should get the following output:

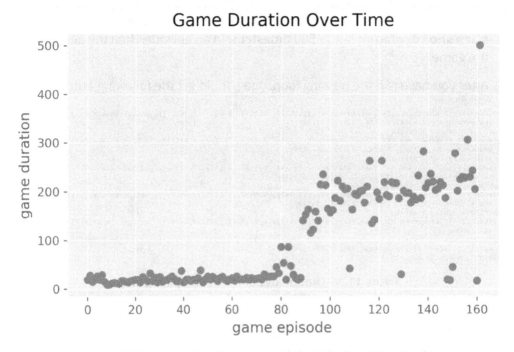

Figure 11.31: Game duration increases to 500 after 175 episodes

We can see that our deep Q-learning algorithm learns to keep the pole balanced for longer and longer as the number of episodes increases. During the first 80 episodes, our model learns very slowly, then it suddenly makes a huge improvement between the 80th and 120th episodes. However, after the 120th episode, the algorithm fails to continue to converge and isn't able to keep the pole balanced for long. After the 160th episode, the algorithm gets back on track and converges on a solution and keeps the pole balanced for more than 500 timestamps. The pattern of convergence is also a result of the epsilon-greedy strategy. Recall the concept of exploration and exploitation: for the first 80 episodes, the algorithm wasn't able to take good actions to keep the pole balanced. This is because the algorithm spent more time on exploration than on exploiting its policy. Later, as its policy improved, the algorithm started to exploit more and explore less. Therefore, we see a convergence after 90 episodes.

> **NOTE**
>
> To access the source code for this specific section, please refer to https://packt.live/2Wh0hTo.

Let's recap what we learned in this exercise. We started it by learning the math and theory behind reinforcement learning for playing a game, looking in particular at the deep Q-learning algorithm. We then went on to implement the deep Q-learning algorithm to play CartPole-v1 from OpenAI Gym. We first defined the DQN model, then we created an agent that implemented an epsilon-greedy strategy and experience replays for model training. We then coded up the main training loop for optimizing our deep Q-learning algorithm. Lastly, we visualized how the algorithm improved over time.

Notice in *Figure 11.31* that our deep Q-learning algorithm doesn't converge consistently. The number of timesteps in one episode fluctuates up and down wildly between the 125th episode and 175th episode. This is probably when the model is overfitting the Q values for actions during that time. We used the same DQN for calculating both the Q values and the expected Q values. The loss from the same function sometimes doesn't converge well, which results in the model overestimating the Q values. The natural question is this: can we improve the algorithm to stabilize the loss function?

The answer is yes. In 2015, Hado van Hasselt, along with two other scientists, proposed a new reinforcement learning algorithm: double Q-learning. The idea behind double Q-learning is simple. Because using the same DQN to calculate the loss is problematic, we can use two different DQN models: one is for action selection, and the other one is for calculating the expected Q values. We call the first one a policy neural network and the second one the target neural network. For more information about double Q-learning, please refer to https://arxiv.org/abs/1509.06461.

In the next activity, we will take what we have learned from this chapter, particularly from the last exercise, and apply it. We will implement a double deep Q-learning algorithm to play CartPole-v1 from OpenAI Gym.

ACTIVITY 11.01: IMPLEMENTING A DOUBLE DEEP Q-LEARNING ALGORITHM TO SOLVE THE CART POLE PROBLEM

In this activity, we will implement a double deep Q-learning algorithm for the cart pole problem. Recall how in *Exercise 11.04, Implementing a Deep Q-Learning Algorithm in PyTorch to Solve the Classic Cart Pole Problem*, we used a single DQN model to calculate the Q-learning loss function. In this activity, we will use two DQN models to calculate the Q-learning loss function. In particular, one DQN will be used for choosing the best action, which is called the policy network. The other DQN will be used for estimating the expected Q value for a given state, which is called the target network.

The reason why we chose to play the same game is that we can compare the two different algorithms at the end of the game with all else being equal. The rules are the same as for the previous exercise.

> **NOTE**
>
> The code and the resulting output for this exercise have been loaded in a Jupyter Notebook that can be found here: https://packt.live/2OzZOrb.

Perform the following steps to complete the activity:

1. Import all necessary modules, such as **numpy, torch, gym, matplotlib. pyplot**, and others. Don't forget to set the seed for all the random number generators that will be used in later steps.

2. Define the model architecture for the DQN for both **policy_network** and **target_network**, which are two different model instances using the same model architecture.

3. Create an agent that implements three methods: **select_action()**, **remember()**, and **experience_replay()**. In the **experience_replay()** method, please use **policy_network** to calculate the Q values while using **target_network** to calculate the expected Q values.

4. Implement the training loop for the double deep Q-learning algorithm. The training loop is similar to the one in the previous exercise. The only difference is that you need to implement the loss function with two DQNs, the policy network, and the target network. The output of the training loop should look similar to the following:

```
[ episode 133 ][ timestamp 476 ] state=tensor([[-2.2825, -0.4745, -0.0851, -0.6469]]), action=tensor([[1]]), reward=
1.0, next_state=tensor([[-2.2920, -0.2783, -0.0980, -0.9651]])
[ Experience replay ] starts
[ episode 133 ][ timestamp 477 ] state=tensor([[-2.2920, -0.2783, -0.0980, -0.9651]]), action=tensor([[1]]), reward=
1.0, next_state=tensor([[-2.2976, -0.0821, -0.1174, -1.2869]])
[ Experience replay ] starts
[ episode 133 ][ timestamp 478 ] state=tensor([[-2.2976, -0.0821, -0.1174, -1.2869]]), action=tensor([[1]]), reward=
1.0, next_state=tensor([[-2.2992,  0.1144, -0.1431, -1.6139]])
[ Experience replay ] starts
[ episode 133 ][ timestamp 479 ] state=tensor([[-2.2992,  0.1144, -0.1431, -1.6139]]), action=tensor([[1]]), reward=
1.0, next_state=tensor([[-2.2970,  0.3108, -0.1754, -1.9476]])
[ Experience replay ] starts
[ episode 133 ][ timestamp 480 ] state=tensor([[-2.2970,  0.3108, -0.1754, -1.9476]]), action=tensor([[1]]), reward=
1.0, next_state=None
[ Experience replay ] starts
[ Ended! ] Episode 133: Exploration_rate=0.05. Score=480.
[ Solved! ] Score is now 480
```

Figure 11.32: Output logs of the main training loop

5. Lastly, create a plot to visualize the algorithm's improvement over the episodes. The plot should look similar to the following:

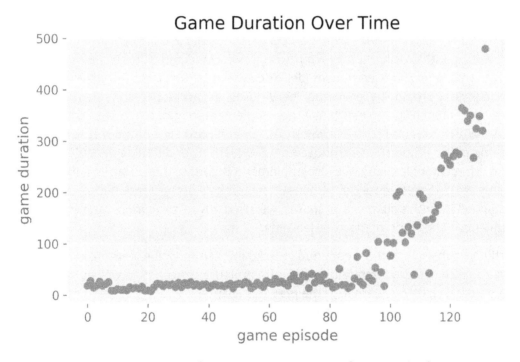

Figure 11.33: Game duration increases to 500 after 133 episodes

With a double Q-learning algorithm, the algorithm can converge faster and in a more consistent fashion.

SUMMARY

This chapter gave you an overview of the theory underpinning machine learning algorithms, looking at constructing a loss function, and using gradient descent. These fundamental concepts will help you better understand a lot of the implementation details of current deep learning practices. They will also help you separate yourself from your peers. The exercises in this chapter focused on hands-on practical skills such as building and training machine learning algorithms for AI from scratch. With the practical skills you learned from this chapter, you will be able to build machine learning models to solve real-world problems.

We started by training a simple linear regression model and implementing a gradient descent algorithm using NumPy from scratch, which helped us better understand how to train a machine learning model. Then we moved on to building training models with PyTorch low-level modules. We also talk about batch gradient descent versus mini-batch SGD in depth. We then implemented mini-batch SGD using high-level PyTorch modules such as **torch.nn**, **torch.optim**, and **torch.utils**. At the end of the chapter, we introduced deep Q-learning and double deep Q-learning for reinforcement learning. We also implemented both of these learning algorithms to play the cart pole game from OpenAI Gym. With all of the hands-on practical skills we gained from this chapter's exercises, we are ready to solve more interesting problems in the world of machine learning.

In the next chapter, you will continue to build on your machine learning knowledge and enhance your understanding of AI infrastructure concepts, as well as developing the practical skills you learned in the previous chapters. You will be learning how to build a production-level AI application.

12

PRODUCTIONIZING YOUR AI APPLICATIONS

OVERVIEW

In this chapter, we'll tie a lot of things together that we have covered in the book. The end goal of an AI application should always be to bring business value. That means all the great models that are created by data scientists should be put into production. It's quite difficult to productionize and maintain an application that contains one or more machine learning models. In this chapter, we'll discuss three major options for productionizing an AI application, so you'll be well equipped to pick a method for your situation.

By the end of this chapter, you will be able to describe the steps it takes to run a machine learning model in production, list a few common options to run models in production, and design and implement a continuous delivery pipeline for models.

INTRODUCTION

In the previous chapters, you have learned ways to set up a data storage environment for AI. In this chapter, we will explore the final step: taking machine learning models into production, so that they can be used in live business applications. There are several methods for productionizing models, and we will elaborate on a few common ones.

Data scientists are trained to wrangle data, pick a machine learning algorithm, do feature engineering, and optimize the models they create. But even an excellent model has no value if it only runs in a machine learning environment or on the laptop of the data scientist; it has to be deployed in a production application. Furthermore, models have to be regularly updated to reflect the latest feedback from customers. Ideally, a model is continuously and automatically refreshed in a feedback loop; we call that reinforcement learning. An example of a system that uses reinforcement learning is a recommendation engine on a video website. Every time the algorithm makes a recommendation to a customer to view a movie, it tracks whether the recommendation is followed. If that is the case, the connection between the customer profile (the features of the algorithm) and the recommendation becomes stronger (*reinforced*), making a similar recommendation more likely in the future.

At this moment, it's important to realize that models don't have to run in an API, a data stream, or in an interactive way at all. It's perfectly acceptable to run a model periodically (say, once per day or even once per month) on a dataset, and write the outcomes to a lookup table. In this way, data is preselected to be used by a live application. A production system only has to find the right records with a simple query on the lookup table. However, more and more systems rely on model execution on demand, simply because the results have to be updated in real-time.

In this chapter, we will describe a few options for running models in production. First, we'll look at ways to export a model from a machine learning environment to run it in production. A common approach is to serialize a model into an intermediate format such as `pickle` and load the resulting files in an API. In this way, your model functions as the core functionality in a microservice.

Next, we'll look at the most popular framework for running containers: Docker. Based on this framework, all public cloud providers offer scalable services that allow customers to run their containers with ease. Kubernetes, OpenShift, and Docker Swarm are examples of services that make use of Docker images.

Finally, we'll describe a method to run models in a streaming data environment. Since performance is an important requirement for streaming data engines, we have to keep the latency low by making our model execution as effective as possible. This can be done by loading the models into an in-memory cache with an intermediate format such as PMML.

Let's start with a basic form of serving models in production: creating an API. We will use the popular frameworks `pickle` and Flask for this.

PICKLE AND FLASK

A machine learning model can "live" in (be part of) many different environments. The choice of environment should depend on the type of application that is being developed, the performance requirements, and the expected frequency of updates. For example, a model that has to predict the weather once per day for a weather analyst has different requirements than a model that makes friend suggestions for millions of people on a social network.

For extreme cases, there are specialized techniques such as streaming models. We'll have a look at them later in this chapter. For now, we'll focus on a method that works for most use cases: running a model as part of an API. In doing so, our model can be part of a microservices architecture, which gives a lot of flexibility and scalability. To build such an API, `pickle` and `joblib` are two popular libraries that can be used when working with Python models. They offer the possibility to *capture* a dataset or a model that was trained in memory, thereby preparing it for transportation to a different environment. As such, this is a good way to share the same model in the development, testing, and production environments. There are also some disadvantages. `pickle` and `joblib` are not language-neutral since they can only be used in Python environments. This gives both data scientists and data engineers a disadvantage since it would be better to have a wider set of technologies to choose from. If you require a cross-platform framework, you could be better off with (for example) Express.js, Spring Boot, or FastAPI for API development and PMML or PFA for model serialization. However, if you're sure that both the machine learning environment and the production environment are running Python, `pickle` and `joblib` are good options for serializing your models. They are well documented, easy to use, high performing, and have a large customer base and community behind them.

We will focus in the first exercise of this chapter on the **pickle** framework. You will learn how to serialize (or **pickle**) a simple model, and how to unserialize or marshal (unpickle) it. Serializing is the process of exporting a model that is built in a notebook. The model lives in the memory of a server in a file format such as JSON, XML, or a binary format. A serialized model can be treated like any other asset in an application; this is needed to transport the model, create versions of it, and deploy it. Then, you will use the model in a Flask API to execute it and get predictions from an input dataset. Let's implement this in the next exercise.

EXERCISE 12.01: CREATING A MACHINE LEARNING MODEL API WITH PICKLE AND FLASK THAT PREDICTS SURVIVORS OF THE TITANIC

In this exercise, we're going to train a simple model and expose it as an API. This exercise aims to create a working API that can be called to get a prediction from a machine learning model. We'll use a dataset of Titanic passengers to build a model that predicts whether a person could have survived the disaster of 15 April 1912. We'll use the **pickle** framework to serialize and deserialize the model, and Flask to expose the API.

pickle is part of the standard Python 3 library, so no installation is needed if you have Python 3 running. For Flask, we'll install it first with **pip** within the exercise.

We will be using a sample dataset that is based on the Titanic dataset. In this famous dataset, all passengers of the first and final trip of the Titanic are listed. The dataset includes details about the persons, such as their family situation during the boat trip, the price they paid for a ticket, and whether they survived the disaster. Predictions can be made such as who is most likely to have survived. The dataset can be found in our GitHub repository at the following location:

https://packt.live/2C72sBN

You need to copy the **Titanic** folder from the GitHub repository.

We'll do this exercise in two parts. The first part consists of building a model and exporting it. In the second part, we'll load the model into an API to get predictions from it.

Perform the following steps to complete the exercise:

1. Create a directory, **Chapter12**, for all the exercises and activities of this chapter. In the **Chapter12** directory, create **Exercise12.01** and **Datasets** directories to store the files for this exercise. Make sure that the **Datasets** directory contains the **Titanic** subdirectory with two files in it: **train.csv** and **test.csv**.

2. Start Jupyter Notebook and create a new Python 3 notebook in the **Exercise12.01** directory. Give the notebook the name **development**.

3. Let's start by installing the Python libraries **pandas** and **sklearn** for model building. Enter the following code in the first cell:

```
!pip install pandas
!pip install sklearn
```

It should give the following output:

```
WARNING: pip is being invoked by an old script wrapper. This will fail in a future version of pip.
Please see https://github.com/pypa/pip/issues/5599 for advice on fixing the underlying issue.
To avoid this problem you can invoke Python with '-m pip' instead of running pip directly.
Collecting pandas
  Downloading pandas-1.0.1-cp37-cp37m-manylinux1_x86_64.whl (10.1 MB)
     |████████████████████████████████| 10.1 MB 4.1 MB/s eta 0:00:01
Requirement already satisfied: python-dateutil>=2.6.1 in /home/bas/anaconda3/lib/python3.7/site-packages (from pan
das) (2.8.0)
Requirement already satisfied: pytz>=2017.2 in /home/bas/anaconda3/lib/python3.7/site-packages (from pandas) (201
9.3)
Requirement already satisfied: numpy>=1.13.3 in /home/bas/anaconda3/lib/python3.7/site-packages (from pandas) (1.1
7.2)
Requirement already satisfied: six>=1.5 in /home/bas/anaconda3/lib/python3.7/site-packages (from python-dateutil>=
2.6.1->pandas) (1.12.0)
Installing collected packages: pandas
Successfully installed pandas-1.0.1
WARNING: pip is being invoked by an old script wrapper. This will fail in a future version of pip.
Please see https://github.com/pypa/pip/issues/5599 for advice on fixing the underlying issue.
To avoid this problem you can invoke Python with '-m pip' instead of running pip directly.
Collecting sklearn
  Downloading sklearn-0.0.tar.gz (1.1 kB)
Requirement already satisfied: scikit-learn in /home/bas/anaconda3/lib/python3.7/site-packages (from sklearn) (0.2
1.3)
Requirement already satisfied: scipy>=0.17.0 in /home/bas/anaconda3/lib/python3.7/site-packages (from scikit-learn
->sklearn) (1.3.1)
Requirement already satisfied: joblib>=0.11 in /home/bas/anaconda3/lib/python3.7/site-packages (from scikit-learn-
>sklearn) (0.13.2)
Requirement already satisfied: numpy>=1.11.0 in /home/bas/anaconda3/lib/python3.7/site-packages (from scikit-learn
->sklearn) (1.17.2)
Building wheels for collected packages: sklearn
  Building wheel for sklearn (setup.py) ... done
  Created wheel for sklearn: filename=sklearn-0.0-py2.py3-none-any.whl size=1316 sha256=b98efd69f9eecee888121087ba
29933b9ee2d9a94ba4dce7f3013ae945cca958
  Stored in directory: /home/bas/.cache/pip/wheels/46/ef/c3/157e41f5ee1372d1be90b09f74f82b10e391eaacca8f22d33e
Successfully built sklearn
Installing collected packages: sklearn
Successfully installed sklearn-0.0
```

Figure 12.1: Installing pandas and sklearn

It will download the libraries and install them within your active Anaconda environment. There is a good chance that both frameworks are already available in your system, as part of Anaconda or from previous installations. If they are already installed, you will have the following output:

```
Requirement already satisfied: pandas in /usr/local/anaconda3/lib/python3.7/site-packages (0.25.1)
Requirement already satisfied: pytz>=2017.2 in /usr/local/anaconda3/lib/python3.7/site-packages (from pand
as) (2019.3)
Requirement already satisfied: numpy>=1.13.3 in /usr/local/anaconda3/lib/python3.7/site-packages (from pan
das) (1.17.2)
Requirement already satisfied: python-dateutil>=2.6.1 in /usr/local/anaconda3/lib/python3.7/site-packages
(from pandas) (2.8.0)
Requirement already satisfied: six>=1.5 in /usr/local/anaconda3/lib/python3.7/site-packages (from python-d
ateutil>=2.6.1->pandas) (1.12.0)
Requirement already satisfied: sklearn in /usr/local/anaconda3/lib/python3.7/site-packages (0.0)
Requirement already satisfied: scikit-learn in /usr/local/anaconda3/lib/python3.7/site-packages (from skle
arn) (0.21.3)
Requirement already satisfied: scipy>=0.17.0 in /usr/local/anaconda3/lib/python3.7/site-packages (from sci
kit-learn->sklearn) (1.3.1)
Requirement already satisfied: numpy>=1.11.0 in /usr/local/anaconda3/lib/python3.7/site-packages (from sci
kit-learn->sklearn) (1.17.2)
Requirement already satisfied: joblib>=0.11 in /usr/local/anaconda3/lib/python3.7/site-packages (from scik
it-learn->sklearn) (0.13.2)
```

Figure 12.2: pandas and sklearn already installed

4. Next, we import the **pickle**, **pandas**, and **sklearn** libraries:

```
import pickle
import pandas as pd
from sklearn.linear_model import LogisticRegression
```

5. Load the training data into a pandas DataFrame object called **train** with the following statement:

```
# load the training dataset
train = pd.read_csv('../../Datasets/Titanic/train.csv')
```

6. Observe the results by using **train.info()**. The output will look as follows:

```
<class 'pandas.core.frame.DataFrame'>
RangeIndex: 891 entries, 0 to 890
Data columns (total 12 columns):
PassengerId   891 non-null int64
Survived      891 non-null int64
Pclass        891 non-null int64
Name          891 non-null object
Sex           891 non-null object
Age           714 non-null float64
SibSp         891 non-null int64
Parch         891 non-null int64
Ticket        891 non-null object
Fare          891 non-null float64
Cabin         204 non-null object
Embarked      889 non-null object
dtypes: float64(2), int64(5), object(5)
memory usage: 83.7+ KB
```

Figure 12.3: Information about the training dataset

What you can see in this output is that the **train** object now holds a dataset of **891** rows, with **12** columns. The names and datatypes of the columns are specified; we can see, for example, that the first column is called **PassengerId** and is of type **int64**. Take a note of the columns that are of type **object**. These are difficult to work within a machine learning model; we should convert them into a numerical datatype.

7. Let's check whether the dataset is loaded correctly. We can explore the dataset a bit with some simple pandas commands. Find the list of people in the training dataset who survived the Titanic disaster by using the following code:

```
# count the survivors
train[train['Survived'] == 0]
```

This will produce the following output, showing that there were **549** survivors:

	PassengerId	Survived	Pclass	Name	Sex	Age	SibSp	Parch	Ticket	Fare	Cabin	Embarked
0	1	0	3	Braund, Mr. Owen Harris	male	22.0	1	0	A/5 21171	7.2500	NaN	S
4	5	0	3	Allen, Mr. William Henry	male	35.0	0	0	373450	8.0500	NaN	S
5	6	0	3	Moran, Mr. James	male	NaN	0	0	330877	8.4583	NaN	Q
6	7	0	1	McCarthy, Mr. Timothy J	male	54.0	0	0	17463	51.8625	E46	S
7	8	0	3	Palsson, Master. Gosta Leonard	male	2.0	3	1	349909	21.0750	NaN	S
...	...	...	...	...	...	...	...	...	...	...	...	...
884	885	0	3	Sutehall, Mr. Henry Jr	male	25.0	0	0	SOTON/OQ 392076	7.0500	NaN	S
885	886	0	3	Rice, Mrs. William (Margaret Norton)	female	39.0	0	5	382652	29.1250	NaN	Q
886	887	0	2	Montvila, Rev. Juozas	male	27.0	0	0	211536	13.0000	NaN	S
888	889	0	3	Johnston, Miss. Catherine Helen "Carrie"	female	NaN	1	2	W./C. 6607	23.4500	NaN	S
890	891	0	3	Dooley, Mr. Patrick	male	32.0	0	0	370376	7.7500	NaN	Q

549 rows × 12 columns

Figure 12.4: Count of the passengers that survived

8. After loading the training dataset, load the testing dataset with the following command:

```
# load the testing dataset
test = pd.read_csv('../../Datasets/Titanic/test.csv')
```

9. Check the result with the **test.info()** command, which will produce the following result:

```
<class 'pandas.core.frame.DataFrame'>
RangeIndex: 418 entries, 0 to 417
Data columns (total 11 columns):
PassengerId    418 non-null int64
Pclass         418 non-null int64
Name           418 non-null object
Sex            418 non-null object
Age            332 non-null float64
SibSp          418 non-null int64
Parch          418 non-null int64
Ticket         418 non-null object
Fare           417 non-null float64
Cabin          91 non-null object
Embarked       418 non-null object
dtypes: float64(2), int64(4), object(5)
memory usage: 36.0+ KB
```

Figure 12.5: Information about the testing dataset

The output shows that there are **418** records in our testing dataset, with the same columns as in the training dataset.

10. We'll now continue to prepare and clean the training dataset as mentioned in *Chapter 3*, *Data Preparation*. Machine learning models work best with numerical datatypes for all columns, so let's convert the **Sex** column to zeros and ones:

```
# prepare the dataset
train.Sex = train.Sex.map({'male':0, 'female':1})
```

We have now transformed the values in the **Sex** column to either **0** (for **male**) or **1** (for **female**).

11. Create the training set, **X**, which doesn't contain the **Survived** output column, using the following code:

```
# use the values in the Survived column as output targets
y = train.Survived.copy()
X = train.drop(['Survived'], axis=1)
```

Since the **Survived** column contains our output value on which we have to train our model (the target values), we have to extract that from the dataset. We create a new dataset for it called **y** and then remove the column from the training dataset. We call the new training set **X**.

Now, let's do some feature engineering. We can be quite certain that a lot of the columns will not hold any predictive value as to whether a person survived. For example, the name of someone and their passenger ID are interchangeable and will not contribute much to the predictive power of the machine learning model.

12. Remove a set of columns that are not needed to predict the surviving passengers:

```
X.drop(['Name'], axis=1, inplace=True)
X.drop(['Embarked'], axis=1, inplace=True)
X.drop(['PassengerId'], axis=1, inplace=True)
X.drop(['Cabin'], axis=1, inplace=True)
X.drop(['Ticket'], axis=1, inplace=True)
```

We have removed the **Name**, **Embarked**, **PassengerId**, **Cabin**, and **Ticket** columns. There is one more thing to do: the **Age** column contains some empty (**null**) values. These could get in the way when training the model.

13. Replace the empty values with the **mean** age of all the passengers in the dataset:

```
X.Age.fillna(X.Age.mean(), inplace=True)
```

14. Let's have a look at the current size and contents of the training dataset:

```
X.info()
X.head()
```

This should give the following output:

```
<class 'pandas.core.frame.DataFrame'>
RangeIndex: 891 entries, 0 to 890
Data columns (total 6 columns):
 #   Column  Non-Null Count  Dtype
---  ------  --------------  -----
 0   Pclass  891 non-null    int64
 1   Sex     891 non-null    int64
 2   Age     891 non-null    float64
 3   SibSp   891 non-null    int64
 4   Parch   891 non-null    int64
 5   Fare    891 non-null    float64
dtypes: float64(2), int64(4)
memory usage: 41.9 KB
```

Out[18]:

	Pclass	Sex	Age	SibSp	Parch	Fare
0	3	0	22.0	1	0	7.2500
1	1	1	38.0	1	0	71.2833
2	3	1	26.0	0	0	7.9250
3	1	1	35.0	1	0	53.1000
4	3	0	35.0	0	0	8.0500

Figure 12.6: Dataset information after feature engineering

As becomes clear from this output, the training dataset still holds all the **891** rows but there are fewer columns. Moreover, all columns are of a numerical type – either **int64** or **float64**. To make even better models, there is a lot that data scientists can do. For example, it's a good option to normalize the columns (get them within the same range) and to do more feature engineering, such as natural language processing, on the non-numerical columns in the source dataset.

15. Let's now train the actual model. We'll create a logistic regression model, which is essentially a mathematical algorithm to separate the survivors from the people who died in the accident:

```
from sklearn.linear_model import LogisticRegression
model = LogisticRegression()
model.fit(X, y)
```

The **model.fit** method takes the training dataset, **X**, and the target values, **y**, and will perform its calculations to make the best fit. This will produce a model, as can be seen in the following output:

```
/home/bas/anaconda3/lib/python3.7/site-packages/sklearn/linear_model/logistic.py:432: FutureWarning: Default solve
r will be changed to 'lbfgs' in 0.22. Specify a solver to silence this warning.
  FutureWarning)
Out[19]: LogisticRegression(C=1.0, class_weight=None, dual=False, fit_intercept=True,
                   intercept_scaling=1, l1_ratio=None, max_iter=100,
                   multi_class='warn', n_jobs=None, penalty='l2',
                   random_state=None, solver='warn', tol=0.0001, verbose=0,
                   warm_start=False)
```

Figure 12.7: Fitting the model

16. To see what the accuracy of the model is, we can run the full set of test data on the model and see in how many cases the algorithm produces a correct outcome using the following code:

```
# evaluate the model
model.score(X, y)
```

You'll get the following output:

```
0.8002244668911336
```

> **NOTE**
>
> The preceding output will vary slightly. For full reproducibility, you can set a random seed.

A score of **0.8** (and something more) means that the model performs accurately in 80% of the cases in the test set.

We can also get some more understanding of how the model works. Enter the following command:

```
train.corr()
```

The resulting correlation graph indicates columns that are closely related:

	PassengerId	Survived	Pclass	Sex	Age	SibSp	Parch	Fare
PassengerId	1.000000	-0.005007	-0.035144	-0.042939	0.036847	-0.057527	-0.001652	0.012658
Survived	-0.005007	1.000000	-0.338481	0.543351	-0.077221	-0.035322	0.081629	0.257307
Pclass	-0.035144	-0.338481	1.000000	-0.131900	-0.369226	0.083081	0.018443	-0.549500
Sex	-0.042939	0.543351	-0.131900	1.000000	-0.093254	0.114631	0.245489	0.182333
Age	0.036847	-0.077221	-0.369226	-0.093254	1.000000	-0.308247	-0.189119	0.096067
SibSp	-0.057527	-0.035322	0.083081	0.114631	-0.308247	1.000000	0.414838	0.159651
Parch	-0.001652	0.081629	0.018443	0.245489	-0.189119	0.414838	1.000000	0.216225
Fare	0.012658	0.257307	-0.549500	0.182333	0.096067	0.159651	0.216225	1.000000

Figure 12.8: Correlation between columns

A correlation value of 0 indicates no relationship. The further away from 0, a value is, toward a minimum of -1 and a maximum of 1, the stronger the relationship between the columns is. In the table, you can see that the data in the columns **Fare** and **Pclass** (the class of the customer) is related since the correlation between them is **-0.549500**. On the contrary, the columns **Age** and **PassengerId** are not related to each other, as expected; since the passenger ID is just an arbitrary number, it would be strange if that was dependent on the age of a person. The column that is most indicative of the survival of a person is **Sex**; the value of **0.543351** indicates that once you know whether a person is **male** or **female**, you can predict whether they survived the disaster with reasonable accuracy. Therefore, the **Sex** column is a good feature for the model.

17. Finally, we have to serialize (export) the model to a file with the **pickle** framework:

```
file = open('model.pkl', 'wb')
pickle.dump(model, file)
file.close()
```

The **pickle.dump** method serializes the model to a **model.pkl** file in the exercise directory. The file has been opened as **wb**, which means that it will write bytes to disk. You can check the file exists in the **Exercise12.01** folder.

The second part of this exercise is to load the model from disk in an API and expose the API.

18. Create a new Python 3 notebook in the **Exercise12.01** directory and give it the name **production**.

> **NOTE**
>
> The **production.ipynb** file can be found here: https://packt.live/2ZtjzH7

19. In the first cell, enter the following line to install Flask:

```
!pip install flask
```

You'll get the following output:

```
WARNING: pip is being invoked by an old script wrapper. This will fail in a future version of pip.
Please see https://github.com/pypa/pip/issues/5599 for advice on fixing the underlying issue.
To avoid this problem you can invoke Python with '-m pip' instead of running pip directly.
Collecting flask
  Downloading Flask-1.1.1-py2.py3-none-any.whl (94 kB)
     |████████████████████████████████| 94 kB 2.2 MB/s eta 0:00:011
Requirement already satisfied: Werkzeug>=0.15 in /home/bas/anaconda3/lib/python3.7/site-packages (from flask) (0.1
6.0)
Requirement already satisfied: click>=5.1 in /home/bas/anaconda3/lib/python3.7/site-packages (from flask) (7.0)
Requirement already satisfied: itsdangerous>=0.24 in /home/bas/anaconda3/lib/python3.7/site-packages (from flask)
(1.1.0)
Requirement already satisfied: Jinja2>=2.10.1 in /home/bas/.local/lib/python3.7/site-packages (from flask) (2.11.
1)
Requirement already satisfied: MarkupSafe>=0.23 in /home/bas/anaconda3/lib/python3.7/site-packages (from Jinja2>=
2.10.1->flask) (1.1.1)
Installing collected packages: flask
Successfully installed flask-1.1.1
```

Figure 12.9: Installing Flask

If Flask is already installed, you will get the following output:

```
Requirement already satisfied: flask in /usr/local/anaconda3/lib/python3.7/site-packages (1.1.1)
Requirement already satisfied: itsdangerous>=0.24 in /usr/local/anaconda3/lib/python3.7/site-packages (fro
m flask) (1.1.0)
Requirement already satisfied: click>=5.1 in /usr/local/anaconda3/lib/python3.7/site-packages (from flask)
(7.0)
Requirement already satisfied: Jinja2>=2.10.1 in /usr/local/anaconda3/lib/python3.7/site-packages (from fl
ask) (2.10.3)
Requirement already satisfied: Werkzeug>=0.15 in /usr/local/anaconda3/lib/python3.7/site-packages (from fl
ask) (0.16.0)
Requirement already satisfied: MarkupSafe>=0.23 in /usr/local/anaconda3/lib/python3.7/site-packages (from
Jinja2>=2.10.1->flask) (1.1.1)
```

Figure 12.10: Flask already installed

20. Now import the required libraries to deserialize the model and to create an API:

```
from flask import Flask, jsonify, request
import pickle
```

21. The model that we trained in the first part was stored in the **model.pkl** file. Let's get that file from disk and deserialize the model into memory:

```
file = open('model.pkl', 'rb')  # read bytes
model = pickle.load(file)
file.close()
```

We now have the same model running in our **production** environment (the **production** Jupyter notebook) as in our **model training** environment (the **development** notebook from part 1 of this exercise). To test the model, we can make a few predictions.

22. Call the **predict** function of the model and give it an array of values that represent a person; the values are in the same order as the columns of the training set (namely, class, sex, age, number of siblings aboard, number of parents aboard, and fare that was paid) as shown in the following code:

```
print(model.predict([[3,0,22.0,1,0,7.2500]]))
print(model.predict([[3,1,22.0,1,0,7.2500]]))
```

You should get the following output:

```
[0]
[1]
```

The values in the output (**0** and **1**) are the predictions of the **Survived** output field. This means that the model predicted that our first test subject, a man in class **3** of age **22** with **1** child and **0** parents on board who paid **7.25** for the ticket, was more likely to have survived than a woman with the same characteristics.

We have now created a machine learning model that predicts whether a passenger on the Titanic survived the disaster of 1912. But, this model can only be executed by calling it within a Python script.

23. Let's continue with productionizing our model. We first must set up a Flask app, which we'll call **Titanic**:

```
app = Flask('Titanic')
```

24. The **app** is now an empty API with no HTTP endpoint exposed. To make things a bit more interesting and test whether it works, add a **GET** method that returns a simple string:

```
@app.route('/hi', methods=['GET'])
def bar():
    result = 'hello!'
    return result
```

After running this cell, test the app by entering and running the following line in a new cell:

```
app.run()
```

This will result in an ever-running cell (like an infinite loop), indicated by an

asterisk (*****), as shown in the following output:

```
* Serving Flask app "Titanic" (lazy loading)
* Environment: production
  WARNING: This is a development server. Do not use it in a production deployment.
  Use a production WSGI server instead.
* Debug mode: off

* Running on http://127.0.0.1:5000/ (Press CTRL+C to quit)
```

Figure 12.11: Testing the app

The bottom of the cell already indicates what to do: let's open a new browser window and go to **http://127.0.0.1:5000/hi**. Since the action that your browser takes is a **GET** request to the API, you'll get the correct response back, as you can see in the following figure:

Figure 12.12: Running the app as seen in the browser window

The API now only returns **hello!** to demonstrate that it's working technically. We can now continue to add business logic that produces useful results.

25. We have to add a new method to the API that executes the model and returns the result as shown in the following code:

```
@app.route('/survived', methods=['POST'])
def survived():
    payload = request.get_json()
    person = [payload['Pclass'], \
             payload['Sex'], payload['Age'], \
             payload['SibSb'], payload['Parch'], \
             payload['Fare']]
    result = model.predict([person])
    print(f'{person} -> {str(result)}')
    return f'I predict that person {person} has \
    {"_not_ " if result == [0] else ""}\
    survived the Titanic\n'
```

These lines define an HTTP **POST** method under the URL **'/survived'**. When called, a **person** object is generated from the JSON payload. The **person** object, which is an array of input parameters for the model, is then passed to the model in the **predict** statement that we've seen before. The result is finally wrapped up in a string and returned to the caller.

Run the app again (stop it by clicking on the interrupt icon (■) or by typing *Ctrl + C* first, if needed). This time, we will test the API with a **curl** statement. cURL is a program that allows you to make an HTTP request across a network in a similar way to how a web browser makes requests, only the result will be just text instead of a graphical interface.

26. Start a Terminal window (or Anaconda prompt) and enter the following command:

```
curl -X POST -H "Content-Type: application/json" -d '{"Pclass":
3, "Sex": 0, "Age": 72, "SibSb": 2, "Parch": 0, "Fare": 8.35}'
http://127.0.0.1:5000/survived
```

NOTE

There is a Jupyter notebook called **validation** in GitHub that contains the same command in a cell. It can be found here:https://packt.live/2Cu9QY3

The string after the **–d** parameter in the **curl** script contains a JSON object with passenger data. The fields for a person have to be named explicitly.

After running this, you'll see the result of your API call that executed the model as follows in your Terminal or Anaconda prompt:

```
I predict that person [3, 0, 72, 2, 0, 8.35] has
not survived the Titanic
```

This output shows that our API works and that the model is evaluated! The API takes the input of one person (the JSON object in the **curl** statement) and provides it as input to the machine learning model that it has loaded from disk, by deserializing the **pickle model.pkl** file. The output is wrapped in a string and transferred back to the Terminal across the network.

NOTE

To access the source code for this specific section, please refer to https://packt.live/306HgE5.

By completing this exercise, you have practiced a lot of useful techniques. You have built and trained a machine learning model, stored your model as a **pickle** file, and used the file in a deserialized form in an API using the Flask framework. This has resulted in an application that can predict whether a passenger of the Titanic survived the disaster of 1912 or not. The API can be used from any other application that can access it across a network. In the following activity, you'll perform the same steps to build an API that predicts in which class a passenger was sitting.

ACTIVITY 12.01: PREDICTING THE CLASS OF A PASSENGER ON THE TITANIC

In this activity, you'll use the same dataset as in the previous exercise. Rather than building a model that predicts whether a person survived the Titanic disaster, we're going to try to predict which class (1, 2, or 3) a person was in based on their details.

> **NOTE**
>
> The code for this activity can be found here:
>
> https://packt.live/38T32iF.

Perform the following steps to complete the activity:

1. Start Jupyter Notebook and create a new Python 3 notebook in the same directory, with the name **development**.

2. Install the **pandas** and **sklearn** Python libraries for model building and import them into your notebook.

3. Load the training data into a Pandas DataFrame object called **train**.

4. Convert the **Sex** column to zeros and ones.

5. Create a new dataset for the output values (in the **Pclass** column). Call that dataset **y** and then remove the **Pclass** column from the training dataset. Call the new training set **X**.

6. Remove the columns that you cannot parse or that don't contain good features for your model.

7. Replace the empty values in the **Age** column with the **fillna** method.

8. Create a logistic regression model with the **sklearn.linearmodel. LogisticRegression** class.

9. Serialize the model to a file with the **pickle.dump** method. Check that the model file exists in the **Activity12.01** folder.

10. Create a new Python 3 notebook in the **Activity12.01** directory and give it a meaningful name such as **production**.

11. In the new notebook, install Flask and import the required libraries to deserialize the model and to create an API.

12. Load the model file from disk and deserialize the model into memory.

13. Set up a Flask app called **ClassPredictor**.

14. Define an HTTP **POST** method under the URL `'/class'`, by making use of the Flask method annotation @**app.route('/class', methods=['POST'])**. When called, a **person** object (an array of input parameters) should be generated from the JSON payload. The result of the model should be wrapped up in a string and returned to the caller.

15. Run the app with an **app.run()** statement and test the API with a **curl** statement. You should get an output similar to the following:

```
I predict that person [1, 0, 72, 2, 0,28.35] was
  in class [2] of the Titanic
```

You have now completed an activity to train a model and build an API for it. In the next part of this chapter, we are going to explore options to productionize these kinds of services in a live application.

> **NOTE**
>
> The solution to this activity can be found on page 654.

DEPLOYING MODELS TO PRODUCTION

After creating an API that contains your machine learning model, it has to be hosted in a production environment. There are several ways to do this, such as the following, for example:

- By copying the API to a (virtual) server

- By containerizing the API and deploying the container to a cluster

- By installing the API in a serverless framework such as Amazon AWS Lambda or Microsoft Azure Functions

We'll focus on the practice that is very common nowadays and still gaining popularity: containerizing the API and model.

DOCKER

AI applications usually work with large datasets. With "big data" comes the requirement for scalability. This means that models in production should scale in line with the data. One way to scale your software services is to distribute them in containers. A container is a small unit of computational power, similar to a virtual machine. There are many other advantages when containerizing your software: deployment becomes easier and more predictable since the containers stay the same in every environment.

The best framework for containerizing applications is Docker. This open-source tool has become the de facto way to containerize applications. Docker works with the concepts of *images* and *containers*. An image is a template of an application that is deployable to an environment. A container is a concrete implementation of an image. What's inside an image (and in its resulting containers) is very flexible; it can range from a simple "hello world" script to a full-blown enterprise application. It's considered good practice to build Docker images for a single purpose and have them communicate with each other via standard protocols such as REST. In this way, it's possible to create infrastructure and software by carefully selecting a set of Docker images and connecting them. For example, one Docker image might contain a database and the data of an application itself. Another Docker image can be created to hold the business logic and expose an API. A third Docker image could then contain a website that is published on the internet. In the next part, about Kubernetes, you will see an example diagram of such an application architecture. In our other exercise and activity in this chapter, we'll use Docker to productionize a machine learning model.

Docker in itself is not so useful in an enterprise; an environment is needed where Docker images can be published and maintained. One of the most popular frameworks to do so is Kubernetes, which we'll discuss next.

KUBERNETES

To create a cluster of many containers, and thus to generate scale in an application, a tool needs to be used that can manage those containers in one environment. Ideally, you would want to deploy many containers with the same software and treat them as if they were one unit. This is exactly the purpose of frameworks such as Kubernetes, Docker Swarm, and OpenShift. They allow software developers to distribute their applications across many "workers," which are, for example, cloud-based nodes. Kubernetes was created by Google to be used in their cloud computing cluster. It was made open source in 2015. Nowadays, it's one of the most popular frameworks for scaling software applications. The large Kubernetes community is very active.

Kubernetes works with the concepts of Pods and Nodes. A Pod is the main abstraction for an application; you always deploy an application in one or more Pods. A Kubernetes cluster consists of Nodes, which are the worker servers that run applications. There is always one Master Node that is responsible for managing the cluster. The following figure gives an overview of this model:

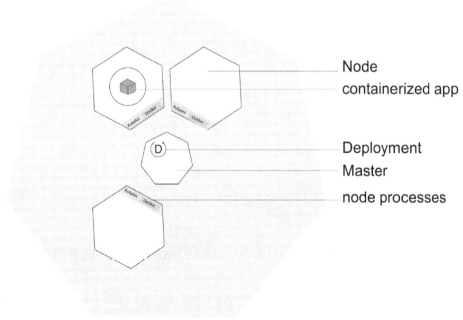

Kubernetes Cluster

Figure 12.13: Kubernetes cluster (redrawn from https://kubernetes.io/docs/)

In the previous figure, you see a Kubernetes cluster with four nodes. One of the nodes has the Master role and contains one Deployment. From the Deployment, an application can be deployed onto the other Nodes. Once an application is deployed to nodes in a Kubernetes cluster, it runs in a Pod. A Pod contains a group of resources such as storage and networking that are needed to run a container. In the previous figure, there are three nodes. On one of them, there is a Pod that contains an application (the containerized app). If we zoom in on a Node, we can see a structure such as in the following figure:

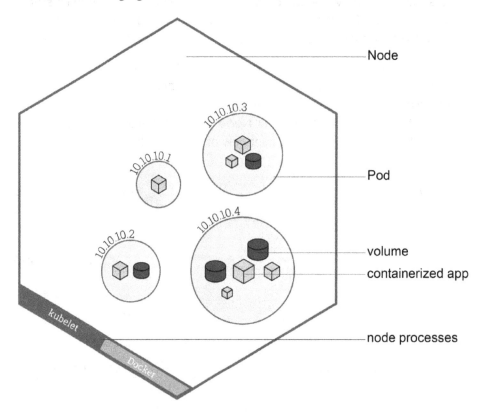

Figure 12.14: Kubernetes Node with Pods (redrawn from https://kubernetes.io/docs/)

In the preceding figure, you can see one Node with four Pods in it. Each Pod has its function and contents; for example, the Pod with ID **10.10.10.4** contains three containerized apps and two storage volumes, which work together toward a business goal such as providing a website. You can see that each Node has two core processes running: **kubelet** for communication, and Docker for running the images. The Kubernetes command-line interface is called **kubectl**.

> **NOTE**
>
> If you want to read more about Kubernetes and Docker, please refer to the following link for the official documentation: https://kubernetes.io/docs/home/

In the next exercise, you will learn how to dockerize an API that contains a machine learning model, and how to deploy it to a Kubernetes cluster.

EXERCISE 12.02: DEPLOYING A DOCKERIZED MACHINE LEARNING API TO A KUBERNETES CLUSTER

In this exercise, we'll store a machine learning API as a Docker image, and deploy an instance (a container) to a Kubernetes cluster. This exercise is a follow-up of the previous exercise, so make sure you have completed that one.

In *Exercise 12.01*, *Creating a Machine Learning Model API with pickle and Flask That Predicts Survivors of the Titanic*, you created a machine learning model that predicts whether a person survived the Titanic disaster in 1912. You created an API that contains the model. We will be using that model in this exercise. The final goal of this exercise is to deploy the same model to a Kubernetes cluster.

We will not work with Jupyter in this exercise; we'll do most of the work from Command Prompt and a text editor (or IDE).

Before you begin, follow the installation instructions in the *Preface* to install Docker and Kubernetes.

This exercise consists of two parts. In the first part, we'll create a Docker image and publish it to a local registry. In the second part, we'll run the image in a Kubernetes cluster.

Perform the following steps to complete the exercise:

1. Create a new directory, **Exercise12.02**, in the **Chapter12** directory to store the files for this exercise.

2. Copy the **model.pkl** file from the **Chapter12/Exercise12.01** directory into the **Exercise12.02** directory.

3. Let's create a new API first. Since we'll work outside of Jupyter on this one, we have to make sure that it runs on our local machine. Create a new file in the **Exercise12.02** folder called **api.py**. Open it with a text editor or an IDE and enter the following Python code:

```python
from flask import Flask, request
import pickle

# load the model from pickle file
file = open('model.pkl', 'rb') # read bytes
model = pickle.load(file)
file.close()

# create an API with Flask
app = Flask('Titanic')

# call this: curl -X POST -H "Content-Type: application/json" \
# -d '{"Pclass": 3, "Sex": 0, "Age": 72, "SibSb": 2, "Parch": 0,
#"Fare": 8.35}' http://127.0.0.1:5000/survived
@app.route('/survived', methods=['POST'])
def survived():
    payload = request.get_json()
    person = [payload['Pclass'], payload['Sex'], \
              payload['Age'], payload['SibSb'], \
              payload['Parch'], payload['Fare']]
    result = model.predict([person])
    return f'I predict that person {person} has \
    {"_not_ " if result == [0] else ""}survived the Titanic\n'

app.run()
```

This code is the same as in the **production** notebook of *Exercise 12.01, Creating a Machine Learning Model API with pickle and Flask That Predicts Survivors of the Titanic*, but is now pulled together in one executable Python file since it's the most efficient way to run a Python program. It uses the machine learning model in the **pickle model.pkl** file to predict whether a person survived the Titanic disaster.

4. Open a new Terminal or Anaconda Prompt and navigate to the **Exercise12.02** folder using the **cd** command. Enter the following command to check whether the API is working:

```
python api.py
```

If all is OK, you should get the message that your API is running on localhost, as seen in the following screenshot:

```
* Serving Flask app "Titanic" (lazy loading)
* Environment: production
  WARNING: This is a development server. Do not use it in a production deployment.
  Use a production WSGI server instead.
* Debug mode: off
* Running on http://127.0.0.1:5000/ (Press CTRL+C to quit)
```

Figure 12.15: Running the API

5. Test the local API by opening a new Terminal or Anaconda Prompt and enter the following code:

```
curl -X POST -H "Content-Type: application/json" -d '{"Pclass":
  3, "Sex": 0, "Age": 72, "SibSb": 2, "Parch": 0, "Fare": 8.35}'
  http://127.0.0.1:5000/survived
```

You'll get the following output in the same Terminal, indicating that the API is working and that we get good predictions from the model:

```
I predict that person [3, 0, 72, 2, 0, 8.35] has
not survived the Titanic
```

6. Add a new file in the **Exercise12.02** folder with the name **requirements. txt**. If we will deploy the API to another environment, it's good practice to indicate which libraries were used. To do so, Python offers the **requirements. txt** file, where you can write the dependencies of an application.

7. Open the **requirements.txt** file in an editor and enter the following lines:

```
Flask
sklearn
pandas
```

This simple line is enough for our API. It indicates that we are depending on the **Flask**, **sklearn**, and **pandas** libraries for everything to work. Since **pickle** is a standard Python library, we don't have to reference it explicitly. The **requirements** file is used in the Docker image once it runs in production.

> **NOTE**
>
> It's possible to specify the exact version, for example, by entering
> `Flask==1.1.1`. To see which version you have in your development
> environment, enter `pip freeze`. To export the current list of
> dependencies and store them as `requirements.txt`, enter `pip
> freeze > requirements.txt`.

Now, let's continue with containerizing the API to make it ready to deploy to a production environment. We need to create a Docker *image*, which is a template for creating the actual Docker *containers* that will be deployed.

8. First, let's verify that Docker is installed correctly on your system. If you have followed the installation instructions, enter the following command in a new Terminal or Anaconda Prompt:

```
sudo docker run hello-world
```

This command will get a basic Docker image called **hello-world** from the central repository (Docker Hub). Based on that image, a local container will be created on your local machine. The output will be as follows:

```
Unable to find image 'hello-world:latest' locally
latest: Pulling from library/hello-world
1b930d010525: Already exists
Digest: sha256:fc6a51919cfeb2e6763f62b6d9e8815acbf7cd2e476ea353743570610737b752
Status: Downloaded newer image for hello-world:latest

Hello from Docker!
This message shows that your installation appears to be working correctly.

To generate this message, Docker took the following steps:
 1. The Docker client contacted the Docker daemon.
 2. The Docker daemon pulled the "hello-world" image from the Docker Hub.
    (amd64)
 3. The Docker daemon created a new container from that image which runs the
    executable that produces the output you are currently reading.
 4. The Docker daemon streamed that output to the Docker client, which sent it
    to your terminal.

To try something more ambitious, you can run an Ubuntu container with:
 $ docker run -it ubuntu bash

Share images, automate workflows, and more with a free Docker ID:
 https://hub.docker.com/

For more examples and ideas, visit:
 https://docs.docker.com/get-started/
```

Figure 12.16: Output of running the 'hello-world' Docker image

9. You can check the images you have stored on your local machine by typing the following:

```
sudo docker image ls
```

This will give an output like the following:

REPOSITORY	TAG	IMAGE ID	CREATED	SIZE
hello-world	latest	fce289e99eb9	14 months ago	1.84kB

Figure 12.17: Checking stored images

At the moment, we only see the **hello-world** image that was created in the previous step.

10. Instead of working with a pre-defined Docker image, we want to set up our own Docker images that are defined in strict files that are named **Dockerfile**. So, let's add a new text file in the same directory and call it **Dockerfile** (without an extension). Open **Dockerfile** in a text editor or IDE, enter the following code, and save the file:

```
FROM python:3.7

RUN mkdir /api
WORKDIR /api
ADD . /api/

RUN pip install -r ./requirements.txt

EXPOSE 5000

ENV PYTHONPATH="$PYTHONPATH:/api"
CMD ["python", "/api/api.py"]
```

Dockerfile is the most important artifact when creating your images. There are five main parts to our file.

First (line 1), a base image is acquired. In our case, this is a Python **3.7** image that can run Python applications.

Next (lines 3 to 5), the contents of the current directory are added to the image; this ensures that the **api.py** and **requirements.txt** files are packaged within the container. We add all files to a directory called **api**.

Next (line 7), the required Python libraries that we marked as dependencies in the **requirements.txt** file are installed with the **pip install** command. In our case, this is just the Flask library.

Next (line 8), we tell Docker to expose network port **5000** to the outside world. If this is omitted, the API cannot be reached from the network.

Finally (lines 11 and 12), we start the API by setting the Python path to the **api** directory and executing the **python api/api.py** command. This is the same command that we tested locally in *Step 3* of this exercise.

11. Open a new Terminal window (or Anaconda Prompt) in the **Exercise12.02** directory and enter the following command to create an image:

```
sudo docker build -t titanic .
```

This command will pull the base image (with Python 3.7) from Docker Hub and then follow the other instructions in the default Dockerfile (called **Dockerfile**). The . at the end of the command points to the current directory. The **-t** parameter specifies the name of our image, **titanic**. The command might take some time to complete. Once the script completes, this is the expected output:

```
Sending build context to Docker daemon  6.144kB
Step 1/8 : FROM python:3.7
3.7: Pulling from library/python
50e431f79093: Pull complete
dd8c6d374ea5: Pull complete
c85513200d84: Pull complete
55769680e827: Pull complete
f5e195d50b88: Pull complete
94cdd3612287: Pull complete
b45109600839: Pull complete
e0c7a90c35ea: Pull complete
97fb33d206b1: Pull complete
Digest: sha256:46e17eae392c2d99e0ac9c26e867807c5022958ec19a9cc46b9668962e0dff78
Status: Downloaded newer image for python:3.7
 ---> f66befd33669
Step 2/8 : RUN mkdir /api
 ---> Running in f268fcc68eab
Removing intermediate container f268fcc68eab
 ---> d949c7351bcc
Step 3/8 : WORKDIR /api
 ---> Running in e6e2e2427634
Removing intermediate container e6e2e2427634
 ---> 6365e3e4c6bc
Step 4/8 : ADD . /api/
 ---> 8a9ec669b1a0
Step 5/8 : RUN pip install -r ./requirements.txt
 ---> Running in 2505a2acc83b
Collecting Flask
  Downloading Flask-1.1.1-py2.py3-none-any.whl (94 kB)
Collecting itsdangerous>=0.24
  Downloading itsdangerous-1.1.0-py2.py3-none-any.whl (16 kB)
Collecting Werkzeug>=0.15
  Downloading Werkzeug-1.0.0-py2.py3-none-any.whl (298 kB)
Collecting Jinja2>=2.10.1
  Downloading Jinja2-2.11.1-py2.py3-none-any.whl (126 kB)
Collecting click>=5.1
  Downloading Click-7.0-py2.py3-none-any.whl (81 kB)
Collecting MarkupSafe>=0.23
  Downloading MarkupSafe-1.1.1-cp37-cp37m-manylinux1_x86_64.whl (27 kB)
Installing collected packages: itsdangerous, Werkzeug, MarkupSafe, Jinja2, click, Flask
Successfully installed Flask-1.1.1 Jinja2-2.11.1 MarkupSafe-1.1.1 Werkzeug-1.0.0 click-7.0 itsdangerous-1.1.0
Removing intermediate container 2505a2acc83b
 ---> 2193044bef15
Step 6/8 : EXPOSE 5000
 ---> Running in 50b6a43517fb
Removing intermediate container 50b6a43517fb
 ---> f8c3dbd5870c
Step 7/8 : ENV PYTHONPATH="$PYTHONPATH:/api"
 ---> Running in 4668e1e57e3a
Removing intermediate container 4668e1e57e3a
 ---> 66ea7a8cfa9e
Step 8/8 : CMD ["python", "/api/api.py"]
 ---> Running in 4165789341b9
Removing intermediate container 4165789341b9
 ---> d4d78bd64165
Successfully built d4d78bd64165
Successfully tagged titanic:latest
```

Figure 12.18: Building a Docker image

In the output, it becomes clear that all steps in our **Dockerfile** have been followed. The Flask library is loaded, port **5000** is exposed, and the API is running.

12. Let's check our local image repository again:

```
sudo docker image ls
```

Next to the **hello-world** image, you'll see the base Python **3.7** image and our Titanic API:

```
REPOSITORY        TAG          IMAGE ID          CREATED           SIZE
titanic           latest       d4d78bd64165      15 minutes ago    928MB
python            3.7          f66befd33669      10 days ago        919MB
hello-world       latest       fce289e99eb9      14 months ago      1.84kB
```

Figure 12.19: Checking stored images

It's great that we have a Docker image now, but that image has to be published to a Docker **registry** when we want to deploy it. Docker Hub is the central repository, but we don't want our Titanic API to end up there.

13. Create a local **registry** and publish our image there, ready to be deployed to Kubernetes in the next part of this exercise. Enter the following command in your Terminal:

```
docker run -d -p 6000:5000 --restart=always --name registry
registry:2
```

This will download the registry libraries and will run a local registry. The output is as follows:

```
--restart=always --name registry registry:2
Unable to find image 'registry:2' locally
2: Pulling from library/registry
486039affc0a: Pull complete
ba51a3b098e6: Pull complete
8bb4c43d6c8e: Pull complete
6f5f453e5f2d: Pull complete
42bc10b72f42: Pull complete
Digest: sha256:7d081088e4bfd632a88e3f3bcd9e007ef44a796fddfe3261407a3f9f04abe1e7
Status: Downloaded newer image for registry:2
98869ab76c12c518e80e86fd8a2c6d19333bc85c2d2fe20d4671577a4f7f1428
```

Figure 12.20: Creating a local registry

14. Tag your Docker image with the following command, to give it a suitable name:

```
docker tag titanic localhost:6000/titanic
```

15. Now, push the **titanic** image to the running local registry with the following command:

```
docker push localhost:6000/titanic
```

This generates the following output:

```
The push refers to repository [localhost:5000/titanic]
0a14e0dbee45: Pushed
6559eb7a04fd: Pushed
f83dca05cf96: Pushed
e77942735734: Pushed
69f872f2a048: Pushed
2de70d7dbfa9: Pushed
3dffd131f01f: Pushed
271910c4c150: Pushed
6670e930ed33: Pushed
c7f27a4eb870: Pushed
e70dfb4c3a48: Pushed
1c76bd0dc325: Pushed
latest: digest: sha256:3c0d9793acccdebb021848cf2ce8f4bd2d3ab2ab6e214ccf479ff68ac82be82a size: 2844
```

Figure 12.21: Pushing a Docker image to a local registry

16. To verify that your image is pushed to the local registry, enter the following command:

```
curl -X GET http://localhost:6000/v2/titanic/tags/list
```

If all is well, you'll see the image with the **latest** tag:

```
{"name":"titanic","tags":["latest"]}
```

17. Double-check that all is OK by entering the **docker image ls** and **docker ps** commands. You'll have a similar output:

```
 ~/Proj/g/The-Artificial-Intelligence-Infrastructure-Workshop/Chapter12/Exercise12.02   master !3 ?5  docker image ls
REPOSITORY                  TAG        IMAGE ID        CREATED        SIZE
192.168.64.1:5000/titanic   latest     bbcdbd7cc564    4 weeks ago    928MB
titanic                     latest     bbcdbd7cc564    4 weeks ago    928MB
localhost:5000/titanic      latest     bbcdbd7cc564    4 weeks ago    928MB
python                      3.7        f66befd33669    6 weeks ago    919MB
registry                    2          708bc6af7e5e    2 months ago   25.8MB
 ~/Proj/g/The-Artificial-Intelligence-Infrastructure-Workshop/Chapter12/Exercise12.02   master !3 ?5  docker ps
CONTAINER ID    IMAGE        COMMAND              CREATED          STATUS        PORTS                      NAMES
98869ab76c12    registry:2   "/entrypoint.sh /etc…"  2 minutes ago  Up 2 minutes  0.0.0.0:5000->5000/tcp     registry
```

Figure 12.22: Viewing the Docker images and containers

As you can see in the output, the **titanic** image is now also available as **localhost:6000/titanic**. The **registry** image is new, and it has a running container called **registry:2**. We have just successfully published our **titanic** image to that registry.

In the second part of this exercise, we'll use the Docker image to host a container on a Kubernetes cluster.

18. We first have to make sure that Minikube, the local version of Kubernetes, is running. Minikube is, in fact, a local virtual machine. Type the following command in a new Terminal or Anaconda Prompt:

```
minikube version
```

If all is OK, this will produce an output like **minikube version: v1.8.1**, along with a **commit** hashtag, as follows:

```
minikube version: v1.8.1
commit:   cbda04cf6bbe65e987ae52bb393c10099ab62014
```

19. Start **minikube** by entering the following command in a new Terminal window or Anaconda Prompt:

```
minikube start --insecure-registry="localhost:6000"
```

If you're running within a virtual machine like VirtualBox, the command will be **minikube start --driver=none**. If all goes well, you'll get the following output:

```
minikube v1.8.1 on Darwin 10.15.3
Using the hyperkit driver based on existing profile
Downloading VM boot image ...
Reconfiguring existing host ...
Starting existing hyperkit VM for "minikube" ...
Preparing Kubernetes v1.17.3 on Docker 19.03.6 ...
Launching Kubernetes ...
Enabling addons: default-storageclass, storage-provisioner
Done! kubectl is now configured to use "minikube"
/usr/local/bin/kubectl is version 1.15.5, and is incompatible with Kubernetes 1.17.3. You will need to update /usr/local/bin/kubectl or use
'minikube kubectl' to connect with this cluster
```

Figure 12.23: Starting Minikube

20. Confirm that Minikube is running by using the following command:

```
minikube status
```

You'll get a status update like the following:

```
host:  Running
kubelet:  Running
apiserver:  Running
kubeconfig:  Configured
```

If you want to look even further into your Kubernetes cluster, start up a dashboard with the following command:

```
minikube dashboard
```

This will open a browser window with a lot of useful information and configuration options:

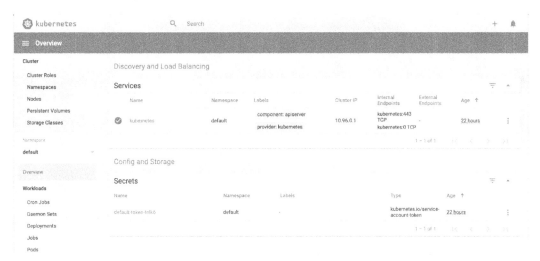

Figure 12.24: The Kubernetes dashboard

21. We have to connect our now-running Minikube Kubernetes cluster to a command-line interface, called **kubectl**. Type in the following command in your Terminal or Anaconda Prompt, as was already suggested in the previous output:

```
minikube kubectl
```

This will produce an output like the following:

```
  > kubectl.sha256: 65 B / 65 B [────────────────────] 100.00% ? p/s 0s
  > kubectl: 47.28 MiB / 47.28 MiB [───────────] 100.00% 1.25 MiB p/s 38s
kubectl controls the Kubernetes cluster manager.

 Find more information at: https://kubernetes.io/docs/reference/kubectl/overview/

Basic Commands (Beginner):
  create       Create a resource from a file or from stdin.
  expose       Take a replication controller, service, deployment or pod and expose it as a new Kubernetes Service
  run          Run a particular image on the cluster
  set          Set specific features on objects

Basic Commands (Intermediate):
  explain      Documentation of resources
  get          Display one or many resources
  edit         Edit a resource on the server
  delete       Delete resources by filenames, stdin, resources and names, or by resources and label selector
```

Figure 12.25: Connecting Minikube to your kubectl command

You can read the **kubectl controls the Kubernetes cluster manager** line, which indicates that we can now use the **kubectl** tool to give commands to our cluster. You can get more information about the cluster with the **kubectl version** and **kubectl get nodes** commands:

```
kubectl version                                              ✓  anaconda3  minikube *  15:02:12
Client Version: version.Info{Major:"1", Minor:"15", GitVersion:"v1.15.5", GitCommit:"20c265fef0741dd71a66480e35bd69f18351daea", Git
TreeState:"clean", BuildDate:"2019-10-15T19:16:51Z", GoVersion:"go1.12.10", Compiler:"gc", Platform:"darwin/amd64"}
Server Version: version.Info{Major:"1", Minor:"17", GitVersion:"v1.17.3", GitCommit:"06ad960bfd03b39c8310aaf92d1e7c12ce618213", Git
TreeState:"clean", BuildDate:"2020-02-11T18:07:13Z", GoVersion:"go1.13.6", Compiler:"gc", Platform:"linux/amd64"}
kubectl get nodes                                            ✓  anaconda3  minikube  15:24:05
NAME   STATUS   ROLES    AGE    VERSION
m01    Ready    master   27m    v1.17.3
```

Figure 12.26: Getting information from Minikube

22. We have already created a Docker image and practiced deploying that to a Docker registry in part 1 of this exercise. We'll rebuild the image now, but with the Docker daemon (process) of the Kubernetes cluster to make sure it lands on the Minikube cluster. Start by entering the following command:

```
eval $(minikube docker-env)
```

This command points our Terminal to use a different **docker** command, namely, the one in the Minikube environment.

23. Let's build the container again but now with Minikube. Enter the following command:

```
docker build -t titanic .
```

You'll get the same output as in part 1 of this exercise.

24. Now all that's left to do is to deploy our Docker image to the cluster and run it as a container. For this, we create a Kubernetes deployment with the following command:

```
kubectl run titanic --image=titanic --image-pull-policy=Never
```

If this is successful, you'll see the following output:

```
deployment.apps/titanic created
```

25. Verify the deployment of the container with the following command:

```
kubectl get deployments
```

This will produce a list of deployed containers:

```
NAME       READY    UP-TO-DATE    AVAILABLE    AGE
titanic    1/1      1             1            3m
```

Figure 12.27: Retrieving a list of deployed containers from Kubernetes

You can also check the Kubernetes dashboard if you have started it, and check whether the deployment and Pod have been created:

Figure 12.28: The Kubernetes dashboard

26. Now that our app is deployed, we want to connect to the Titanic API. By default, the running containers can only be accessed by other resources (containers) in the same Kubernetes Pod. There are several ways to connect to our app from the outside world, such as an Ingress and Load Balancer. To keep it simple, we will just instruct our cluster to forward the network traffic on port **5000** to our running Pod in the Minikube cluster:

```
kubectl port-forward titanic-6d8f58fc8b-znmx9 5000:5000
```

> **NOTE**
>
> In the preceding code, replace the name of the pod with your own; you can find it in the Kubernetes dashboard or by entering **kubectl get pods**

This will create a local task that forwards network traffic to the **titanic** Pod. You will have the following output:

```
Forwarding from 127.0.0.1:5000 -> 5000
Forwarding from [::1]:5000 -> 5000
```

27. To test the API, enter the following command in a new terminal window (while keeping the port forwarding running):

```
curl -X POST -H "Content-Type: application/json" -d '{"Pclass":
  2, "Sex": 1, "Age": 34, "SibSb": 1, "Parch": 1, "Fare": 5.99}'
  http://127.0.0.1:5000/survived
```

This sends a JSON string through the proxy to the running container in our Minikube Kubernetes cluster. If all goes well, you'll get the output of a prediction in the familiar form, in the same Terminal where you executed the **curl** command:

```
I predict that person [2, 1, 34, 1, 1, 5.99] has
survived the Titanic
```

> **NOTE**
>
> To access the source code for this specific section, please refer to https://packt.live/32s2PC3.

By completing this exercise, you have containerized a machine learning API and have published it to a Docker registry. You have also gained experience with deploying Docker images to a Kubernetes cluster. Together, this gives you the skills to publish a machine learning model in production.

In the next activity, you'll deploy a machine learning model to a Kubernetes cluster that predicts the class of a Titanic passenger.

ACTIVITY 12.02: DEPLOYING A MACHINE LEARNING MODEL TO A KUBERNETES CLUSTER TO PREDICT THE CLASS OF TITANIC PASSENGERS

In this activity, you will deploy the machine learning model that you created in *Activity 12.01, Predicting the Class of a Passenger on the Titanic*, to predict the passenger class of a person on board the Titanic to a Kubernetes cluster.

> **NOTE**
>
> The code for this activity can be found here:
>
> https://packt.live/3fDoH0E.

Perform the following steps to complete the activity:

1. To start, copy the **model.pkl** file from the **Chapter12/Activity12.01** directory into a new **Activity12.02** directory.

2. Create a new file in the **Activity12.02** folder called **api.py**.

3. Open a new Terminal or Anaconda Prompt and navigate to the **Activity12.02** folder. Enter the following command to check whether the API is working:

```
python api.py
```

If all is OK, you should get the message that your API is running on localhost.

4. Test the local API by opening a new Terminal or Anaconda Prompt and enter the following command:

```
curl -X POST -H "Content-Type: application/json" -d '{"Survived":
  0, "Sex": 1, "Age": 52, "SibSb": 1, "Parch": 0, "Fare": 82.35}'
  http://127.0.0.1:5000/class
```

5. Add a new **requirements.txt** file, open it in an editor, and enter the lines to import the **Flask**, **sklearn**, and **pandas** libraries.

6. Add a new file in the same directory and call it **Dockerfile**.

7. Enter the **docker build** command to create an image with the right tag. In the output, it should become clear that all steps in our **Dockerfile** have been followed. The Flask library is loaded, port **5000** is exposed, and the API is running.

8. Create a local Docker registry and publish the **titanic** image.

9. Tag your Docker image.

10. Push the **titanic** image to the running local registry.

11. To verify that your image is pushed to the local registry, enter the following command:

```
curl -X GET http://localhost:6000/v2/titanic/tags/list
```

12. Start and confirm **minikube** in a new Terminal window or Anaconda Prompt.

13. Startup a **minikube** dashboard.

14. Point your Terminal to the **docker** command of the Minikube cluster.

15. Build the container again but now with the Minikube daemon.

16. Create a Kubernetes **deployment**. If this is successful, you'll see the **deployment.apps/titanic created** statement.

17. Verify the deployment of the container by checking the Kubernetes dashboard.

18. Create a network connection that allows communication (via port forwarding) between the outside world and the resources in your Pod on the Minikube cluster.

19. Test the API using the following command:

```
curl -X POST -H "Content-Type: application/json" \
# -d '{"Survived": 1, "Sex": 0, "Age": 72, "SibSb": 2, "Parch":
  0, "Fare": 68.35}' http://127.0.0.1:5000/class
```

This will produce the following output:

```
I predict that person [1, 0, 72, 2, 0, 68.35] was
in passenger class [1] of the Titanic
```

You have now successfully dockerized an API with a machine learning model, and deployed the Docker image to a Kubernetes cluster. In the real world, this is a common approach to productionize software in a cloud environment.

> **NOTE**
>
> The solution to this activity can be found on page 661.

In the next section, we'll explore how to deploy machine learning models to an ever-running streaming application.

MODEL EXECUTION IN STREAMING DATA APPLICATIONS

In the first part of this chapter, you learned how to export models to the **pickle** format, to be used in an API. That is a good way to productionize models since the resulting microservices architecture is flexible and robust. However, calling an API across a network might not be the best-performing way to get a forecast. As we learned in *Chapter 2*, *Artificial Intelligence Storage Requirements*, latency is always an issue when working with high loads of event data. If you're processing thousands of events per second and have to execute a machine learning model for each event, your network and **pickle** file that's stored on disk might not be able to handle the load. So, in a similar way to how we cache data, we should cache models in memory as close to the data stream as possible. That way, we can reduce or even eliminate the network traffic and disk I/O. This technique is often used in high-velocity stream processing applications, for example, fraud detection in banks and real-time recommendation systems for websites.

There are several methods of caching models in memory. Some platforms offer built-in capabilities. For example, the H2O.ai platform has the option to export models to a POJO/MOJO binary object. Spark has its machine learning library called SparkML, which is quite extensive and easy to use. All these methods have one disadvantage: they require a lock-in with the platform. It's not possible to distribute a model from H2O.ai to DataBricks, or from Spark to DataIku. To enable this kind of flexibility, an intermediate format has to be picked as the "glue" between data scientists and data engineers that gives both practitioners the freedom to choose the tools they want. PMML is such a format, and we'll discuss it in the next section.

PMML

As this book is focused on open source standards, we have picked a popular intermediate model format that we can load in memory – PMML, short for Predictive Model Markup Language. A PMML file is an XML-based file that contains the input parameters, calculations of the algorithms, and output field of a model. Exporting a model to PMML is essentially a way of serializing a model, similar to how exporting to **pickle** works. A PMML file can be read quite easily by humans, as can be seen in the following figure:

```xml
<?xml version="1.0" encoding="UTF-8" standalone="yes"?>
<PMML xmlns="http://www.dmg.org/PMML-4_3" xmlns:data="http://jpmml.org/jpmml-model/InlineTable" version="4.3">
    <Header>
        <Application name="JPMML-SkLearn" version="1.5.32"/>
        <Timestamp>2020-03-08T14:37:37Z</Timestamp>
    </Header>
    <DataDictionary>
        <DataField name="y" optype="categorical" dataType="integer">
            <Value value="0"/>
            <Value value="1"/>
        </DataField>
        <DataField name="x1" optype="continuous" dataType="double"/>
        <DataField name="x2" optype="continuous" dataType="double"/>
        <DataField name="x3" optype="continuous" dataType="double"/>
        <DataField name="x4" optype="continuous" dataType="double"/>
        <DataField name="x5" optype="continuous" dataType="double"/>
        <DataField name="x6" optype="continuous" dataType="double"/>
    </DataDictionary>
    <RegressionModel functionName="classification" normalizationMethod="logit">
        <MiningSchema>
            <MiningField name="y" usageType="target"/>
            <MiningField name="x1"/>
            <MiningField name="x2"/>
            <MiningField name="x3"/>
            <MiningField name="x4"/>
            <MiningField name="x5"/>
            <MiningField name="x6"/>
        </MiningSchema>
        <Output>
            <OutputField name="probability(0)" optype="continuous" dataType="double" feature="probability" value="0"/>
            <OutputField name="probability(1)" optype="continuous" dataType="double" feature="probability" value="1"/>
        </Output>
        <RegressionTable intercept="1.7352745880935379" targetCategory="1">
            <NumericPredictor name="x1" coefficient="-0.9542728798036803"/>
            <NumericPredictor name="x2" coefficient="2.6453670664181863"/>
            <NumericPredictor name="x3" coefficient="-0.0342878148625247"/>
            <NumericPredictor name="x4" coefficient="-0.3292077163727886"/>
            <NumericPredictor name="x5" coefficient="-0.10430754400356529"/>
            <NumericPredictor name="x6" coefficient="0.003925079235425489"/>
        </RegressionTable>
        <RegressionTable intercept="0.0" targetCategory="0"/>
    </RegressionModel>
</PMML>
```

Figure 12.29: Example of a PMML file with a simple logistic regression model

The PMML format is maintained by the Data Mining Group, which is an independent, non-profit consortium of organizations. There is also a JSON-based format for models called **Portable Format for Analytics** (**PFA**). Since that format is still emerging and thus less mature than PMML, we will not discuss it further in this book. The next paragraph contains a short introduction to a popular stream processing framework, Apache Flink.

APACHE FLINK

Apache Flink is one of the most popular streaming engines, and for a good reason. It offers low latency and high throughput for streaming data processing and gives a lot of power to developers. Compared to Spark Structured Streaming, another popular stream processing framework, it offers more features and better performance. A Flink job is a Java application that can be executed on a local machine or within a cluster.

In the next exercise, you'll practice with PMML and Flink by creating a real-time stream processing application that includes a machine learning model.

EXERCISE 12.03: EXPORTING A MODEL TO PMML AND LOADING IT IN THE FLINK STREAM PROCESSING ENGINE FOR REAL-TIME EXECUTION

In this exercise, you'll create a simple machine learning model, export it to PMML, and load it in memory to be used in a data stream.

Before you begin, follow the installation instructions in the *Preface* to install Java, Maven, Netcat, and a suitable IDE (IntelliJ IDEA or Eclipse).

Perform the following steps to complete the exercise:

1. In the **Chapter12** directory, create the **Exercise12.03** directory to store the files for this exercise.

2. Open Jupyter Notebook and create a new Python 3 notebook called **export_ pmml** in the **Exercise12.03** folder.

3. Copy the **model.pkl** file from the **Exercise 12.01** directory to the **Exercise12.03** directory. This can be done manually, or with the following statement in a notebook:

   ```
   !cp ../Exercise12.01/model.pkl .
   ```

4. Enter the following line in the first cell of the new notebook:

   ```
   !pip install sklearn2pmml
   ```

This will install the required library, which can export our models to the PMML format:

```
WARNING: pip is being invoked by an old script wrapper. This will fail in a future version of pip.
Please see https://github.com/pypa/pip/issues/5599 for advice on fixing the underlying issue.
To avoid this problem you can invoke Python with '-m pip' instead of running pip directly.
Collecting sklearn2pmml
  Downloading sklearn2pmml-0.55.1.tar.gz (5.6 MB)
     |████████████████████████████████| 5.6 MB 4.1 MB/s eta 0:00:01
Requirement already satisfied: joblib>=0.13.0 in /home/bas/anaconda3/lib/python3.7/site-packages (from sklearn2pmm
l) (0.13.2)
Requirement already satisfied: scikit-learn>=0.18.0 in /home/bas/anaconda3/lib/python3.7/site-packages (from sklea
rn2pmml) (0.21.3)
Collecting sklearn-pandas>=0.0.10
  Downloading sklearn_pandas-1.8.0-py2.py3-none-any.whl (12 kB)
Requirement already satisfied: numpy>=1.11.0 in /home/bas/anaconda3/lib/python3.7/site-packages (from scikit-learn
>=0.18.0->sklearn2pmml) (1.17.2)
Requirement already satisfied: scipy>=0.17.0 in /home/bas/anaconda3/lib/python3.7/site-packages (from scikit-learn
>=0.18.0->sklearn2pmml) (1.3.1)
Requirement already satisfied: pandas>=0.11.0 in /home/bas/anaconda3/lib/python3.7/site-packages (from sklearn-pan
das>=0.0.10->sklearn2pmml) (1.0.1)
Requirement already satisfied: python-dateutil>=2.6.1 in /home/bas/anaconda3/lib/python3.7/site-packages (from pan
das>=0.11.0->sklearn-pandas>=0.0.10->sklearn2pmml) (2.8.0)
Requirement already satisfied: pytz>=2017.2 in /home/bas/anaconda3/lib/python3.7/site-packages (from pandas>=0.11.
0->sklearn-pandas>=0.0.10->sklearn2pmml) (2019.3)
Requirement already satisfied: six>=1.5 in /home/bas/anaconda3/lib/python3.7/site-packages (from python-dateutil>=
2.6.1->pandas>=0.11.0->sklearn-pandas>=0.0.10->sklearn2pmml) (1.12.0)
Building wheels for collected packages: sklearn2pmml
  Building wheel for sklearn2pmml (setup.py) ... done
  Created wheel for sklearn2pmml: filename=sklearn2pmml-0.55.1-py3-none-any.whl size=5636073 sha256=467d42831a6f06
ba4c2825110a63e7c4458c2fceffb284733e9fc88c26f7bd11
  Stored in directory: /home/bas/.cache/pip/wheels/d1/43/46/407e4a36a26f43a599e7ca185b894237be91355dbc8aa1ebde
Successfully built sklearn2pmml
Installing collected packages: sklearn-pandas, sklearn2pmml
Successfully installed sklearn-pandas-1.8.0 sklearn2pmml-0.55.1
```

Figure 12.30: Installation of sklearn2pmml

If **sklearn2pmml** is already installed, you will have the following output:

```
Requirement already satisfied: sklearn2pmml in /usr/local/anaconda3/lib/python3.7/site-packages (0.55.4)
Requirement already satisfied: joblib>=0.13.0 in /usr/local/anaconda3/lib/python3.7/site-packages (from sk
learn2pmml) (0.13.2)
Requirement already satisfied: scikit-learn>=0.18.0 in /usr/local/anaconda3/lib/python3.7/site-packages (f
rom sklearn2pmml) (0.21.3)
Requirement already satisfied: sklearn-pandas>=0.0.10 in /usr/local/anaconda3/lib/python3.7/site-packages
(from sklearn2pmml) (1.8.0)
Requirement already satisfied: numpy>=1.11.0 in /usr/local/anaconda3/lib/python3.7/site-packages (from sci
kit-learn>=0.18.0->sklearn2pmml) (1.17.2)
Requirement already satisfied: scipy>=0.17.0 in /usr/local/anaconda3/lib/python3.7/site-packages (from sci
kit-learn>=0.18.0->sklearn2pmml) (1.3.1)
Requirement already satisfied: pandas>=0.11.0 in /usr/local/anaconda3/lib/python3.7/site-packages (from sk
learn-pandas>=0.0.10->sklearn2pmml) (0.25.1)
Requirement already satisfied: python-dateutil>=2.6.1 in /usr/local/anaconda3/lib/python3.7/site-packages
(from pandas>=0.11.0->sklearn-pandas>=0.0.10->sklearn2pmml) (2.8.0)
Requirement already satisfied: pytz>=2017.2 in /usr/local/anaconda3/lib/python3.7/site-packages (from pand
as>=0.11.0->sklearn-pandas>=0.0.10->sklearn2pmml) (2019.3)
Requirement already satisfied: six>=1.5 in /usr/local/anaconda3/lib/python3.7/site-packages (from python-d
ateutil>=2.6.1->pandas>=0.11.0->sklearn-pandas>=0.0.10->sklearn2pmml) (1.12.0)
```

Figure 12.31: sklearn2pmml already installed

5. Create a new cell and enter the following lines to start using the **pickle** and **sklearn2pmml** libraries:

```
from sklearn2pmml import sklearn2pmml, make_pmml_pipeline
import pickle
```

6. Deserialize the pickled model by executing the following code in a new cell:

```
file = open('model.pkl', 'rb')   # read bytes
model = pickle.load(file)
file.close()
```

Now we can export the model to the PMML format.

7. Enter the following lines in a new cell to create a **pipeline**, export the method, and write it to the PMML file:

```
pmml_pipeline = make_pmml_pipeline(model)
sklearn2pmml(pmml_pipeline, 'titanic.pmml')
```

First, we create a **pipeline** model, which is another representation of the model. Then, we call the **sklearn2pmml** method that performs the export and writes the resulting PMML file (**titanic.pmml**) to disk.

8. Check the contents of the generated **titanic.pmml** file by entering the following line in a new cell:

```
!cat titanic.pmml
```

This will produce the following output:

```xml
<?xml version="1.0" encoding="UTF-8" standalone="yes"?>
<PMML xmlns="http://www.dmg.org/PMML-4_3" xmlns:data="http://jpmml.org/jpmml-model/InlineTable" version="4.3">
    <Header>
        <Application name="JPMML-SkLearn" version="1.5.32"/>
        <Timestamp>2020-03-08T14:31:43Z</Timestamp>
    </Header>
    <DataDictionary>
        <DataField name="y" optype="categorical" dataType="integer">
            <Value value="0"/>
            <Value value="1"/>
        </DataField>
        <DataField name="x1" optype="continuous" dataType="double"/>
        <DataField name="x2" optype="continuous" dataType="double"/>
        <DataField name="x3" optype="continuous" dataType="double"/>
        <DataField name="x4" optype="continuous" dataType="double"/>
        <DataField name="x5" optype="continuous" dataType="double"/>
        <DataField name="x6" optype="continuous" dataType="double"/>
    </DataDictionary>
    <RegressionModel functionName="classification" normalizationMethod="logit">
        <MiningSchema>
            <MiningField name="y" usageType="target"/>
            <MiningField name="x1"/>
            <MiningField name="x2"/>
            <MiningField name="x3"/>
            <MiningField name="x4"/>
            <MiningField name="x5"/>
            <MiningField name="x6"/>
        </MiningSchema>
        <Output>
            <OutputField name="probability(0)" optype="continuous" dataType="double" feature="probabil
ity" value="0"/>
            <OutputField name="probability(1)" optype="continuous" dataType="double" feature="probabil
ity" value="1"/>
        </Output>
        <RegressionTable intercept="1.7352745880935379" targetCategory="1">
            <NumericPredictor name="x1" coefficient="-0.9542728798036803"/>
            <NumericPredictor name="x2" coefficient="2.64536706664181863"/>
            <NumericPredictor name="x3" coefficient="-0.0342878148625247"/>
            <NumericPredictor name="x4" coefficient="-0.3292077163727886"/>
            <NumericPredictor name="x5" coefficient="-0.10430754400356529"/>
            <NumericPredictor name="x6" coefficient="0.003925079235425489"/>
        </RegressionTable>
        <RegressionTable intercept="0.0" targetCategory="0"/>
    </RegressionModel>
</PMML>
```

Figure 12.32: Viewing the contents of a PMML model

As you can see, the PMML file is an XML-structured file that is quite easy to read. The input and output fields are listed, and the type of model becomes clear from the **<RegressionModel>** tag.

Now let's create a streaming job. We'll use Apache Flink for this. We'll write the code with an example from IntelliJ IDEA. You can also choose another IDE. It's also possible (though not recommended) to use a plain text editor and run the code from a Terminal or command line.

9. Open a new Terminal window in the **Exercise12.03** directory and enter the following lines to set up Apache Flink:

```
mvn archetype:generate                                      \
        -DarchetypeGroupId=org.apache.flink                 \
        -DarchetypeArtifactId=flink-quickstart-java         \
        -DarchetypeVersion=1.10.0
```

10. During the installation, enter the following values:

```
groupId: com
artifactId: titanic
version: 0.0.1
package: packt
```

This will generate a project from a template.

11. Open your favorite IDE (IntelliJ IDEA, Eclipse, or another) and import the generated files by selecting the **pom.xml** file (if you use Eclipse, the m2e plugin allows you to import Maven projects) as shown in the following figure:

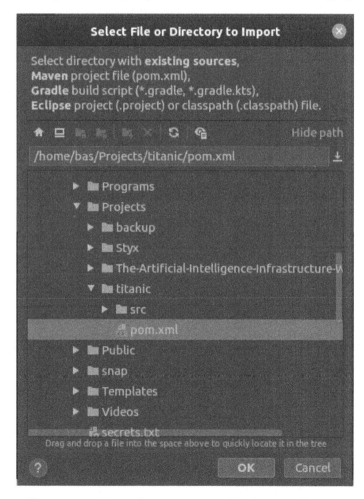

Figure 12.33: Importing a Maven project into IntelliJ IDEA

12. The Flink project that was generated by running the **mvn archetype:generate** command contains two Flink jobs: one for streaming data, and one for batch processing. Since we don't need the one for batch processing in this exercise, remove the **src/main/java/packt/BatchJob.java** file. Copy the **titanic.pmml** file that we created in *Step 8* to the **titanic/src/resources** directory. You'll see the following file structure:

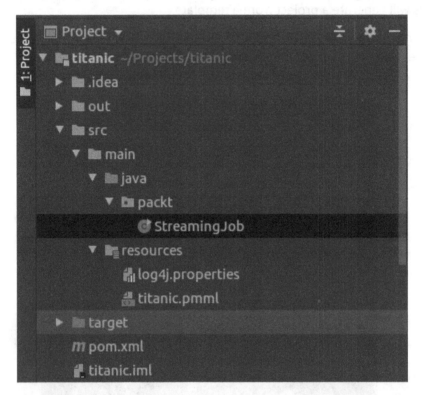

Figure 12.34: File structure of the generated Flink project

13. If you installed JDK and Maven, you should be able to compile the code. Type the following in your Terminal to compile the code and package it into a JAR file:

```
mvn clean package
```

This produces the following output:

```
[INFO] Scanning for projects...
[INFO]
[INFO] ------------------------< com:titanic >------------------------
[INFO] Building Flink Quickstart Job 0.0.1
[INFO] --------------------------[ jar ]-------------------------------
[INFO]
[INFO] --- maven-clean-plugin:2.5:clean (default-clean) @ titanic ---
[INFO] Deleting /home/bas/Projects/titanic/target
[INFO]
[INFO] --- maven-resources-plugin:2.6:resources (default-resources) @ titanic ---
[INFO] Using 'UTF-8' encoding to copy filtered resources.
[INFO] Copying 1 resource
[INFO]
[INFO] --- maven-compiler-plugin:3.1:compile (default-compile) @ titanic ---
[INFO] Changes detected - recompiling the module!
[INFO] Compiling 2 source files to /home/bas/Projects/titanic/target/classes
[INFO]
[INFO] --- maven-resources-plugin:2.6:testResources (default-testResources) @ titanic ---
[INFO] Using 'UTF-8' encoding to copy filtered resources.
[INFO] skip non existing resourceDirectory /home/bas/Projects/titanic/src/test/resources
[INFO]
[INFO] --- maven-compiler-plugin:3.1:testCompile (default-testCompile) @ titanic ---
[INFO] No sources to compile
[INFO]
[INFO] --- maven-surefire-plugin:2.12.4:test (default-test) @ titanic ---
[INFO] No tests to run.
[INFO]
[INFO] --- maven-jar-plugin:2.4:jar (default-jar) @ titanic ---
[INFO] Building jar: /home/bas/Projects/titanic/target/titanic-0.0.1.jar
[INFO]
[INFO] --- maven-shade-plugin:3.1.1:shade (default) @ titanic ---
[INFO] Excluding org.slf4j:slf4j-api:jar:1.7.15 from the shaded jar.
[INFO] Excluding org.slf4j:slf4j-log4j12:jar:1.7.7 from the shaded jar.
[INFO] Excluding log4j:log4j:jar:1.2.17 from the shaded jar.
[INFO] Replacing original artifact with shaded artifact.
[INFO] Replacing /home/bas/Projects/titanic/target/titanic-0.0.1.jar with /home/bas/Projects/titanic/target/titanic-0.0.1-shaded.jar
[INFO] -----------------------------------------------------------------
[INFO] BUILD SUCCESS
[INFO] -----------------------------------------------------------------
[INFO] Total time:  3.976 s
[INFO] Finished at: 2020-03-08T16:26:34+01:00
[INFO] -----------------------------------------------------------------
```

Figure 12.35: Output of the Maven package command

In the output, you'll see **BUILD SUCCESS**, which indicates that you now have a working Java program.

Let's start by testing the job – first, the Maven template-generated code that can be deployed to a Flink cluster.

14. Since we want to test locally first, let's change the configuration a bit. Open the **pom.xml** file and remove the **<provided>** tags in the dependencies of **flink-java** and **flink-streaming-java**; these tags are single lines within the **<dependency>** elements. The file will look like the following:

```
<dependencies>
    <!-- Apache Flink dependencies -->
    <!-- These dependencies are provided, because they should not be packaged into the JAR file. -->
    <dependency>
        <groupId>org.apache.flink</groupId>
        <artifactId>flink-java</artifactId>
        <version>${flink.version}</version>
    </dependency>
    <dependency>
        <groupId>org.apache.flink</groupId>
        <artifactId>flink-streaming-java_${scala.binary.version}</artifactId>
        <version>${flink.version}</version>
    </dependency>

    <!-- Add connector dependencies here. They must be in the default scope (compile). -->
```

Figure 12.36: The pom.xml file with provided tags removed for Flink dependencies

Save the file, then import the Maven changes.

15. Open the **StreamingJob.java** file in your IDE and add the following lines at the top:

```
import org.apache.flink.api.common.\
        serialization.SimpleStringEncoder;
import org.apache.flink.core.fs.Path;
import org.apache.flink.streaming.api.\
        datastream.DataStream;
import org.apache.flink.api.common.\
        functions.MapFunction;
import org.apache.flink.streaming.api.datastream.\
        SingleOutputStreamOperator;
import org.apache.flink.streaming.api.environment.\
        StreamExecutionEnvironment;
import org.apache.flink.streaming.api.functions.\
        sink.filesystem.StreamingFileSink;
```

These lines will add the necessary Flink libraries to our **class** file.

16. In the middle part of the file, within the **main** function, add the following lines:

```
DataStream<String> dataStream = env.socketTextStream(\
                        "localhost", 1234, "\n");

StreamingFileSink<String> sink = StreamingFileSink
    .forRowFormat(new Path("out"), \
                new SimpleStringEncoder<String>("UTF-8"))
    .build();

dataStream.addSink(sink);
```

This code sets up a data stream that listens to a local socket on port **1234**. It takes the lines and writes (sinks) the lines to a file in the **out** directory.

17. To write to a local socket, we use the Netcat tool. Test the simple code by opening a Terminal and typing the following command:

```
nc -l -p 1234
```

You get a prompt to enter lines. Leave it open for now.

18. Type the following in a Terminal to build and run the code:

```
mvn clean package
mvn exec:java -Dexec.mainClass="packt.StreamingJob"
```

These lines are the Maven instructions to compile the code, package it into a JAR file, and run the file with the entry point in the **packt.StreamingJob** class that contains our **main** function.

19. In the Terminal that's still running Netcat, type the following lines of text followed by *Enter* for each line:

```
hello!
this
is a test
to check
if Flink is writing
data to a file
```

You should get the following output:

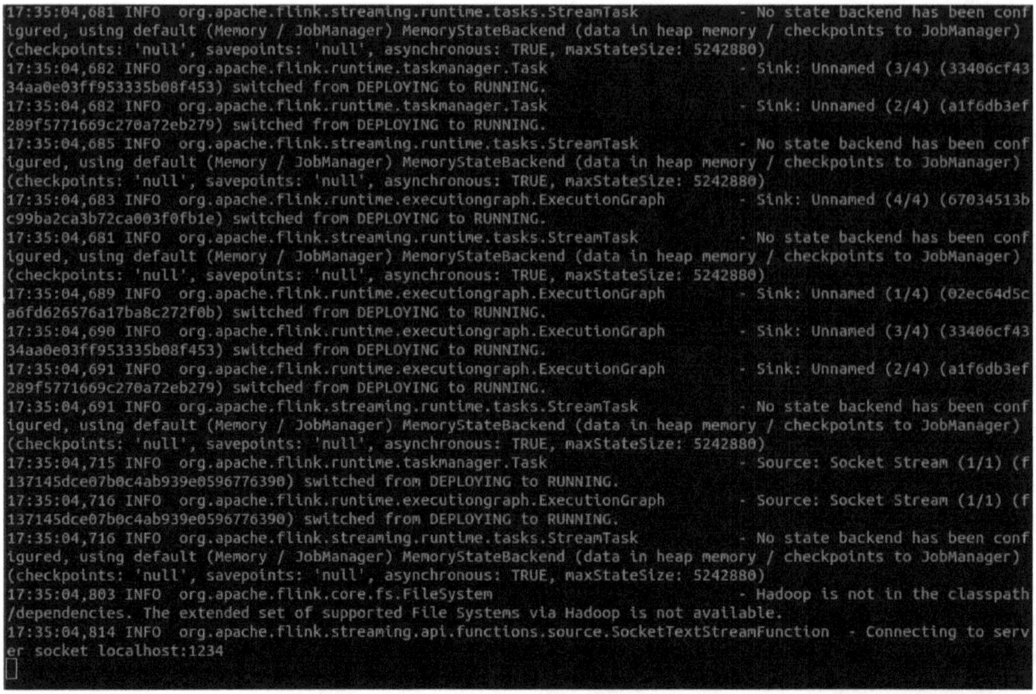

Figure 12.37: Running a Flink job that processes input from a local socket

20. Now check the output in the **titanic_results** directory. You should see a subfolder with a date and time, and a file for each line that was entered in the input socket:

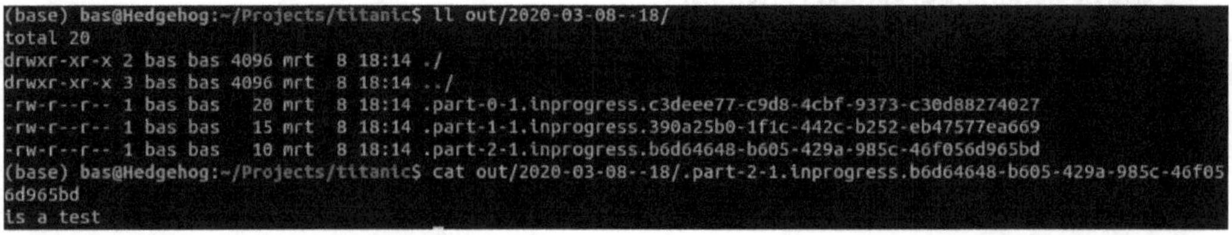

Figure 12.38: Checking the file contents that were produced by the Flink job

21. Let's make our application a bit more interesting by using input for our machine learning model. We have to add some dependencies to our project for this. Add the following lines to **pom.xml** and import the Maven changes:

```
<dependency>
    <groupId>org.jpmml</groupId>
    <artifactId>pmml-evaluator</artifactId>
```

```
    <version>1.4.15</version>
</dependency>
<dependency>
    <groupId>org.jpmml</groupId>
    <artifactId>pmml-evaluator-extension</artifactId>
    <version>1.4.15</version>
</dependency>
```

22. Edit the **StreamingJob.java** file again and add the following lines at the top of the file:

```
import org.dmg.pmml.FieldName;
import org.jpmml.evaluator.*;
import org.jpmml.evaluator.\
        visitors.DefaultVisitorBattery;
import java.io.File;
import java.util.LinkedHashMap;
import java.util.List;
import java.util.Map;
```

These lines import the required PMML and common Java libraries for working with PMML files.

23. Enter the following lines at the beginning of the **main** method:

```
// prepare PMML evaluation
ClassLoader classloader= StreamingJob.class\
                        .getClassLoader();

Evaluator evaluator= new LoadingModelEvaluatorBuilder()\
                    .setLocatable(false)
    .setVisitors(new DefaultVisitorBattery())\
                .load(newFile(classLoader.getResource(\
                "titanic.pmml").getFile())).build();

List<? extends InputField> inputFields = \
                        evaluator.getInputFields();
```

This code deserializes the model from PMML, loading the model in memory ready to be executed. The list of input fields will come in handy in the next step.

We are now ready to parse incoming messages as persons that can be evaluated according to whether they survived the Titanic disaster. The input values will be in a comma-separated string with the values "class, sex, age, number of siblings on board, number of parents on board, and fare paid". For example, "3,0,28,2,0,9.45" indicates a man of 28 years old who entered the ship with his two children.

24. The PMML evaluator expects a map of **FieldName** value objects, so we first have to convert the input strings. To do so, add the following lines to the class:

```
SingleOutputStreamOperator<String> mapped = \
    dataStream.map(new MapFunction<String, String>() {
  @Override
  public String map(String s) throws Exception {
      System.out.println("EVENT: " + s);

      Map<FieldName, FieldValue> arguments = new
        LinkedHashMap<>();
      String[] values = s.split(",");


      // prepare model evaluation
      for (int i = 0; i < values.length; i++) {
          FieldName inputName = inputFields.get(i).getName();
          FieldValue inputValue = inputFields.get(i)\
                                  .prepare(values[i]);
          arguments.put(inputName, inputValue);
      }


      // execute the model
      Map<FieldName, ?> results = evaluator.\
                                  evaluate(arguments);


      // Decoupling results from the
      //JPMML-Evaluator runtime environment
      Map<String, ?> resultRecord =
        EvaluatorUtil.decodeAll(results);

      System.out.println(resultRecord);
      return s;
  }
});
```

This code adds a **map** method to the stream processing job that splits the input string and uses the resulting string values to build a set of arguments for the machine learning model.

25. Now, if you run this code together with Netcat for the input socket, you can get real-time predictions from the model. Enter the following set of comma-separated values in the **nc** prompt that is already running, followed by *Enter*, and see what predictions are made:

```
3,0,28,2,0,9.45
1,1,12,0,2,3.25
2,1,72,1,1,4.62
```

You should get the following output:

```
EVENT: 3,0,28,2,0,9.45
{y=0, probability(0)=0.9375513143323562, probability(1)=0.0624486856676438}
EVENT: 1,1,12,0,2,3.25
{y=1, probability(0)=0.05630304805805353, probability(1)=0.9436969519419465}
EVENT: 2,1,72,1,1,4.62
{y=0, probability(0)=0.6015693023326556, probability(1)=0.39843069766734446}
```

Figure 12.39: Executing a machine learning model that was imported
in a Flink job from PMML

The Flink job that you have created and that now runs on your local machine is an example of a stream processing application with a machine learning model inside. The machine learning model is imported from a static PMML file and resides in the memory of the streaming job. This is an efficient way to work with models in stream processing.

> **NOTE**
>
> To access the source code for this specific section, please refer to https://packt.live/38SssNr.

By completing this exercise, you have created a stream processing job that evaluates a machine learning model in real time. The model is deserialized from the PMML format and sinks output to a local filesystem. The resulting application can predict in real time whether passengers of the Titanic would have survived the disaster of 1912. Although this real-time use case might not be so useful, you have now gained substantial practice with setting upstream processing software that can execute machine learning models.

In the next activity, you'll build a similar stream processing application that predicts the class of Titanic passengers.

ACTIVITY 12.03: PREDICTING THE CLASS OF TITANIC PASSENGERS IN REAL TIME

In this activity, we'll create a streaming job that processes events from a local socket and produces a prediction from a machine learning model.

> **NOTE**
>
> The code for this activity can be found here:
>
> https://packt.live/2Cu8wV4.

Perform the following steps:

1. Copy the model that you created in **Activity12.01** (**model.pkl**) to the **Activity12.03** directory.

2. Open Jupyter and create a new Python 3 notebook in the **Activity12.03** folder.

3. In the first cell of the new notebook, install the **sklearn2pmml** library with the **pip** tool.

4. Create a new cell and enter the **import** lines to start using the **pickle** and **sklearn2pmml** libraries.

5. Deserialize the pickled model by using the **open** and **pickle.load** functions in a new cell.

6. Export the model to the PMML format by creating a PMML pipeline, followed by a call to the **sklearn2pmml** method, which performs the export and writes the resulting PMML file to disk in a file called **titanic_class.pmml**.

7. Check the contents of the generated **titanic_class.pmml** file by entering the following line in a new cell:

```
!cat titanic_class.pmml
```

This will produce the following output:

```
<?xml version="1.0" encoding="UTF-8" standalone="yes"?>
<PMML xmlns="http://www.dmg.org/PMML-4_3" xmlns:data="http://jpmml.org/jpmml-model/InlineTable" version="4.3">
    <Header>
            <Application name="JPMML-SkLearn" version="1.5.32"/>
            <Timestamp>2020-03-09T20:11:55Z</Timestamp>
    </Header>
    <DataDictionary>
            <DataField name="y" optype="categorical" dataType="integer">
                    <Value value="1"/>
                    <Value value="2"/>
                    <Value value="3"/>
            </DataField>
            <DataField name="x1" optype="continuous" dataType="double"/>
            <DataField name="x2" optype="continuous" dataType="double"/>
            <DataField name="x3" optype="continuous" dataType="double"/>
            <DataField name="x4" optype="continuous" dataType="double"/>
            <DataField name="x5" optype="continuous" dataType="double"/>
            <DataField name="x6" optype="continuous" dataType="double"/>
    </DataDictionary>
```

Figure 12.40: Viewing a PMML model in a Jupyter Notebook

8. Open a new Terminal window and set up Apache Flink.

9. During the installation, enter the following values:

```
groupId: com
artifactId: titanic_class
version: 0.0.1
package: packt
```

This will generate a project from a template.

10. Open your favorite IDE (IntelliJ IDEA, Eclipse, or another) and import the generated files by selecting **pom.xml**.

11. Remove the **BatchJob.java** file since we don't need it.

12. Copy the **titanic_class.pmml** file that we created in *Step 8* to the **titanic_class/src/resources** directory.

13. Compile and build the code using the **maven** command. This produces an output that ends with **BUILD SUCCESS**.

14. Open the **pom.xml** file and remove the **<provided>** tags in the dependencies of **flink-java** and **flink-streaming-java**. Save the file, then import the Maven changes.

15. Open the **StreamingJob.java** file and add the **import** code for Flink.

16. In the middle part of the file, within the **main** function, add code to connect the Flink job to a local socket on port **1234**. Use the **socketTextStream** function of the existing **env** object (the Flink **StreamExecutionEnvironment** context).

17. Next, add the code just after that to write (sink) the events to a file in the **out** directory.

18. To write to a local socket, we use the Netcat tool. Test the simple code by opening a Terminal and typing the following command:

```
nc -l -p 1234
```

19. Go back to your IDE and run the **StreamingJob.main()** method. Alternatively, run the code from the Terminal.

20. In the Terminal that still runs Netcat, type a few lines of text.

21. Now check the output in the **out** directory. You should see a subfolder with a date and time, and a file for each line that was entered in the input socket.

22. Modify the **pom.xml** file to add the **JPMML** dependencies to the project, then import the Maven changes.

23. Edit the **StreamingJob.java** file again and import **jpmml.evaluator**.

24. Write the code to deserialize the model from PMML, loading the model in memory ready to be executed. Use the **LoadingModelEvaluatorBuilder** class of the **JPMML** library for this.

25. Add a **map** method to the stream processing job that splits the input string, and uses the resulting string values to build a set of arguments for the machine learning model.

26. Build and run the code of the Flink job together with the Netcat tool for the input socket. Enter a set of comma-separated values in the **nc** prompt and see which predictions are made:

```
1,1,13,1,56.91
0,0,81,0,0,120.96
1,0,41,1,1,18.11
```

You should get the following output:

```
EVENT: 1,1,13,3,1,56.91
{y=2, probability(1)=0.05043057638902439, probability(2)=0.6529580945632668, probability(3)=0.29661132904770876}
EVENT: 0,0,81,0,0,120.96
{y=1, probability(1)=0.9528716249898004, probability(2)=0.047128334498293026, probability(3)=4.051190659187697E-8}
EVENT: 1,0,41,1,1,18.11
{y=3, probability(1)=0.02728000899651477, probability(2)=0.32434910004056755, probability(3)=0.6483708909629177}
```

Figure 12.41: Output of the Flink job that predicts the class of Titanic passengers

By completing this activity, you have built a streaming job with Flink that can generate predictions in real time.

> **NOTE**
>
> The solution to this activity can be found on page 670.

SUMMARY

With this chapter, you have completed the entire book. That means you now have a thorough understanding of the infrastructure of AI systems. We have covered a tremendous number of topics in the book and supplied a great number of exercises and activities for you to follow. Let's have a short recap of all the topics in the book.

In *Chapter 1, Data Storage Fundamentals*, we started with the basics – the chapter covered data storage fundamentals. You learned about AI and machine learning in general, and we used text classification as an example of a machine learning model. *Chapter 2, Artificial Intelligence Storage Requirements*, was about requirements and covered a great number of concepts in depth. For every data storage layer in a data lake, you learned about the specific requirements and methods to store and retrieve data at scale. We addressed security, scalability, and various other aspects of building great data-driven systems. In *Chapter 3, Data Preparation*, the data preparation, and processing techniques that are needed to transform data were evaluated. You learned about ETL and ELT, data cleaning, filtering, aggregating, and feature engineering. You also practiced streaming event data processing using Apache Spark.

Chapter 4, Ethics of AI Data Storage, was a less technical chapter, but perhaps even more important than the solely technical ones. The main topic was the ethics of AI storage. We explored a few famous case studies where the ethics of AI were under discussion. You learned about bias and other prejudice in data and models, which gave you a good basis to start any conversation on these topics.

In *Chapter 5, Data Stores: SQL and NoSQL Databases*, we did a deep dive into databases. You learned about SQL and NoSQL databases; the differences, use cases, best practices, and query languages. By doing some hands-on exercises with technologies such as MySQL, Cassandra, and MongoDB, you learned how to store data in any type of database in the historical and analytics data layers of your data lake. *Chapter 6, Big Data File Formats*, followed up on this by exploring the file format for big data. You practiced with CSV, JSON, Parquet, and Avro to get a broad perspective on data formats.

In *Chapter 7, Introduction to Analytics Engine (Spark) for Big Data*, we moved from storing data to the analysis of data. You learned a lot about Apache Spark, one of the most popular data processing engines available. This knowledge comes in handy when discussing the design of data systems, as we did in *Chapter 8, Data System Design Examples*. Starting from a historical perspective and following up on the requirements of *Chapter 2, Artificial Intelligence Storage Requirements*, this chapter explored the components of system design and addressed hardware, architecture, data pipelines, security, scaling, and much more. By completing this chapter, you gained more experience in the architecture and design of AI systems.

Chapter 9, Workflow Management for AI, contained an in-depth overview of workflow management. You practiced with several techniques, from simple Python and Bash scripts to Apache Airflow for sophisticated workflow management systems.

In *Chapter 10, Introduction to Data Storage on Cloud Services (AWS)*, we moved on to cloud-based storage for AI systems. We used AWS to explain the concepts and technology that come with storing data in the cloud. *Chapter 11, Building an Artificial Intelligence Algorithm*, was a real hands-on chapter where you could practice model building and training, and finally, *Chapter 12, Productionizing Your AI Applications*, explored some techniques to put a machine learning model into production: building an API, running a Docker image in Kubernetes, and serializing the model to PMML to use in a data stream.

We hope that you learned a lot by reading this book. We aimed to provide a good mixture of reading content, hands-on exercises, and fun. We believe that by completing this book we have provided a solid basis for anyone who wants to build AI systems. We have drawn from our own experience, the open-source community, countless colleagues, and other people who have shared their knowledge. We hope you have enjoyed it. Please let us know about your experience with this book; we're eager to hear from you and we'll be happy to receive any feedback. Finally, we would like to thank you – thanks for staying with us for all these chapters, thanks for your attention, and thanks for sharing your feedback. We wish you all the best in your careers and good luck with building awesome AI systems.

APPENDIX

CHAPTER 1: DATA STORAGE FUNDAMENTALS

ACTIVITY 1.01: CREATING A TEXT CLASSIFIER FOR MOVIE REVIEWS

Solution

1. Create a new directory, **Activity01.01**, in the **Chapter01** directory to store the files for this activity.

2. Move the **aclImdb** folder to the **Datasets** directory.

3. Open your Terminal (macOS or Linux) or Command Prompt (Windows), navigate to the **Chapter01** directory, and type **jupyter notebook**.

4. In the Jupyter notebook, click the **Activity01.01** directory and create a new notebook file with a Python3 kernel.

5. Import the **os** library and a **random** library, and define where our training and test data is stored using four variables, as shown in the following code:

```
import os
import random


dataset_train_pos_path = "../Datasets/aclImdb/train/pos/"
dataset_train_neg_path = "../Datasets/aclImdb/train/neg/"

dataset_test_pos_path = "../Datasets/aclImdb/test/pos/"
dataset_test_neg_path = "../Datasets/aclImdb/test/neg/"
```

We have four variables: one for **training_positive**, one for **training_negative**, one for **test_positive**, and one for **test_negative**, each pointing at the respective dataset subdirectory.

6. Define a **read_dataset** function that reads the contents of each file in the given directory and adds these contents into a data structure, as shown in the following code:

```
contents_labels = [('this is the text from one '\
                   'of the files', 'pos'), \
                   ('this is another text', 'pos')]
def read_dataset(dataset_path, label):
    contents_labels = []
    files = os.listdir(dataset_path)
```

```
    for fn in files:
        path = os.path.join(dataset_path, fn)
        with open(path) as f:
            s = f.read()
            contents_labels.append((s, label))
    return contents_labels
```

The **read_dataset** function takes a path to a dataset and a label (either **pos** or **neg**), reads the contents of each file in the given directory, and adds these contents into a data structure that is a list of tuples. Each tuple contains both the text of the file, and the label **pos** or **neg**.

7. Use the **read_dataset** function to read each dataset into its own variable, as shown in the following code:

```
train_pos = read_dataset(dataset_train_pos_path, "pos")
train_neg = read_dataset(dataset_train_neg_path, "neg")

test_pos = read_dataset(dataset_test_pos_path, "pos")
test_neg = read_dataset(dataset_test_neg_path, "neg")
```

We have four variables in total: **train_pos**, **train_neg**, **test_pos**, and **test_neg**, each one of which is a list of tuples containing the relative text and labels.

8. Combine the **train_pos** with **train_neg** datasets and **test_pos** with **test_neg**, as shown in the following code:

```
train = train_pos + train_neg
test = test_pos + test_neg
```

We combined our positive and negative examples into a single dataset so that we can train an algorithm to discriminate between the two classes.

9. Use the **random.shuffle** function to shuffle the train and test datasets separately:

```
random.shuffle(train)
random.shuffle(test)
```

This gives us datasets where the training data is mixed up, instead of feeding all the positive and then all the negative examples to the classifier in order.

10. Split each of the train and test datasets back into data and labels respectively, as shown in the following code:

```
train_data, y_train = zip(*train)
test_data, y_test = zip(*test)
```

You should have four variables again called **train_data**, **y_train**, **test_data**, and **y_test**, where the **y** prefix indicates that the respective array contains labels.

11. Vectorize the training and test sets using **TfidfVectorizer** and output the dimensions of the vectors along with the vectorization time, as shown in the following code:

```
%%time
from sklearn.feature_extraction.text \
import TfidfVectorizer

vectorizer = TfidfVectorizer()
X_train = vectorizer.fit_transform(train_data)
X_test = vectorizer.transform(test_data)
print("The dimensions of our vectors:")
print(X_train.shape)
print(«- - -»)
```

You should get the following output:

```
The dimensions of our vectors:
(25000, 74849)

- - -

CPU times: user 13.4 s, sys: 440 ms, total: 13.8 s
Wall time: 14.7 s
```

We import **TfidfVectorizer** from **sklearn**, initialize an instance of it, fit the vectorizer on the training data, and vectorize both the training and testing data into **X_train** and **X_test** variables respectively. Lastly, we calculate the time it takes and prints out the shape of the training vectors at the end.

12. Initialize a **LinearSVC** model and fit it to the training data:

```
%%time

from sklearn.svm import LinearSVC

svm_classifier = LinearSVC()
svm_classifier.fit(X_train, y_train)

predictions = svm_classifier.predict(X_test)
```

You should get the following output:

```
CPU times: user 799 ms, sys: 63 ms, total: 862 ms
Wall time: 1.17 s
```

We imported the **LinearSVC** classifier and trained it on our training data. Then we generated predictions for every label in our test set.

13. Import **accuracy_score** and **classification_report** from **sklearn** and calculate the results of your predictions using each:

```
from sklearn.metrics import accuracy_score,
   classification_report

print("Accuracy: {}\n"\
      .format(accuracy_score(y_test, predictions)))
print(classification_report(y_test, predictions))
```

You should get the following output:

```
Accuracy: 0.8772

              precision    recall  f1-score   support

         neg       0.87      0.89      0.88     12500
         pos       0.89      0.87      0.88     12500

    accuracy                           0.88     25000
   macro avg       0.88      0.88      0.88     25000
weighted avg       0.88      0.88      0.88     25000
```

Figure 1.12: Results – accuracy and full report

Now, let's see how the classifier performs when we feed it with data on different topics.

14. Create two restaurant reviews as shown in the following code:

```
good_review = "The restaurant was really great! "\
             "I ate wonderful food and had "\
             "a very good time"
bad_review = "The restaurant was awful. "\
             "The staff were rude and the "\
             "food was horrible. I hated it"
```

15. Now vectorize each using the same vectorizer and generate predictions for whether each one is negative or positive:

```
restuarant_reviews = [good_review, bad_review]
vectors = vectorizer.transform(restuarant_reviews)
print(svm_classifier.predict(vectors))
```

You should get the following output:

```
['pos' 'neg']
```

> **NOTE**
>
> To access the source code for this specific section, please refer to https://packt.live/3fq2Hqg.

CHAPTER 2: ARTIFICIAL INTELLIGENCE STORAGE REQUIREMENTS

ACTIVITY 2.01: REQUIREMENTS ENGINEERING FOR A DATA-DRIVEN APPLICATION

Solution

1. Taxi data (GPS locations, current rides), HR system data (drivers), map data, phone calls, and email records with customer interaction, website, and app input (queries for rides, page visits).

   The layers for the solution are as follows:

Layer	Use of the Layer
Raw data layer	Copies of the data files from source systems such as HR data.
Historical data layer	Archived data in a historical overview, such as past rides and contracts.
Analytics data layer	Views and flattened data tables for reports and APIs.
Streaming data layer	Storage and analysis of even data such as GPS coordinates from taxis.
Model development and training environment	Providing data scientists with the opportunity to build predictive models.

Figure 2.16: Layers in a data-driven application

2. There are daily updates from source systems: raw -> historical -> analytics layer.

   There is a streaming data pipeline for events from taxis.

3. The minimum set of metadata to capture is the source, owner, date, type, and the transformations that have been applied to the data records. This metadata can be used for auditing, security and consent management, and lineage reports.

4. The solution will probably receive an AIC rating of 323 or higher since it contains sensitive and private data (personnel records, GPS locations, and so on). Therefore, **security** measures should be top-level, such as strong passwords, multi-factor authentication, and role-based access to all data.

For **scalability**, the raw and historical layers should be highly scalable since new data will flood into the system periodically. The analytics layer should be scalable in use since the number of concurrent users will probably grow in time. The streaming data layer should be scalable to some extent in order to keep up with incoming event data for when the number of taxis increases. The model development environment should be scalable in various aspects: data, number of users, and performance.

The **availability** of the system should be high (99.9% or better) for the streaming data layer since running the new system (and therefore the business strategy) depends on the analysis of that data. Other layers can have lower availability rates since they are less mission-critical.

5. **Time travel** is needed for historical reports and audits, and for historical analysis by data analysts and data scientists.

    Retention is important in the raw and historical layers to comply with laws and regulations, in the analytics layer to retain good performance, and in the streaming data layer to limit the amount of data that is being processed at any given moment.

    The locality of the data becomes important once the taxi company expands to other regions/countries.

6. There is a trade-off between costs and quality. A certain minimum level of quality should be in place; for example, software development principles that ensure a high level of maintainability.

7. The model development layer should import data from the raw data layer, the historical data layer, the analytics layer, and the streaming data layer. It must have enough processing power (memory and processing in the form of CPUs and/or GPUs) to train the models for forecasting.

CHAPTER 3: DATA PREPARATION

ACTIVITY 3.01: USING PYSPARK FOR A SIMPLE ETL JOB TO FIND NETFLIX SHOWS FOR ALL AGES

Solution

1. Create a directory called **Activity03.01** in the **Chapter03** directory to store the files for this activity.

2. Open your Terminal (macOS or Linux) or Command Prompt (Windows), navigate to the **Chapter03** directory, and type **jupyter notebook**.

3. Select the **Activity03.01** directory, then click **New** -> **Python3** to create a new Python 3 notebook.

4. If you have done *Exercise 3.02*, *Building an ETL Job Using Spark*, PySpark is already installed on your local machine. If not, install PySpark with the following lines:

```
import sys
!conda install --yes --prefix {sys.prefix} \
        -c conda-forge pyspark
```

5. Connect to a Spark cluster or a local instance using the following code:

```
from pyspark.sql import SparkSession
from pyspark.sql.functions import col, split, size
spark = SparkSession.builder.appName("Packt").getOrCreate()
```

6. Load and show the contents of the dataset using the following code:

```
data = spark.read.csv(\
        '../../Datasets/netflix_titles_nov_2019.csv', \
        header='true')
data.show()
```

> **NOTE**
>
> We read the data from a relative file path. You can also copy the CSV file to the same **Activity03.01** directory and use **netflix_titles_nov_2019.csv** as the path.

You should get the following output:

```
+--------+-------------------+--------------------+-------------------+-------------------+-------------------+------------------+-----
-------+------+--------+--------------------+-------------------+-------------------+-------+
| show_id|              title|            director|               cast|            country|       date_added|relea
se_year|rating| duration|           listed_in|        description|   type|
+--------+-------------------+--------------------+-------------------+-------------------+-------------------+------------------+-----
-------+------+--------+--------------------+-------------------+-------------------+-------+
|81193313|          Chocolate|                null|Ha Ji-won, Yoon K...|        South Korea|November 30, 2019|
2019| TV-14| 1 Season|International TV ...|Brought together ...|TV Show|
|81197050|Guatemala: Heart ...|Luis Ara, Ignacio...|   Christian Morales|               null|November 30, 2019|
2019|  TV-G|   67 min|Documentaries, In...|From Sierra de la...|  Movie|
|81213894|     The Zoya Factor|     Abhishek Sharma|Sonam Kapoor, Dul...|              India|November 30, 2019|
2019| TV-14|  135 min|Comedies, Dramas,...|A goofy copywrite...|  Movie|
|81082007|           Atlantics|           Mati Diop|Mama Sane, Amadou...|France, Senegal, ...|November 29, 2019|
2019| TV-14|  106 min|Dramas, Independe...|Arranged to marry...|  Movie|
|80213643|     Chip and Potato|                null|Abigail Oliver, A...|Canada, United Ki...|             null|
2019|  TV-Y|2 Seasons|         Kids' TV|Lovable pug Chip ...|TV Show|
|81172754|        Crazy people|        Moses Inwang|Ramsey Nouah, Chi...|            Nigeria|November 29, 2019|
2018| TV-14|  107 min|Comedies, Interna...|Nollywood star Ra...|  Movie|
|81120982|      I Lost My Body|      Jérémy Clapin|Hakim Faris, Vict...|             France|November 29, 2019|
```

Figure 3.23: The contents of the CSV file

7. Apply the **data.filter()** function to filter the shows with a rating of **TV-G** and **TV-Y**, as shown in the following code:

```
movies = data.filter((col('type') == 'TV Show') & \
        ((col('rating') == 'TV-G') | \
        (col('rating') == 'TV-Y')))
movies.show()
```

You should get the following output:

```
---+---------+--------------------+--------------------+-------+
| show_id|               title|director|              cast| type|           country|         date_added|release_year|rat
ing| duration|          listed_in|         description| type|
+--------+--------------------+--------+-------------------+-------+--------------------+-------------------+------------+---
---+---------+--------------------+--------------------+-------+
|80213643|     Chip and Potato|    null|Abigail Oliver, A...|Canada, United Ki...|               null|        2019|  T
V-Y|2 Seasons|           Kids' TV|Lovable pug Chip ...|TV Show|
|80117560|Trolls: The Beat ...|    null|Amanda Leighton, ...|     United States|               null|        2019|  T
V-G|8 Seasons|Kids' TV, TV Come...|As Queen Poppy we...|TV Show|
|80115338|         Llama Llama|    null|Jennifer Garner, ...|     United States|               null|        2019|  T
V-Y|2 Seasons|           Kids' TV|Beloved children'...|TV Show|
|80045925|Bella and the Bul...|    null|Brec Bassinger, C...|     United States|  November 2, 2019|        2015|  T
V-G| 1 Season|Kids' TV, TV Come...|The life of cheer...|TV Show|
|70172485|          Victorious|    null|Victoria Justice,...|     United States|               null|        2013|  T
V-G|3 Seasons|Kids' TV, TV Come...|When aspiring sin...|TV Show|
|81184735|Barbie Dreamhouse...|    null|America Young, Ki...|               null|  November 1, 2019|        2019|  T
V-Y| 1 Season|           Kids' TV|As the Roberts fa...|TV Show|
|80227818|         Hello Ninja|    null|Lukas Engel, Zoey...|               null|  November 1, 2019|        2019|  T
V-Y| 1 Season|           Kids' TV|BFFs Wesley and G...|TV Show|
|80991060|   Flavorful Origins|    null|          Yang Chen|             China|               null|        2019|  T
V-G|2 Seasons|Docuseries, Inter...|Delve into the de...|TV Show|
|81034099|           Jeopardy!|    null|        Alex Trebek|     United States|               null|        2019|  T
V-G|5 Seasons|          Reality TV|Alex Trebek hosts...|TV Show|
|81185502|A Little Thing Ca...|    null|Lai Kuan-lin, Zha...|             China|  October 26, 2019|        2019|  T
V-G| 1 Season|International TV ...|A shy college stu...|TV Show|
|81192130|ChuChu TV Kids So...|    null|               null|               null|               null|        2019|  T
V-Y|2 Seasons|           Kids' TV|This educational ...|TV Show|
|81011059|               Booba|    null|        Roman Karev|            Russia|               null|        2019|  T
V-Y|3 Seasons|Kids' TV, TV Come...|The world is a my...|TV Show|
|80176872|   Little Baby Bum: ...|    null|Chloe Marsden, Aa...|               null|               null|        2019|  T
V-Y|2 Seasons|British TV Shows,...|Twinkle, Mia, Jac...|TV Show|
|81020066|Mighty Little Bhe...|    null|Samriddhi Shukla,...|             India|  October 18, 2019|        2019|  T
V-Y| 1 Season|Kids' TV, TV Come...|From decorating h...|TV Show|
|81094271|Spirit Riding Fre...|    null|Amber Frank, Bail...|     United States|               null|        2019|  T
V-Y|2 Seasons|           Kids' TV|"Find the fun and...|TV Show|
|81154549|       Magical Andes|    null|               null|               null|  October 15, 2019|        2019|  T
V-G| 1 Season|Docuseries, Inter...|From Argentina to...|TV Show|
|80212481| YooHoo to the Rescue|    null|Kira Buckland, Ry...|       South Korea|               null|        2019|  T
V-Y|2 Seasons|Kids' TV, Korean ...|In a series of ma...|TV Show|
|80124711|      Super Monsters|    null|Elyse Maloway, Vi...|               null|               null|        2019|  T
V-Y|3 Seasons|           Kids' TV|Preschool kids wh...|TV Show|
|80218107|Dragons: Rescue R...|    null|Nicolas Cantu, Br...|               null|September 27, 2019|        2019|  T
V-Y| 1 Season|Kids' TV, TV Come...|Twins Dak and Ley...|TV Show|
```

Figure 3.24: The contents of the file with TV shows filtered by rating

8. Add the **count_lists** column, which contains the number of lists that are in the **listed_in** column, as shown in the following code:

```
transformed = movies.withColumn('count_lists', \
            size(split(movies['listed_in'], ',')))
```

9. Select a subset of columns using the following code:

```
selected = transformed.select('title', 'cast', \
            'rating', 'release_year', 'duration', \
            'count_lists', 'listed_in', 'description')
```

10. View the data in the selected column using the following code:

```
selected.show()
```

You should get the following output:

```
+--------------------+--------------------+------+------------+---------+-----------+--------------------+--------
------------+
|               title|                cast|rating|release_year| duration|count_lists|           listed_in|
description|
+--------------------+--------------------+------+------------+---------+-----------+--------------------+--------
------------+
|     Chip and Potato|Abigail Oliver, A...| TV-Y|        2019|2 Seasons|          1|          Kids' TV|Lovable
pug Chip ...|
|Trolls: The Beat ...|Amanda Leighton, ...| TV-G|        2019|8 Seasons|          2|Kids' TV, TV Come...|As Queen
Poppy we...|
|         Llama Llama|Jennifer Garner, ...| TV-Y|        2019|2 Seasons|          1|          Kids' TV|Beloved
children'...|
|Bella and the Bul...|Brec Bassinger, C...| TV-G|        2015| 1 Season|          2|Kids' TV, TV Come...|The life
of cheer...|
|          Victorious|Victoria Justice,...| TV-G|        2013|3 Seasons|          2|Kids' TV, TV Come...|When asp
iring sin...|
|Barbie Dreamhouse...|America Young, Ki...| TV-Y|        2019| 1 Season|          1|          Kids' TV|As the R
oberts fa...|
|         Hello Ninja|Lukas Engel, Zoey...| TV-Y|        2019| 1 Season|          1|          Kids' TV|BFFs Wes
ley and G...|
|   Flavorful Origins|          Yang Chen| TV-G|        2019|2 Seasons|          2|Docuseries, Inter...|Delve in
to the de...|
|           Jeopardy!|         Alex Trebek| TV-G|        2019|5 Seasons|          1|          Reality TV|Alex Tre
bek hosts...|
|A Little Thing Ca...|Lai Kuan-lin, Zha...| TV-G|        2019| 1 Season|          3|International TV ...|A shy co
llege stu...|
|ChuChu TV Kids So...|                null| TV-Y|        2019|2 Seasons|          1|          Kids' TV|This edu
cational ...|
|               Booba|         Roman Karev| TV-Y|        2019|3 Seasons|          2|Kids' TV, TV Come...|The worl
d is a my...|
|Little Baby Bum: ...|Chloe Marsden, Aa...| TV-Y|        2019|2 Seasons|          2|British TV Shows,...|Twinkle,
Mia, Jac...|
|Mighty Little Bhe...|Samriddhi Shukla,...| TV-Y|        2019| 1 Season|          2|Kids' TV, TV Come...|From dec
orating h...|
|Spirit Riding Fre...|Amber Frank, Bail...| TV-Y|        2019|2 Seasons|          1|          Kids' TV|"Find th
e fun and...|
|       Magical Andes|                null| TV-G|        2019| 1 Season|          3|Docuseries, Inter...|From Arg
entina to...|
|YooHoo to the Rescue|Kira Buckland, Ry...| TV-Y|        2019|2 Seasons|          2|Kids' TV, Korean ...|In a ser
ies of ma...|
|      Super Monsters|Elyse Maloway, Vi...| TV-Y|        2019|3 Seasons|          1|          Kids' TV|Preschoo
l kids wh...|
|Dragons: Rescue R...|Nicolas Cantu, Br...| TV-Y|        2019| 1 Season|          2|Kids' TV, TV Come...|Twins Da
k and Ley...|
|Loo Loo Kids: Joh...|                null| TV-Y|        2016| 1 Season|          1|          Kids' TV|Music-lo
ving baby...|
+--------------------+--------------------+------+------------+---------+-----------+--------------------+--------
------------+
only showing top 20 rows
```

Figure 3.25: The contents of the file with the selected columns

11. Write the contents of our still in-memory DataFrame to the **transformed2** directory using the following code:

```
selected.write.csv('transformed2', header='true')
```

> **NOTE**
>
> Alternatively, we can add the complete code of the ETL process so far in a Python script and run it through the Terminal (macOS or Linux) or Command Prompt (Windows). We have created the same **spark_etl. py** Python script at the following location: https://packt.live/2ZqQxb9.

12. Open the CSV file in the **transformed2** directory in Jupyter Notebook using the following command:

```
# note: the filename ('part-....') will differ
#in your local machine


!head transformed2/part-00000-a9837c96-549d-4a8c-981a-\
cae147e36801-c000.csv
```

You should get the following output:

```
title,cast,rating,release_year,duration,count_lists,listed_in,description
Chip and Potato,"Abigail Oliver, Andrea Libman, Briana Buckmaster, Brian Dobson, Chance Hurstfield, Dominic Good,
Emma Jayne Maas, Evan Byarushengo, Scotia Anderson, Alessandro Juliani",TV-Y,2019,2 Seasons,1,Kids' TV,"Lovable pu
g Chip starts kindergarten, makes new friends and tries new things — with a little help from Potato, her secret mo
use pal."
Trolls: The Beat Goes On!,"Amanda Leighton, Skylar Astin, Ron Funches, David Fynn, David Koechner, David Kaye, Sea
n T. Krishnan, Sam Lerner, Patrick Pinney, Kevin Michael Richardson, Kari Wahlgren, Fryda Wolff",TV-G,2019,8 Seaso
ns,2,"Kids' TV, TV Comedies","As Queen Poppy welcomes a new time of peace in Troll Village with parties, sports an
d holiday celebrations, Branch tries to be more fun-loving."
Llama Llama,"Jennifer Garner, Shayle Simons, Vania Gill, Islie Hirvonen, Brendon Sunderland, Austin Abell, Evans J
ohnson, Kathleen Barr, David Hoole",TV-Y,2019,2 Seasons,1,Kids' TV,"Beloved children's book character Llama Llama
springs to life in this heartwarming series about family, friendship and learning new things."
Bella and the Bulldogs,"Brec Bassinger, Coy Stewart, Jackie Radinsky, Buddy Handleson, Lilimar, Haley Tju, Dorien
Wilson, Rio Mangini, Annie Tedesco",TV-G,2015,1 Season,2,"Kids' TV, TV Comedies","The life of cheerleader Bella Da
wson turns upside down when, in a twist of fate, she becomes quarterback for her middle school's football team."
Victorious,"Victoria Justice, Leon Thomas III, Matt Bennett, Elizabeth Gillies, Ariana Grande, Avan Jogia, Daniell
a Monet, Michael Eric Reid, Jake Farrow, Eric Lange",TV-G,2013,3 Seasons,2,"Kids' TV, TV Comedies","When aspiring
singer Tori Vega joins the eccentric students at Hollywood Arts High, she struggles to fit in with the amazingly t
alented teens."
Barbie Dreamhouse Adventures: Go Team Roberts,"America Young, Kirsten Day, Cassandra Morris, Cassidy Naber, Emma G
alvin, Stephanie Sheh, Desirae Whitfield, Cristina Milizia, Lisa Fuson, Greg Chun, Ritesh Rajan, Eamon Brennan",TV
-Y,2019,1 Season,1,Kids' TV,"As the Roberts family heads to Costa Rica to investigate a mermaid legend, Barbie tak
es on a summer job at a water park run by a devious boss."
Hello Ninja,"Lukas Engel, Zoey Siewert, Sam Vincent, Mayumi Yoshida",TV-Y,2019,1 Season,1,Kids' TV,"BFFs Wesley an
d Georgie and their silly cat sidekick Pretzel transform into ninjas and enter a magic world, where they solve pro
blems and save the day."
Flavorful Origins,Yang Chen,TV-G,2019,2 Seasons,2,"Docuseries, International TV Shows","Delve into the delectable
world of Chaoshan cuisine, explore its unique ingredients and hear the stories of the people behind its creation."
Jeopardy!,Alex Trebek,TV-G,2019,5 Seasons,1,Reality TV,"Alex Trebek hosts one of TV's longest-running game shows,
where a trio of players buzz in with their knowledge of history, arts, pop culture and more."
```

Figure 3.26: CSV file in the transformed2 directory

> **NOTE**
>
> To access the source code for this specific section, please refer to https://packt.live/3iXSinP.

ACTIVITY 3.02: COUNTING THE WORDS IN A TWITTER DATA STREAM TO DETERMINE THE TRENDING TOPICS

Solution

1. Create a new Python file in your favorite editor (for example, PyCharm or VS Code) or a Jupyter Notebook and name it **spark_twitter.py** or **spark_ twitter.ipynb**.

2. If you have done *Exercise 3.02*, *Building an ETL Job Using Spark*, PySpark is already installed on your local machine. If not, install PySpark with the following lines:

```
import sys
!conda install --yes --prefix {sys.prefix} \
-c conda-forge pyspark
```

3. We first have to connect to a Spark cluster or a local instance. Enter the following lines in the file, notebook, or Python shell:

```
from pyspark.sql import SparkSession
from pyspark.sql.functions import from_json, window,\
     to_timestamp, explode, split, col
from pyspark.sql.types import StructType, StructField, StringType
```

4. Enter the following line to create a Spark session:

```
spark = SparkSession.builder.appName('Packt').getOrCreate()
```

5. To connect to **socket** on localhost, enter the following line:

```
raw_stream = spark.readStream.format('socket').option(\
'host', 'localhost').option('port', 1234).load()
```

6. We'll define the JSON **schema** and add the string format that Twitter uses for its timestamps:

```
tweet_datetime_format = 'EEE MMM dd HH:mm:ss ZZZZ yyyy'
schema = StructType([StructField('created_at', \
                      StringType(), True),
                      StructField('text', StringType(), True)])
```

7. We can now convert the JSON strings with the **from_json** PySpark function:

```
tweet_stream = raw_stream.select(from_json('value',
  schema).alias('tweet'))
```

8. Convert the field that contains the event time to a timestamp with the **to_timestamp** function and split the text of the tweets in words by using the **explode** and **split** functions as shown in the following code:

```
timed_stream = tweet_stream.select(
    to_timestamp('tweet.created_at',
  tweet_datetime_format).alias('timestamp'),
    explode(
        split('tweet.text', ' ')
    ).alias('word'))
```

9. Create a tumbling window of **10 minutes** with **groupBy(window(…))** and add a watermark that ensures that we have a slack of 1 minute before the window evaluates. Make sure to group the tweets in two fields – the window, and the words of the tweets:

```
windowed = timed_stream \
    .withWatermark('timestamp', '1 minute') \
    .groupBy(window('timestamp', '10 minutes'), 'word')
```

10. Specify the evaluation function of the window that is a count of all the words in the window as shown in the following code:

```
counts_per_window = windowed.count().orderBy(['window', 'count'],
    ascending=[0, 1])
```

11. Send the output of the stream to the console and start executing the stream with the **awaitTermination** function:

```
query = counts_per_window.writeStream.outputMode('complete')\
        .format('console').option("truncate", False).start()
query.awaitTermination()
```

You should get the following output:

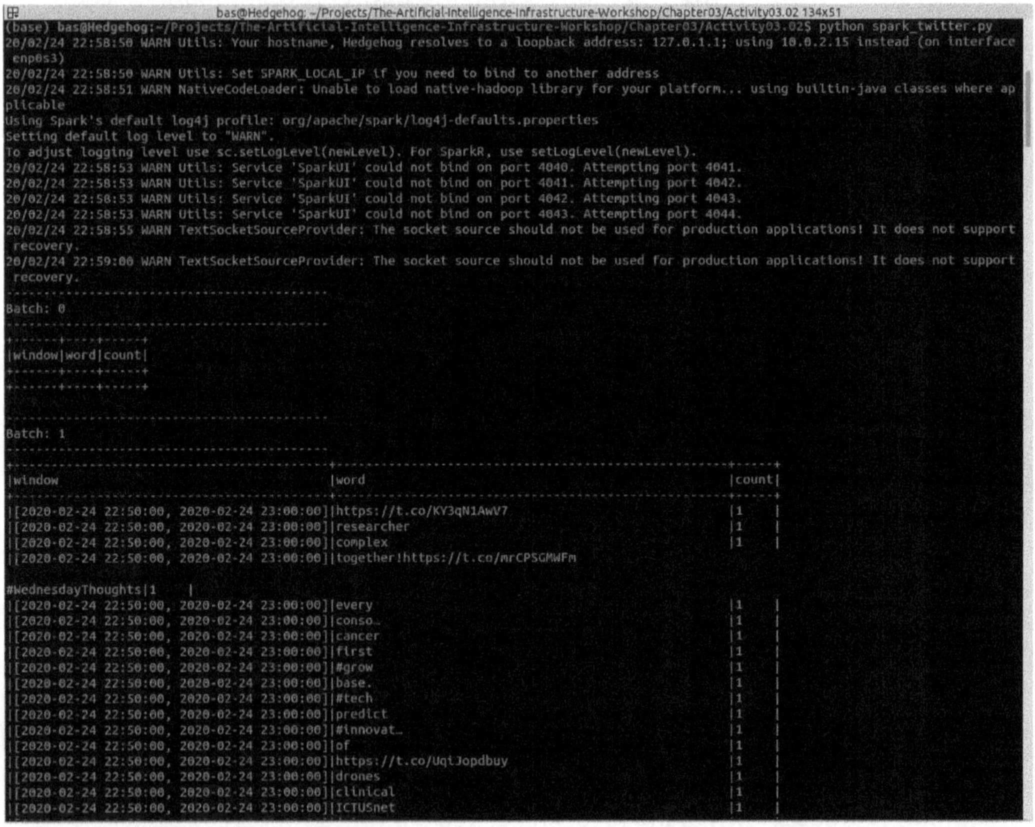

Figure 3.27: Spark Structured Streaming job that connects to Twitter

---

NOTE

To access the source code for this specific section, please refer to https://packt.live/3iX0ODx.

CHAPTER 4: ETHICS OF AI DATA STORAGE

ACTIVITY 4.01: FINDING MORE LATENT PREJUDICES

Solution

1. Create the **Activity04.01** directory in the **Chapter04** directory to store the files for this activity.

2. Open your Terminal (macOS or Linux) or Command Prompt (Windows), navigate to the **Chapter04** directory, and type **jupyter notebook**.

3. In the Jupyter Notebook, click the **Activity04.01** directory and create a new notebook file with the **Python3** kernel.

4. Create a list of at least 16 words that you think might have a positive or negative prejudice:

```
words = """sporty
nerdy
employed
unemployed
clever
stupid
latino
asian
caucasian
disabled
pregnant
introvert
extrovert
politician
florist
CEO"""
```

In the previous code, we added 16 words to our list.

5. Define the same classification model that we used in previous exercises:

```
import spacy
nlp = spacy.load('en_core_web_lg')


def polarity_good_vs_bad(word):
```

```
"""Returns a positive number if a word is closer to good
    than it is to bad, or a negative number if vice versa
IN: word (str): the word to compare
OUT: diff (float): positive if the word is closer to good,
    otherwise negative
"""


good = nlp("good")
bad = nlp("bad")
word = nlp(word)
if word and word.vector_norm:
    sim_good = word.similarity(good)
    sim_bad = word.similarity(bad)
    diff = sim_good - sim_bad
    diff = round(diff * 100, 2)
    return diff
else:
    return None
```

In the preceding code, we defined our basic sentiment classification model again. It takes in a word and checks whether it is closer to **good** or closer to **bad**.

6. Before running the code, guess whether each of the words you chose will be classified as a positive or negative word:

```
# Guesses
"""sporty : POS
nerdy : NEG
employed : POS
unemployed : NEG
clever : POS
stupid : NEG
latino : NEG
asian : NEG
caucasian : POS
disabled : NEG
pregnant : NEG
introvert : NEG
extrovert : POS
politician : NEG
florist : POS
CEO: NEG"""
```

In the preceding code, we added **POS** (positive) or **NEG** (negative) next to the words in our list, guessing their polarity.

7. Run the classifier on the word list and see how close your predictions were, as shown in the following code:

```
for word in words.split("\n"):
    print(word, polarity_good_vs_bad(word))
```

You should get the following output:

```
sporty 13.26
nerdy -6.96
employed 4.46
unemployed -9.36
clever 7.27
stupid -24.37
latino -5.41
asian -4.76
caucasian 1.08
disabled -8.82
pregnant -8.14
introvert -0.36
extrovert 3.92
politician -5.98
florist 6.96
CEO -7.16
```

Figure 4.17: The polarity scores for each word in our new list

In the preceding code, we looped through all the words in our list and calculated their polarity. We can see that our predictions were right in all cases.

> **NOTE**
>
> To access the source code for this specific section, please refer to https://packt.live/2Zqk84p.

CHAPTER 5: DATA STORES: SQL AND NOSQL DATABASES

ACTIVITY 5.01: MANAGING THE INVENTORY OF AN E-COMMERCE WEBSITE USING A MYSQL QUERY

Solution

1. Open a Terminal and run the MySQL client using the following command, based on your OS:

   Windows:

   ```
   mysql
   ```

   Linux:

   ```
   sudo mysql
   ```

   macOS:

   ```
   mysql
   ```

2. Create and select the **PacktFashion** database using the following commands:

   ```
   Create database PacktFashion;
   use PacktFashion;
   ```

   You should get the following output:

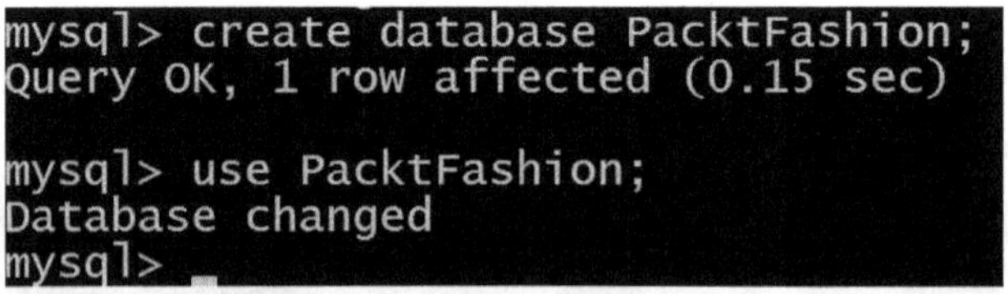

   Figure 5.71: Created and selected database for operation

   Next, we will create the tables as per the data model.

3. Create the **manufacturer** table based on the data model, as shown in the following query:

   ```
   CREATE TABLE manufacturer (m_id INT,
   m_name TEXT,
   m_created_at TIMESTAMP,
   PRIMARY KEY (m_id)
   );
   ```

4. Create the **products** table based on the data model, as shown in the following query:

```
CREATE TABLE products (p_id INT,
p_name TEXT,
p_buy_price FLOAT,
p_manufacturer_id INT,
p_created_at TIMESTAMP,
PRIMARY KEY (p_id),
FOREIGN KEY (p_manufacturer_id)
    REFERENCES manufacturer(m_id)
    ON DELETE CASCADE
);
```

Declaring the **ON DELETE CASCADE** command will force MySQL to delete the data from the child table when any parent key is deleted. For example, if any **m_id** entry is deleted from the **manufacturer** table, then all the rows matching the **p_manufacturer_id** column will also be deleted from the **products** table.

5. Create the **sales** table based on the data model, as shown in the following query:

```
CREATE TABLE sales (s_id INT,
p_id INT,
s_sale_price FLOAT,
s_profit FLOAT,
s_created_at TIMESTAMP,
PRIMARY KEY (s_id),
FOREIGN KEY (p_id)
    REFERENCES products(p_id)
    ON DELETE CASCADE
);
```

6. Create the **location** table based on the data model, as shown in the following query:

```
CREATE TABLE location (loc_id INT,
loc_name TEXT,
loc_created_at TIMESTAMP,
PRIMARY KEY (loc_id)
);
```

7. Create the **status** table based on the data model, as shown in the following query:

```
CREATE TABLE status (status_id INT,
status_name TEXT,
status_created_at TIMESTAMP,
PRIMARY KEY (status_id)
);
```

8. Create the **inventory** table based on the data model, as shown in the following query:

```
CREATE TABLE inventory (inv_id INT,
inv_loc_id INT,
inv_p_id INT,
inv_status_id INT,
inv_created_at TIMESTAMP,
PRIMARY KEY (inv_id),
FOREIGN KEY (inv_loc_id)
    REFERENCES location(loc_id)
    ON DELETE CASCADE,
FOREIGN KEY (inv_p_id)
    REFERENCES products(p_id)
    ON DELETE CASCADE,
FOREIGN KEY (inv_status_id )
    REFERENCES status(status_id)
    ON DELETE CASCADE
);
```

Now that the structure is ready, we will insert data into the tables.

9. Insert the necessary data into the **manufacturer** table using the **INSERT INTO** command, as shown in the following query:

```
INSERT INTO manufacturer(m_id, m_name, m_created_at)
VALUES
(1,"Z-1", now()),
(2,"XIMO", now()),
(3,"NY", now());
```

10. Insert the necessary data into the **products** table using the **INSERT INTO** command, as shown in the following query:

```
INSERT INTO products(p_id, p_name, p_buy_price, p_manufacturer_id,
p_created_at)
VALUES
(1, 'Z-1 Running shoe', 34, 1, now()),
(2, 'XIMO Trek shirt', 15, 2, now()),
(3, 'XIMO Trek shorts', 18, 2, now()),
(4, 'NY cap', 18, 3, now());
```

11. Insert the necessary data into the **sales** table using the **INSERT INTO** command, as shown in the following query:

```
INSERT INTO sales(s_id, p_id, s_sale_price, s_profit, s_created_at)
VALUES
(1,2,18,3,now()),
(2,3,20,2,now()),
(3,3,19,1,now()),
(4,1,40,6,now()),
(5,1,34,0,now());
```

12. Insert the necessary data into the **location** table using the **INSERT INTO** command, as shown in the following query:

```
INSERT INTO location(loc_id, loc_name, loc_created_at)
VALUES
(1, 'California', now()),
(2, 'London', now()),
(3, 'Prague', now());
```

13. Insert the necessary data into the **status** table using the **INSERT INTO** command, as shown in the following query:

```
INSERT INTO status(status_id, status_name, status_created_at)
VALUES
(1, 'IN', now()),
(2, 'OUT', now());
```

14. Insert the necessary data into the **inventory** table using the **INSERT INTO** command, as shown in the following query:

```
INSERT INTO inventory(inv_id,
inv_loc_id,
inv_p_id,
```

```
inv_status_id,
inv_created_at)
VALUES
(1,1,3,1,now()),
(2,3,4,1,now()),
(3,2,2,2,now()),
(4,3,2,2,now()),
(5,1,1,2,now());
```

15. View the **manufacturer** table's data using the **SELECT** query:

```
SELECT * FROM manufacturer;
```

You should get the following output:

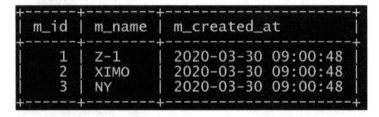

```
+------+--------+---------------------+
| m_id | m_name | m_created_at        |
+------+--------+---------------------+
|    1 | Z-1    | 2020-03-30 09:00:48 |
|    2 | XIMO   | 2020-03-30 09:00:48 |
|    3 | NY     | 2020-03-30 09:00:48 |
+------+--------+---------------------+
```

Figure 5.72: SELECT query output from the manufacturer table

16. View the **products** table's data using the **SELECT** query:

```
SELECT * FROM products;
```

You should get the following output:

```
+------+------------------+--------------+------------------+---------------------+
| p_id | p_name           | p_buy_price  | p_manufacturer_id| p_created_at        |
+------+------------------+--------------+------------------+---------------------+
|    1 | Z-1 Running shoe |           34 |                1 | 2020-03-30 09:00:49 |
|    2 | XIMO Trek shirt  |           15 |                2 | 2020-03-30 09:00:49 |
|    3 | XIMO Trek shorts |           18 |                2 | 2020-03-30 09:00:49 |
|    4 | NY cap           |           18 |                3 | 2020-03-30 09:00:49 |
+------+------------------+--------------+------------------+---------------------+
```

Figure 5.73: SELECT query output from the products table

17. View the **sales** table's data using the **SELECT** query:

```
SELECT * FROM sales;
```

You should get the following output:

```
+------+------+--------------+----------+---------------------+
| s_id | p_id | s_sale_price | s_profit | s_created_at        |
+------+------+--------------+----------+---------------------+
|    1 |    2 |           18 |        3 | 2020-03-30 09:00:49 |
|    2 |    3 |           20 |        2 | 2020-03-30 09:00:49 |
|    3 |    3 |           19 |        1 | 2020-03-30 09:00:49 |
|    4 |    1 |           40 |        6 | 2020-03-30 09:00:49 |
|    5 |    1 |           34 |        0 | 2020-03-30 09:00:49 |
+------+------+--------------+----------+---------------------+
```

Figure 5.74: SELECT query output from the sales table

The formula that's used for calculating profit is (sale price – buy price). In each record, we can see that the sales price is different for the same product. It is assumed that the selling prices are different based on various sale offers.

18. View the **location** table's data using the **SELECT** query:

```
SELECT * FROM location;
```

You should get the following output:

```
+--------+------------+---------------------+
| loc_id | loc_name   | loc_created_at      |
+--------+------------+---------------------+
|      1 | California | 2020-03-30 09:00:49 |
|      2 | London     | 2020-03-30 09:00:49 |
|      3 | Prague     | 2020-03-30 09:00:49 |
+--------+------------+---------------------+
```

Figure 5.75: SELECT query output from the location table

19. View the **status** table's data using the **SELECT** query:

```
SELECT * FROM status;
```

You should get the following output:

```
+-----------+-------------+---------------------+
| status_id | status_name | status_created_at   |
+-----------+-------------+---------------------+
|         1 | IN          | 2020-03-30 09:00:50 |
|         2 | OUT         | 2020-03-30 09:00:50 |
+-----------+-------------+---------------------+
```

Figure 5.76: SELECT query output from the status table

20. View the **inventory** table's data using the **SELECT** query:

```
SELECT * FROM inventory;
```

You should get the following output:

```
+--------+------------+----------+---------------+---------------------+
| inv_id | inv_loc_id | inv_p_id | inv_status_id | inv_created_at      |
+--------+------------+----------+---------------+---------------------+
|      1 |          1 |        3 |             1 | 2020-03-30 09:00:53 |
|      2 |          3 |        4 |             1 | 2020-03-30 09:00:53 |
|      3 |          2 |        2 |             2 | 2020-03-30 09:00:53 |
|      4 |          3 |        2 |             2 | 2020-03-30 09:00:53 |
|      5 |          1 |        1 |             2 | 2020-03-30 09:00:53 |
+--------+------------+----------+---------------+---------------------+
```

Figure 5.77: SELECT query output from the inventory table

This inventory table defines the product inventory where **inv_p_id** refers to **p_id** from the **products** table, **inv_status_id** refers to **status_id** from the **status** table, and **inv_loc_id** refers to **loc_id** from the **location** table. Let's understand this better using a third row where **inv_id** (inventory ID) is **3**. This can be declassified as stating that the **XIMO Trek Shirt** product (that is, **inv_p_id** is **2**) is not available (that is, **inv_status_id** is **2**) in the London warehouse (**inv_loc_id** is **2**).

21. Find the total number of products in the **inventory** table using the **count** function and a **JOIN** clause, as shown in the following query:

```
SELECT count(inventory.inv_p_id) as total_in_stock_products
FROM inventory
JOIN status
ON status.status_id=inventory.inv_status_id
WHERE status.status_name='IN';
```

You should get the following output:

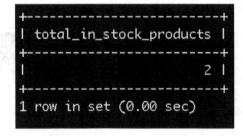

Figure 5.78: Table showing the total of in-stock products

We need to join the **inventory** and **status** tables on **inv_status_id** and **status_id** and count the total products available with **Status_name='IN'** in the **products** table.

22. Find the total number of products not in the inventory using the **count** function and a **JOIN** clause, as shown in the following query:

```
SELECT count(inventory.inv_p_id) as total_in_stock_products
FROM inventory
JOIN status
ON status.status_id=inventory.inv_status_id
WHERE status.status_name='OUT';
```

You should get the following output:

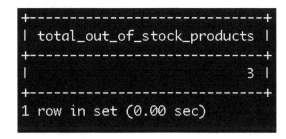

Figure 5.79: Table showing the total of out-of-stock products

This scenario is similar to the previous step. In this step, we are filtering for the **'OUT'** status.

23. Find the status of the **XIMO Trek shirt** product for the **Prague** location, as shown in the following query:

```
SELECT status.status_name as status
FROM inventory, status, products, location
WHERE status.status_id=inventory.inv_status_id
AND products.p_id = inventory.inv_p_id
AND location.loc_id = inventory.inv_loc_id
AND products.p_name='XIMO Trek shirt'
AND location.loc_name='Prague';
```

You should get the following output:

```
+--------+
| status |
+--------+
| OUT    |
+--------+
1 row in set (0.00 sec)
```

Figure 5.80: Table showing the status of the product

We select **status_name** from the **status** table and apply conditions to other tables, that is, **inventory**, **products**, and **location**. Let's join on the primary key and foreign keys from the **inventory**, **status**, **products**, and **location** tables. Then, we'll filter for the location as **'Prague'** and the product name as **'XIMO Trek shirt'**.

> **NOTE**
>
> To access the source code for this specific section, please refer to https://packt.live/2ZpQpbz.

ACTIVITY 5.02: DATA MODEL TO CAPTURE USER INFORMATION

Solution

1. Open a Terminal and start a MongoDB shell:

```
$mongo
```

2. Create a database called PacktFashion through the **use** query:

```
use PacktFashion
```

You should get the following output:

```
switched to db PacktFashion
```

In MongoDB, you can start using the database directly without creating it. It will be created when you create a collection in it.

3. Create a **products** collection, as shown in the following query:

```
db.createCollection("products")
```

You should get the following output:

```
{ "ok" : 1 }
```

4. Insert the data into the **products** collection, as shown in the following query:

```
todayDate=new Date()

products=[{
"p_name": "XIMO Trek shirt",
"p_manufacturer": "XIMO",
"p_buy_price": 15,
"p_created_at": todayDate,
"sales": [
  {
    "s_sale_price": 30,
    "s_profit": 15,
    "p_created_at": todayDate,
  },
  {
    "s_sale_price": 18,
    "s_profit": 3,
    "p_created_at": todayDate
  },
  {
    "s_sale_price": 20,
    "s_profit": 5,
    "p_created_at": todayDate
  },
  {
    "s_sale_price": 15,
    "s_profit": 0,
    "p_created_at": todayDate
  }
  ]
},
{
"p_name": "XIMO Trek shorts",
```

```
    "p_manufacturer": "XIMO",
    "p_buy_price": 18,
    "p_created_at": todayDate,
    "sales": [
      {
        "s_sale_price": 22,
        "s_profit": 4,
        "p_created_at": todayDate,
      },
      {
        "s_sale_price": 18,
        "s_profit": 0,
        "p_created_at": todayDate
      },
      {
        "s_sale_price": 20,
        "s_profit": 2,
        "p_created_at": todayDate
      }
    ]
  },
  {
    "p_name": "NY cap",
    "p_manufacturer": "NY",
    "p_buy_price": 18,
    "p_created_at": todayDate,
    "sales": [
      {
        "s_sale_price": 20,
        "s_profit": 2,
        "p_created_at": todayDate,
      },
      {
        "s_sale_price": 21,
        "s_profit": 3,
        "p_created_at": todayDate
      },
      {
        "s_sale_price": 19,
        "s_profit": 1,
```

```
      "p_created_at": todayDate
    }
  ]
}
]
```

5. Insert the products object as a document into the collection, as shown in the following query:

```
db.products.insert(products)
```

You should get the following output:

```
WriteResult({ "nInserted" : 1 })
```

We have successfully inserted the **product** and **sales** data into the database. Now, we would like to track the user's actions. To do this, we will need two collections, **users** and **user_logs**, as per the data model.

6. Create and insert data into the **users** collection, as shown in the following query:

```
users=[
  {
    "name":"Max",
    "u_created_at":todayDate
  },
  {
    "name":"John Doe",
    "u_created_at":todayDate
  },
  {
    "name":"Roger smith",
    "u_created_at":todayDate
  },
];
```

7. Create and enter the necessary logs into the **user_logs** collection, as shown in the following query:

```
user_logs=[
  {
    user_id:"Max",
    product_id:"XIMO Trek shirt",
    action:"bought",
```

```
        ul_crated_at:todayDate
    },
    {
      user_id:"John Doe",
      product_id:"NY cap",
      action:"bought",
      ul_crated_at:todayDate


    },
    {
      user_id:"Roger smith",
      product_id: "XIMO Trek shorts",
      action:"bought",
      ul_crated_at:todayDate


    }
];
```

8. Insert the **users** object as a document into the collection, as shown in the following query:

```
db.users.insert(users);
```

You should get the following output:

```
WriteResult({ "nInserted" : 1 })
```

9. Insert the **user_logs** object as a document into the collection, as shown in the following query:

```
db.user_logs.insert(user_logs);
```

You should get the following output:

```
WriteResult({ "nInserted" : 1 })
```

10. Join the **users** and **user_logs** collections using an aggregation pipeline, as shown in the following query:

```
var user_logs_aggregate_pipeline = [
    { $lookup:
        {
            from: "users",
            localField: "user_id",
            foreignField: "name",
            as: "users"
        }
    },
    {
      $lookup:
        {
            from: "products",
            localField: "product_id",
            foreignField: "p_name",
            as: "products"
        }
    },
    {
        "$unwind":"$users"
    },
    {
        "$project": {
            "user":"$users.name",
            "product":"$products.p_name",
            "action":"$action"

        }
    }
];
```

We have joined three collections of **products**, **users**, and **user_logs** using the **$lookup** operator of the aggregate pipeline. Once the joining is done, we exploded the **products** and **users** arrays using the **$unwind** operator. You can perform various types of aggregation just to get comfortable with it before diving into more complex usage.

11. Generate the report of user logs using the JavaScript array object with the **aggregate** function, as shown in the following query:

```
db.user_logs.aggregate(user_logs_aggregate_pipeline).pretty()
```

You should get the following output:

```
> db.user_logs.aggregate(user_logs_aggregate_pipeline).pretty()
{
        "_id" : ObjectId("5e817b22b292090c5cf3139f"),
        "user" : "Max",
        "product" : [
                "XIMO Trek shirt"
        ],
        "action" : "bought"
}
{
        "_id" : ObjectId("5e817b22b292090c5cf313a0"),
        "user" : "John Doe",
        "product" : [
                "NY cap"
        ],
        "action" : "bought"
}
{
        "_id" : ObjectId("5e817b22b292090c5cf313a1"),
        "user" : "Roger smith",
        "product" : [
                "XIMO Trek shorts"
        ],
        "action" : "bought"
}
```

Figure 5.81: Using $lookup to join the collections

NOTE

To access the source code for this specific section, please refer to https://packt.live/38T4b9X.

ACTIVITY 5.03: MANAGING CUSTOMER FEEDBACK USING CASSANDRA

Solution

1. Launch the Cassandra CLI based on your OS, as follows:

   Windows:

   ```
   Open Cassandra CLI application
   ```

   Linux:

   ```
   root@ubuntu: -$ cqlsh
   ```

   macOS:

   ```
   MyMac:~ root$ cqlsh
   ```

   You should get the following output:

   ```
   cqlsh>
   ```

2. Create and select the **fashionmart** keyspace, as shown in the following query:

   ```
   CREATE KEYSPACE fashionmart
   WITH replication = {'class':'SimpleStrategy', 'replication_factor' :
   3};

   use fashionmart;
   ```

3. Create a **COLUMNFAMILY** called **feedback_logs**, as shown in the following query:

   ```
   CREATE COLUMNFAMILY feedback_logs(
   fl_id int PRIMARY KEY,
   fl_feedback text,
   fl_location text,
   user_id int,
   fl_created_at timestamp);
   ```

4. Check whether the **feedback_logs** column family was created in the **fashionmart** keyspace using the following query:

   ```
   describe tables;
   ```

   You should get the following output:

   ```
   feedback_logs     user
   ```

5. Insert the records as batches into the **feedback_logs** column family, as shown in the following query:

```
BEGIN BATCH
INSERT INTO feedback_logs(fl_id, fl_feedback, fl_location, user_id, fl_
created_at)
VALUES(1, 'Great website', 'London', 3, '2019-10-30 12:05:00+0000');
INSERT INTO feedback_logs(fl_id, fl_feedback, fl_location, user_id, fl_
created_at)
VALUES(2, 'Good work', 'Seattle', 2, '2019-10-03 12:05:00+0000');
INSERT INTO feedback_logs(fl_id, fl_feedback, fl_location, user_
id, fl_created_at) VALUES(3, 'Amazing', 'Seattle', 2, '2019-11-04
11:05:00+0000');
INSERT INTO feedback_logs(fl_id, fl_feedback, fl_location, user_id,
fl_created_at) VALUES(4, 'Not so good', 'Hong Kong', 1, '2019-11-04
11:05:00+0000');
INSERT INTO feedback_logs(fl_id, fl_feedback, fl_location, user_id, fl_
created_at) VALUES(, 'Informative website', 'Shanghai', 1, '2018-11-
04 11:05:00+0000');
INSERT INTO feedback_logs(fl_id, fl_feedback, fl_location, user_id,
fl_created_at) VALUES(5, 'Great website', 'London', 1, '2019-10-30
12:05:00+0000');
INSERT INTO feedback_logs(fl_id, fl_feedback, fl_location, user_id,
fl_created_at) VALUES(6, 'Good work', 'Seattle', 4, '2019-10-03
12:05:00+0000');
INSERT INTO feedback_logs(fl_id, fl_feedback, fl_location, user_id, fl_
created_at) VALUES(7, 'Informative website', 'Shanghai', 2, '2018-11-
04 11:05:00+0000');
APPLY BATCH;
```

6. View the **feedback_logs** column family's data using the **SELECT** query:

```
SELECT * FROM feedback_logs;
```

You should get the following output:

```
fl_id | fl_created_at                     | fl_feedback         | fl_location | user_id
------+-----------------------------------+---------------------+-------------+--------
    5 | 2019-10-30 12:05:00.000000+0000 |       Great website |      London |       1
    1 | 2019-10-30 12:05:00.000000+0000 |       Great website |      London |       3
    2 | 2019-10-03 12:05:00.000000+0000 |           Good work |     Seattle |       2
    4 | 2019-11-04 11:05:00.000000+0000 |         Not so good |   Hong Kong |       1
    7 | 2018-11-04 11:05:00.000000+0000 | Informative website |    Shanghai |       2
    6 | 2019-10-03 12:05:00.000000+0000 |           Good work |     Seattle |       4
    3 | 2019-11-04 11:05:00.000000+0000 |             Amazing |     Seattle |       2

(7 rows)
```

Figure 5.82: The feedback_logs column family data

7. Calculate the total amount of feedback, as shown in the following query:

```
SELECT COUNT(fl_id) AS total_feedback
FROM feedback_logs;
```

You should get the following output:

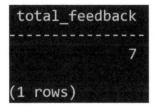

Figure 5.83: Showing total visits

> **NOTE**
>
> To access the source code for this specific section, please refer to https://packt.live/2Ok942r.

CHAPTER 6: BIG DATA FILE FORMATS

ACTIVITY 6.01: SELECTING AN APPROPRIATE BIG DATA FILE FORMAT FOR GAME LOGS

Solution

1. In the **Chapter06** directory, create the **Activity06.01** directory to store the files for this activity.

2. Move the **session_log** file into the **Chapter06/Data** directory.

3. Open your Terminal (macOS or Linux) or Command Prompt window (Windows), move to the installation directory, and open the Spark shell in it using the following command:

```
spark-shell --packages org.apache.spark:spark-avro_2.11:2.4.5
```

You should get the following output:

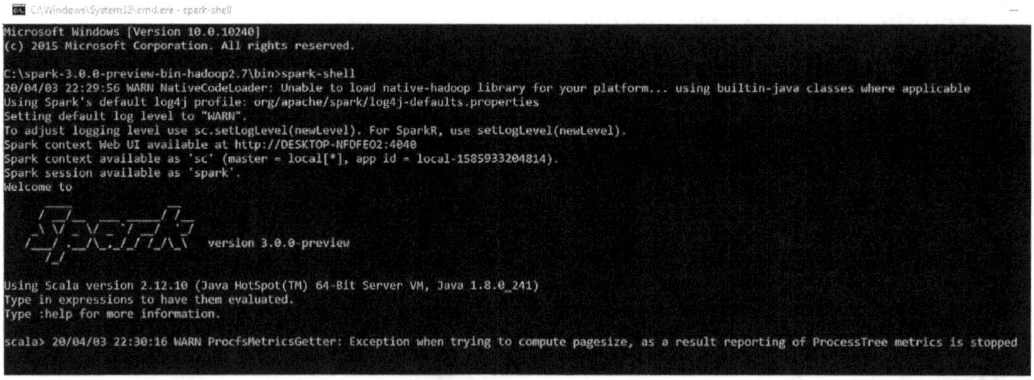

Figure 6.27: Spark shell

By using this command, the Spark shell will be launched and we will now load the dataset from the CSV file.

4. Load the **session_log.csv** dataset:

```
val df_ses_log_csv = spark.read.options(Map("inferSchema"-
  >"true","delimiter"->",","header"-
  >"true")).csv("F:/Chapter06/Data/session_log.csv")
```

> **NOTE**
>
> Update the input path of the file according to your local file path throughout the exercise.

You should get the following output:

```
df_ses_csv: org.apache.spark.sql.DataFrame = [user_id: string, event_
date:
    string … 3 more fields]
```

Here, we have loaded the dataset from the CSV file with an explicit option of a **delimiter** (or value separator), and the header argument set to **true** because the dataset contains the schema or column names. This will result in a DataFrame.

5. Show the CSV file using the following command:

```
df_ses_log_csv.show(false)
```

> **NOTE**
>
> Owing to the size of the data, the commands in this activity can take up to a few minutes to execute.

You should get the following output:

```
+------------------------------------+----------+-------------------+-----------+----------+
|user_id                             |event_date|event_time         |Level__ID  |session_nb|
+------------------------------------+----------+-------------------+-----------+----------+
|6BED1A04-43A3-431D-B908-5CD3E3843B5C|14-10-2015|14-10-2015 10:28|18         |4         |
|FEAE206C-02E5-4610-B2A5-0EC5CBC73BBD|14-10-2015|14-10-2015 10:01|13         |9         |
|DC155AB7-3728-4E8D-A933-1BD7A0BC993A|14-10-2015|14-10-2015 09:24|1          |3         |
|6BED1A04-43A3-431D-B908-5CD3E3843B5C|14-10-2015|14-10-2015 09:16|15         |3         |
|6BED1A04-43A3-431D-B908-5CD3E3843B5C|14-10-2015|14-10-2015 09:17|15         |3         |
|6BED1A04-43A3-431D-B908-5CD3E3843B5C|14-10-2015|14-10-2015 08:47|14         |3         |
|FEAE206C-02E5-4610-B2A5-0EC5CBC73BBD|14-10-2015|14-10-2015 08:26|12         |8         |
|C67C6D4C-53E9-4C77-873D-8A70F9A9C9ED|14-10-2015|14-10-2015 08:05|1          |1         |
|7F5E8F9B-269B-41BF-9DD5-F91EBFB1332D|15-10-2015|15-10-2015 09:03|20         |6         |
|7F5E8F9B-269B-41BF-9DD5-F91EBFB1332D|15-10-2015|15-10-2015 09:23|21         |6         |
|7F5E8F9B-269B-41BF-9DD5-F91EBFB1332D|15-10-2015|15-10-2015 09:24|21         |6         |
|FEAE206C-02E5-4610-B2A5-0EC5CBC73BBD|15-10-2015|15-10-2015 07:58|1          |5         |
|6ED1B75C-F35B-462C-924E-EACA43550000|13-10-2015|13-10-2015 12:14|3          |1         |
|C47A6FCE-6C76-41FE-9A6A-173AC7F01315|13-10-2015|13-10-2015 08:41|1          |1         |
|5E9FA0C9-8363-4F75-9432-4608C495291B|12-10-2015|12-10-2015 10:26|1          |1         |
|494CB6CD-33D7-49C3-8414-35CC5D4A86DD|12-10-2015|12-10-2015 08:35|1          |3         |
|9F39B934-01E4-49A1-AD1F-04A48D4930B7|09-10-2015|09-10-2015 04:56|1          |1         |
|6ED1B75C-F35B-462C-924E-EACA43550000|08-10-2015|08-10-2015 06:29|2          |3         |
|8BF191CF-567A-4977-B2C9-90EC756EFD41|07-10-2015|07-10-2015 10:46|1          |1         |
|8BF191CF-567A-4977-B2C9-90EC756EFD41|07-10-2015|07-10-2015 10:39|1          |1         |
+------------------------------------+----------+-------------------+-----------+----------+
only showing top 20 rows
```

Figure 6.28: Displaying the DataFrame

This will display the first **20** rows from the DataFrame. The **false** argument represents the width of the columns shown to match the complete raw value.

6. Convert the CSV file into the Avro file format and verify the file size as shown in the following code:

```
df_ses_log_csv.write.format("avro").save("F:/Chapter06/Data/session_
log.avro")
```

A file is created of the following size:

```
Output file size: 454 MB
```

This command will convert the DataFrame(that loaded the CSV file) and save it into the Avro file format. The time taken to execute this command will vary as per the size of the data. Let's convert the data to the ORC format now.

7. Convert the CSV file into the ORC file format and verify the file size as shown in the following code:

```
df_ses_log_csv.write.orc("F:/Chapter06/Data/session_log.orc")
```

A file is created of the following size:

```
Output file size: 182 MB
```

This command will convert the DataFrame(that loaded the CSV file) and save it in the ORC file format. The time taken to execute this command will vary as per the size of the data. Let's convert the data into the Parquet format now.

8. Convert the CSV file into the Parquet file format and verify the file size as shown in the following code:

```
df_ses_log_csv.write.parquet(«F:/Chapter06/Data/session_log.parquet»)
```

A file is created of the following size:

```
[Output file size: 115 MB]
```

This command will convert the DataFrame(that loaded the CSV file) and save it in the Parquet file format. The time taken to execute this command will vary as per the size of the data. Now, let's measure the query performance.

9. Read the Parquet file through the DataFrame as shown in the following code:

```
var df_ses_log_parquet = \
spark.read.parquet(«F:/Chapter06/Data/session_log.parquet»)
```

You should get the following output:

```
df_ses_log_parquet: org.apache.spark.sql.DataFrame = \
[user_id: string, event_date: string … 3 more fields]
```

10. Read the ORC file through the DataFrame as shown in the following code:

```
var df_ses_log_orc = spark.read.orc("F:/Chapter06/Data/session_log.
orc")
```

You should get the following output:

```
df_ses_log_orc: org.apache.spark.sql.DataFrame = [user_id: string,
  event_date: string … 3 more fields]
```

11. Read the Avro file through the DataFrame as shown in the following code:

```
var df_ses_log_avro =
    spark.read.format("avro").load("F:/Chapter06/Data/session_log.
avro")
```

You should get the following output:

```
df_ses_log_avro: org.apache.spark.sql.DataFrame = [user_id: string,
  event_date: string … 3 more fields]
```

We have created three DataFrames by loading the data from our Parquet, Avro, and ORC files. Now we will create a function to measure the execution time.

12. Create a function to measure the execution time as follows:

```
// Function for time consumption

  def time[A](f: => A) = {
        val s = System.nanoTime
        val ret = f
        println("Time: "+(System.nanoTime-s)/1e6+" ms")
        ret

  }
```

In this step, we have create a time-measuring function that will take the query that needs to be executed as an input. Let's run the performance query to measure the efficiency of each file format.

13. Execute the **count** query within the **time** function for all three DataFrames as shown in the following code:

```
time{df_ses_log_avro.groupBy("session_nb").count()}
time{df_ses_log_parquet.groupBy("session_nb").count()}
time{df_ses_log_orc.groupBy("session_nb").count()}
```

You should get the following output:

```
scala> time{df_ses_log_avro.groupBy("session_nb").count()}
time: 6.050187ms
res51: org.apache.spark.sql.DataFrame = [session_nb: int, count: bigint]

scala> time{df_ses_log_parquet.groupBy("session_nb").count()}
time: 5.024889ms
res52: org.apache.spark.sql.DataFrame = [session_nb: int, count: bigint]

scala> time{df_ses_log_orc.groupBy("session_nb").count()}
time: 5.129231ms
res53: org.apache.spark.sql.DataFrame = [session_nb: int, count: bigint]
```

Figure 6.29: Time consumed for count queries by each format

In this step, we have executed the **count** query over the DataFrames created in the previous steps. We can observe that the ORC and Parquet files have nearly the same time-efficiency over this query. Let's execute a few complex queries.

14. Create a table from the DataFrames as shown in the following code:

```
//Parquet
df_ses_log_parquet.createOrReplaceTempView(«session_log_parquet»)
//ORC
df_ses_log_orc.createOrReplaceTempView(«session_log_orc»)
//AVRO
df_ses_log_avro.createOrReplaceTempView(«session_log_avro»)
```

By creating a table from the DataFrames, we can execute SQL syntax queries over the same dataset. Let's now execute the queries.

15. Execute a **GROUP BY** query over the created tables as follows:

```
// Defining sql query as a variable

val p_yr_query = "Select count(Year),Year from (Select
  SUBSTRING(event_date,7,10) as Year from session_log_parquet) GROUP
BY
  Year"

val o_yr_query = "Select count(Year),Year from (Select
  SUBSTRING(event_date,7,10) as Year from session_log_orc) GROUP BY
Year"

val a_yr_query = "Select count(Year),Year from (Select
  SUBSTRING(event_date,7,10) as Year from session_log_avro) GROUP BY
Year"

// Executing the query inside the time function
time{spark.sql(p_yr_query)}
time{spark.sql(o_yr_query)}
time{spark.sql(a_yr_query)}
```

You should get the following output:

```
scala> time{spark.sql(p_yr_query)}
time: 11.249209ms
res57: org.apache.spark.sql.DataFrame = [count(Year): bigint, Year: string]

scala> time{spark.sql(o_yr_query)}
time: 9.789433ms
res58: org.apache.spark.sql.DataFrame = [count(Year): bigint, Year: string]

scala> time{spark.sql(a_yr_query)}
time: 19.105728ms
res59: org.apache.spark.sql.DataFrame = [count(Year): bigint, Year: string]
```

Figure 6.30: Time consumed by a GROUP BY query

In this step, we have executed a complex **GROUP BY** query with a **SUBSTRING** function over all the DataFrames. And again, the ORC and Parquet files have nearly the same time-efficiency over this query.

We can conclude from the results that the Parquet file format has the highest compression for our dataset, and ORC has the quickest query response. However, the difference in operational efficiency does not have a high delta.

Along with technical specifications, there are a few other considerations before finalizing your choice of file format. These include scopes such as the current environment or platform, the cost and time of adopting the new environment, and the impact of such a change on the business. Let's include the technology stack of the company in our decision making.

16. Consolidating these factors into a single selection, and taking into consideration the existing technology stack, which is Spark and Cloudera, which have seamless support for Parquet, your report will likely contain a preference for **Parquet** over the ORC file format.

NOTE

To access the source code for this specific section, please refer to https://packt.live/3gU6yMu.

CHAPTER 7: INTRODUCTION TO ANALYTICS ENGINE (SPARK) FOR BIG DATA

ACTIVITY 7.01: EXPLORING AND PROCESSING A MOVIE LOCATIONS DATABASE BY USING SPARK'S TRANSFORMATIONS AND ACTIONS

Solution

1. The first step involves logging in to the **COMMUNITY EDITION** of Databricks.

2. Upload the file you have downloaded, **Film_Locations_in_San_Francisco.csv**, into Databricks:

Figure 7.33: Uploading the file

3. Read the CSV file to a DataFrame:

```
from pyspark.sql.functions import desc
# File location and type
file_location =
    "/FileStore/tables/Film_Locations_in_San_Francisco.csv"
file_type = "csv"


# The applied options are for CSV files. For other file types,
    these will be ignored.
dataTable = spark.read.format(file_type) \
    .option("inferSchema", "true") \
    .option("header", "true") \
    .option("sep", ",") \
    .load(file_location)


display(dataTable)
dataTable.printSchema()
```

You should get the following output:

Title	Release Year	Locations	Fun Facts	Production Company	Distributor	Director	Writer	Actor 1	Actor 2	Actor 3
180	2011	Epic Roasthouse (399 Embarcadero)	null	SPI Cinemas	null	Jayendra	Umarji Anuradha, Jayendra, Aarthi Sriram, & Suba	Siddarth	Nithya Menon	Priya Anand
180	2011	Mason & California Streets (Nob Hill)	null	SPI Cinemas	null	Jayendra	Umarji Anuradha, Jayendra, Aarthi	Siddarth	Nithya Menon	Priya Anand

Showing the first 1000 rows.

Figure 7.34: View of the data read

> **NOTE**
>
> As the output was long, the view is truncated for ease of representation.

We read the CSV file into a DataFrame called **dataTable** by using the **spark. read.format()** function. The function takes in several arguments, we specify the delimiter as a comma (**,**), specify the file path, specify the first row as the header of the DataFrame using **header=true**, and the data types can be inferred from the samples of the DataFrame by setting **inferSchema = true**.

4. Rename the columns with no spaces between the words, as shown in the following code:

```
newcolnames = ['Title','ReleaseYear', 'Locations', \
               'FunFacts', 'ProductionCompany', \
               'Distributor', 'Director', 'Writer', \
               'Actor1', 'Actor2', 'Actor3']
for old,new in zip(dataTable.columns, newcolnames):
    dataTable = dataTable.withColumnRenamed(old,new)
display(dataTable)
```

You should get the following output:

Title	ReleaseYear	Locations	FunFacts	ProductionCompany	Distributor	Director	Writer	Actor1	Actor2
180	2011	Epic Roasthouse (399 Embarcadero)	null	SPI Cinemas	null	Jayendra	Umarji Anuradha, Jayendra, Aarthi Sriram, & Suba	Siddarth	Nithya Menon
180	2011	Mason & California Streets (Nob Hill)	null	SPI Cinemas	null	Jayendra	Umarji Anuradha, Jayendra, Aarthi Sriram, &	Siddarth	Nithya Menon

Figure 7.35: Snapshot of the view created with the renamed column

NOTE

As the output was long, the view is truncated for ease of representation.

We remove the space in the column names by renaming them using the **withColumnRenamed** function. We do this by creating a list of new column names, **newcolnames**, and using the new list to rename the old list of columns.

5. Find the recent movies released in or after **2015** using the following code:

```
#finding the recent movies released after 2015
recentMovies= dataTable.filter(dataTable.ReleaseYear >= 2015)
display(recentMovies)
```

You should get the following output:

Title	ReleaseYear	Locations	FunFacts	ProductionCompany	Distributor	Director	Writer	Actor1	Actor2	Actor3
Age of Adaline	2015	Pier 50- end of the pier	null	Lionsgate / Sidney Kimmel Entertainment / Lakeshore Entertainment	null	Lee Toland Krieger	J. Mills Goodloe	Blake Lively	Harrison Ford	Ellen Burstyn
Age of Adaline	2015	California @ Montgomery	null	Lionsgate / Sidney Kimmel Entertainment /	null	Lee Toland Krieger	J. Mills Goodloe	Blake Lively	Harrison Ford	Ellen Burstyn

Showing the first 1000 rows.

Figure 7.36: Recently released movies (released in 2015 or later) shot in SFO

> **NOTE**
>
> As the output was long, the view is truncated for ease of representation.

We use the **filter** function to find the list of movies released in the year 2015 or later.

6. Find and show the popular locations of recently released movies shot in SFO using the following code:

```
#group by locations and find the count
recentLocations = recentMovies.groupby('Locations')\
                    .count().sort(desc("count"))
display(recentLocations)
```

You should get the following output:

Locations	count
60 Leavenworth St.	10
770 Haight Street	8
1246 Folsom Street	8
Treasure Island	8
State Garage, 818 Leavenworth	6
1890 Clay Street	6
Fairmont Hotel (950 Mason Street, Nob Hill)	5
San Francisco City Hall	5
Willy's Barber shop, 3227 22nd Street	4

Figure 7.37: Popular locations of recently released movies shot in SFO

We find the popular location of recently released movies by aggregating on a writer by using the **groupby** and **count** functions. We then sort in descending order with the help of the **sort(desc("count"))** function.

7. Find and show the top three popular locations in recent movies shot in SFO using the following code:

```
display(recentLocations.take(3))
```

You should get the following output:

Locations	count
60 Leavenworth St.	10
1246 Folsom Street	8
Treasure Island	8

Figure 7.38: Top three popular locations of recently released movies shot in SFO

We find and display the top three rows from the **recentLocations** DataFrame by using the **take** function, with an argument of **3**.

8. Find and show the most popular location for the recent movies shot in SFO using the following code:

```
recentLocations.first()
display(recentLocations.take(1))
```

You should get the following output:

Locations	count
60 Leavenworth St.	10

Figure 7.39: The most popular location of recently released movies shot in SFO

We find and display the topmost row from the **recentLocations** DataFrame by using the **first()** function.

9. Show the popular writers of recently released movies shot in SFO using the following code:

```
recentWriters =
    recentMovies.groupby('Writer').count().sort(desc("count"))
display(recentWriters)
```

You should get the following output:

Writer	count
Michael Lannan	168
Alexandra Cunningham	71
Eric Lodal	66
Sarah Bellwood, Michal Lali Kagan, Michael McGowan Amital Stern, Jose Luis Sampedro	56
Linda Demetrick	50
James Cameron	48
The Wachowskis	46
Kay Cannon	42
Tom Brown	38

Figure 7.40: Popular writers of recently released movies shot in SFO

> **NOTE**
>
> As the output was long, the view is truncated for ease of representation.

We find the popular writers of recently released movies by aggregating on a writer by using the **groupby** and **count** functions. We then sort in descending order with the help of the **sort(desc("count"))** function.

10. Show the top three popular writers of recently released movies shot in SFO using the following code:

```
display(recentWriters.take(3))
```

You should get the following output:

Writer	count
Michael Lannan	168
Alexandra Cunningham	71
Eric Lodal	66

Figure 7.41: Top three popular writers of recently released movies shot in SFO

We find and display the top three rows from the **recentWriters** DataFrame by using the **take** function, with an argument of **3**.

11. Find and show the most popular writer of recently released movies shot in SFO using the following code:

```
recentWriters.first()
display(recentWriters.take(1))
```

You should get the following output:

Writer	count
Michael Lannan	168

Figure 7.42: The most popular writer of recently released movies shot in SFO

We find and display the topmost row from the **recentWriters** DataFrame by using the **first()** function.

> **NOTE**
>
> To access the source code for this specific section, please refer to https://packt.live/2OoDtN2.

CHAPTER 8: DATA SYSTEM DESIGN EXAMPLES

ACTIVITY 8.01: BUILDING THE COMPLETE SYSTEM WITH PIPELINES AND QUEUES

Solution

1. Import the **random** and **time** standard libraries, as well as the **Queue** and **Thread** classes from their respective modules:

```
from queue import Queue
from threading import Thread
import random
import time
```

We imported the modules that we will use to design our next mock system.

2. Initialize the mock dataset and put it into a queue, as shown in the following query:

```
urls = ['url1-', 'url1-', 'url2-', 'url3-', 'url4-', \
'url5-', 'url6-', 'url7-', 'url8-', 'url9-', 'url10-']
seen = set()

url_queue = Queue()
for url in urls:
    url_queue.put(url)
```

We created 11 mock URLs and a **seen** set to find duplicates. We then created a queue for our URLs and added each URL to the queue.

3. Set up queues for each of the components, as shown in the following query:

```
scraped_queue = Queue()
cleaned_queue = Queue()
deduplicated_queue = Queue()
insights_queue = Queue()
decisions_queue = Queue()
```

We initialized **Queue()** objects for each component to push to when done.

4. Define the **scraper**, **cleaner**, and **deduplicator** modules, just like we did in *Exercise 8.02, Adding Queues to a System to Make It Highly Available*:

```
def scraper():
    while True:
        time.sleep(random.randrange(0,2))
        url = url_queue.get()
        print("Scraping {}".format(url))
        scraped_queue.put(url[3:])

def cleaner():
    while True:
        time.sleep(random.randrange(2,4))
        raw = scraped_queue.get()
        print("Cleaning {}".format(raw))
        cleaned_queue.put(raw.replace("-", ""))

def deduplicator():
    while True:
        time.sleep(random.randrange(4,6))
        cleaned = cleaned_queue.get()
        print("Deduplicating {}".format(cleaned))
        if cleaned not in seen:
            deduplicated_queue.put(cleaned)
            seen.add(cleaned)
```

We defined each of these functions in the same way we did previously.

5. Define the **analyzer** module, as shown in the following query:

```
def analyzer():
    while True:
        time.sleep(random.randrange(1,4))
        unique = deduplicated_queue.get()
        print(«Analyzing {}».format(unique))
        n = int(unique)
        if n % 2 == 0:
            insights_queue.put(-n)
        else:
            insights_queue.put(n)
```

Here, we defined the **analyzer** module, which works in a similar way. It has an **if** statement to put either **-n** or **n** on the next queue, depending on whether the clean URL it fetches is odd or even.

6. Define the **decision_maker** module, as shown in the following query:

```
def decision_maker():
    while True:
        time.sleep(random.randrange(1,4))
        insight = insights_queue.get()
        print("Deciding {}".format(insight))
        if insight > 0:
            decisions_queue.put("Buy {}".format(insight))
        else:
            decisions_queue.put("Sell {}".format(-insight))
```

Here, we created the **decision_maker** module, which is similar to the **analyzer**. It puts a **Buy** decision on the next queue for positive numbers and a **Sell** decision for negative ones.

7. Define the **trader** module, as shown in the following query:

```
def trader():
    while True:
        time.sleep(random.randrange(1,4))
        trade = decisions_queue.get()
        print("Trading {}".format(trade))
        print(trade)
```

The **trader** module is the simplest module. It simply fetches items off the decision queue and prints them out.

8. Initialize the threads for each component, as shown in the following query:

```
scraper_worker = Thread(target=scraper)
cleaner_worker = Thread(target=cleaner)
deduplicator_worker = Thread(target=deduplicator)
analyzer_worker = Thread(target=analyzer)
decision_maker_worker = Thread(target=decision_maker)
trader_worker = Thread(target=trader)
```

Here, we created three threads, one for each component using the **Thread** class, and passed the respective functions in using the **target** parameter.

9. Add all the components to an array and start each thread, as shown in the following query:

```
threads = [
    scraper_worker, cleaner_worker, deduplicator_worker, \
    analyzer_worker, decision_maker_worker, trader_worker
]


[t.start() for t in threads]
```

You should get the following output:

```
[None, None, None, None, None, None]

Scraping url1-
Scraping url2-
Cleaning 1-
Scraping url3-
Scraping url4-
Scraping url5-
Deduplicating 1Scraping url6-

Analyzing 1
Deciding 1
Trading Buy 1
Buy 1
Cleaning 1-
Scraping url7-
Scraping url8-
Cleaning 2-
Scraping url9-
Scraping url10-
```

Figure 8.7: Excerpt of the output from the full system

We added each thread to an array and started each thread. Again, we saw the different components work at different speeds.

> **NOTE**
>
> To access the source code for this specific section, please refer to https://packt.live/3ftzYR9.

CHAPTER 9: WORKFLOW MANAGEMENT FOR AI

ACTIVITY 9.01: CREATING A DAG IN AIRFLOW TO CALCULATE THE RATIO OF LIKES-DISLIKES FOR EACH CATEGORY

Solution

1. Create an **Activity09.01** directory in the **Chapter09** directory to store the files for this activity.

2. Open your Terminal (macOS or Linux) or Command Prompt (Windows), navigate to the **Chapter09** directory, and type **jupyter notebook**. The Jupyter Notebook should resemble what you can see in the following screenshot:

Figure 9.42: The Jupyter Notebook launched in the Chapter09 directory

3. In the Jupyter Notebook, click the **Activity09.01** directory, create a notebook file with the **Python 3** kernel, and add the following code:

```
import json
import pandas as pd


# read video data
df_vids = pd.read_csv('../Data/USvideos.csv.zip',
    compression='zip')


# read category data
data_cats = json.load(open('../Data/US_category_id.json', 'r'))
df_cat = pd.DataFrame(data_cats)
df_cat['category'] = df_cat['items'].apply(lambda x:
    x['snippet']['title'])
```

```
df_cat['id'] = df_cat['items'].apply(lambda x: int(x['id']))
df_cat_drop = df_cat.drop(columns=['kind', 'etag', 'items'])


# join data
df_join = df_vids.merge(df_cat_drop, left_on='category_id', \
                        right_on='id')


df_join.head()
```

You should get the following output:

	video_id	trending_date	title	channel_title	category_id	publish_time	tags	views	likes
0	2kyS6SvSYSE	17.14.11	WE WANT TO TALK ABOUT OUR MARRIAGE	CaseyNeistat	22	2017-11-13T17:13:01.000Z	SHANtell martin	748374	57527
1	0mINzVSJrT0	17.14.11	Me-O Cats Commercial	Nobrand	22	2017-04-21T06:47:32.000Z	cute|"cats"|"thai"|"eggs"	98966	2486
2	STI2fi7sKMo	17.14.11	AFFAIRS, EX BOYFRIENDS, $18MILLION NET WORTH -...	Shawn Johnson East	22	2017-11-11T15:00:03.000Z	shawn johnson|"andrew east"|"shawn east"|"shaw...	321053	4451
3	KODzih-pYIU	17.14.11	BLIND(folded) CAKE DECORATING CONTEST (with Mo...	Grace Helbig	22	2017-11-11T18:08:04.000Z	itsgrace|"funny"|"comedy"|"vlog"|"grace"|"helb...	197062	7250
4	8mhTWqWIQzU	17.14.11	Wearing Online Dollar Store Makeup For A Week	Safiya Nygaard	22	2017-11-11T01:19:33.000Z	wearing online dollar store makeup for a week|...	2744430	115426

Figure 9.43: YouTube video data with category information

To get the video data with category information, we reused a lot of the code we developed in the previous exercises. It's important to write reusable and modular code so that we can simply use it without writing code from scratch.

4. Calculate the likes-dislikes ratio for each category by adding the following code snippet in the next cell of the Jupyter Notebook:

```
# aggregate likes and dislikes by category
df_agg = df_join[['category', 'likes',
  'dislikes']].groupby('category').sum()


# calculate ratio
df_agg['ratio_likes_dislikes'] = df_agg['likes'] / \
                                 df_agg['dislikes']


df_agg.head()
```

You should get the following output:

category	likes	dislikes	ratio_likes_dislikes
Autos & Vehicles	4245656	243010	17.471116
Comedy	216346746	7230391	29.921860
Education	49257772	1351972	36.434018
Entertainment	530516491	42987663	12.341134
Film & Animation	165997476	6075148	27.324022

Figure 9.44: Ratio of likes to dislikes for each category

When we perform a **groupby** operation, it requires an aggregation action for each group. In this case, our aggregation action is **sum**. We are summing the likes and dislikes for each category.

In the **pandas** API, **df_agg['likes'] / df_agg['dislikes']** returns a **pandas** series, each entry of which is the likes-dislikes ratio for each category. We assign the returned **pandas** series to the **df_agg** DataFrame's new column, **['rato_likes_dislikes']**.

5. Create a **ratio_dag.py** DAG script and add the following code to the script:

ratio_dag.py

```
1 import json
2 import os
3 import shutil
4 import sys
5 from datetime import datetime
6
7 import pandas as pd
8 from airflow import DAG
9 from airflow.operators.python_operator import PythonOperator
10
11
12 def filter_data(**kwargs):
13     # read data
14     path_vids = kwargs['dag_run'].conf['path_vids']
15     date = str(kwargs['dag_run'].conf['date'])
16
17     print(os.getcwd())
18     print(path_vids)
19
20     df_vids = pd.read_csv(path_vids, compression='zip') \
21         .query('trending_date==@date')
22     # cache
23     try:
24         df_vids.to_csv('./tmp/data_vids.csv', index=False)
25     except FileNotFoundError:
26         os.mkdir('./tmp')
27         df_vids.to_csv('./tmp/data_vids.csv', index=False)
```

The full code is available at https://packt.live/2C8bqyP.

At a glance, you will notice that this DAG script is almost the same as the **top_ cat_dag.py** DAG script from *Exercise 9.06, Creating a DAG for Our Data Pipeline Using Airflow*, except that there is one more operator and its corresponding Python callable, as shown in the following code:

```
def calc_ratio(**kwargs):
    try:
        df_join = pd.read_csv('./tmp/data_joined.csv')
    except Exception as e:
        print('>>>>>>>>>>>> Error: {}'.format(e))
        sys.exit(1)
    # aggreate likes and dislikes by category
    df_agg = df_join[['category', 'likes',
        'dislikes']].groupby('category').sum()
    # calculate ratio
    df_agg['ratio_likes_dislikes'] = df_agg['likes'] / df_
agg['dislikes']
    df_agg.reset_index('category')\
                        .to_csv('./tmp/data_ratio.csv', \
                        index=False)


op4 = PythonOperator(
    task_id='calc_ratio',
    python_callable=calc_ratio,
    dag=dag)
```

This is the advantage of writing reusable and modular code. You can reuse most of the functions from previous exercises. Taking advantage of what you have already built is a very common practice in software engineering. One of the most often spoken phrases in the industry is, "Don't reinvent the wheel."

6. Register the DAG by copying the **ratio_dag.py** DAG file to the Airflow home directory by running the following command in your Terminal:

```
cp ./ratio_dag.py ~/airflow/dags/
```

> **NOTE:**
>
> The Airflow webserver and scheduler should be stopped when you copy the DAG script to **~/airflow**. We can stop them using **Ctrl + C** in the Command Prompt for Windows/Linux or **Cmd + C** for macOS. After the DAG script has been moved to the Airflow home directory, **~/airflow**, we can launch the Airflow scheduler and webserver again. When the Airflow scheduler is launched again, it will register new DAGs from the DAG scripts in its home directory, **~/airflow**. So, when we issue the **airflow list_dags** command, we can see new DAGs that are newly added.

7. Verify whether the **ratio_dag.py** DAG is successfully registered in the Airflow system by running the following command:

```
airflow list_dags
```

You should get the following output:

```
[2020-01-26 17:56:33,695] {__init__.py:51} INFO - Using executor SequentialExecutor
[2020-01-26 17:56:33,695] {dagbag.py:403} INFO - Filling up the DagBag from /Users/Kevin/airflow/dags


-------------------------------------------------------------------
DAGS
-------------------------------------------------------------------
example_bash_operator
example_branch_dop_operator_v3
example_branch_operator
example_http_operator
example_passing_params_via_test_command
example_pig_operator
example_python_operator
example_short_circuit_operator
example_skip_dag
example_subdag_operator
example_subdag_operator.section-1
example_subdag_operator.section-2
example_trigger_controller_dag
example_trigger_target_dag
example_xcom
latest_only
latest_only_with_trigger
ratio_dag
test_utils
top_cat_dag
tutorial
```

Figure 9.45: List of DAGs registered in the Airflow system

The fourth entry from the bottom is our **ratio_dag** DAG. If your DAG is registered, you will see your DAG in the list returned by the **airflow list_dags** command.

8. Launch the Airflow **scheduler** by running the following command in your Terminal:

```
airflow scheduler
```

You should get the following output:

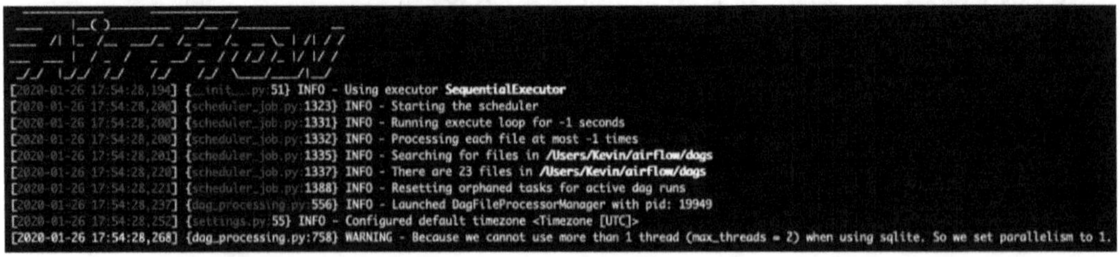

Figure 9.46: Logs of the Airflow scheduler running in the foreground

> **NOTE**
>
> Launch the Airflow scheduler in the **Activity09.01** directory. We are assuming that **Activity09.01** is the working directory for the Airflow system.
>
> When you launch an Airflow scheduler and it throws an error that reads **attempt to write a readonly database**, it means the user who launched the Airflow scheduler does not have write permission to the Airflow SQLite database file, **airflow.db**. One way to resolve the permission issue is to issue **sudo chmod 644 ~/airflow/airflow.db** in your Terminal and restart the Airflow scheduler so that the Airflow scheduler has write access to the database file.

9. Trigger the DAG by running the following command:

```
airflow trigger_dag -c '
{
    "path_vids": "../Data/USvideos.csv.zip",
    "path_cats": "../Data/US_category_id.json",
```

```
    "date": "17.14.11",
    "path_output": "../Data/Ratio_Likes_Dislikes.csv"
}' 'ratio_dag'
```

You should get the following output:

```
[2020-01-26 17:59:23,903] {__init__.py:51} INFO - Using executor SequentialExecutor
[2020-01-26 17:59:23,903] {dagbag.py:403} INFO - Filling up the DagBag from /Users/Kevin/airflow/dags/ratio_dag.py
Created <DagRun ratio_dag @ 2020-01-27 01:59:24+00:00: manual__2020-01-27T01:59:24+00:00, externally triggered: True>
```

Figure 9.47: ratio_dag is successfully scheduled to be executed

The **trigger_dag** action will trigger the DAG to run immediately. The CLI requires the DAG ID, **ratio_dag**. However, there is also a **-c** positional argument, which is used for passing parameters into each operator in our DAG.

10. Launch the Airflow UI to monitor our DAG using the following command in a new Terminal:

```
airflow webserver
```

11. Open your browser and go to http://localhost:8080/admin/ to open the UI of the Airflow system. Look for the **ratio_dag** DAG and click the **Graph View** icon as shown in the following figure:

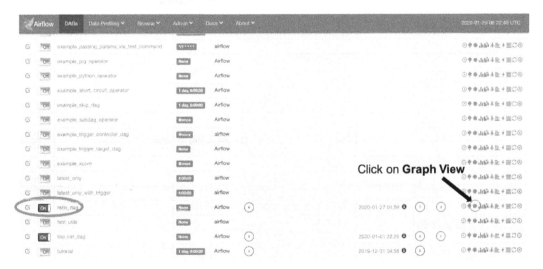

Figure 9.48: Airflow UI dashboard

12. Clicking the **Graph View** icon will redirect you to the following page:

Figure 9.49: Monitoring a running pipeline

As we can see from this dashboard, every step in the pipeline is marked with a dark green border, which means the step completed successfully. All tiles have a dark green border, which means our data pipeline successfully finished.

13. List the files in **../Data/** by running the following command in the Terminal:

```
ls ../Data/
```

You should get the following output:

```
Ratio_Likes_Dislikes.csv      US_category_id.json
   USvideos.csv.zip      top_10_trendy_cats.csv
   top_10_trendy_vids.csv
```

If you see the first file as **Ratio_Likes_Dislikes.csv**, it means the pipeline is successfully running.

14. Open a new **Python 3** Jupyter Notebook, import **pandas**, and use the **read_csv** function to read the **Ratio_Likes_Dislikes.csv** file, as shown in the following code:

```
import pandas as pd
pd.read_csv('../Data/Ratio_Likes_Dislikes.csv ')
```

You should get the following output:

	category	likes	dislikes	ratio_likes_dislikes
0	Shows	1082639	24508	44.174922
1	Pets & Animals	19370702	527379	36.730135
2	Education	49257772	1351972	36.434018
3	Comedy	216346746	7230391	29.921860
4	Howto & Style	162880075	5473899	29.755769
5	Music	1416838584	51179008	27.683979
6	Film & Animation	165997476	6075148	27.324022
7	Sports	98621211	5133551	19.211110
8	People & Blogs	186615999	10187901	18.317414
9	Science & Technology	82532638	4548402	18.145414
10	Autos & Vehicles	4245656	243010	17.471116
11	Travel & Events	4836246	340427	14.206411
12	Entertainment	530516491	42987663	12.341134
13	Gaming	69038284	9184466	7.516853
14	Nonprofits & Activism	14815646	3310381	4.475511
15	News & Politics	18151033	4180049	4.342301

Figure 9.50: Ratio of likes to dislikes for each category

NOTE

To access the source code for this specific section, please refer to https://packt.live/32puldZ.

CHAPTER 10: INTRODUCTION TO DATA STORAGE ON CLOUD SERVICES (AWS)

ACTIVITY 10.01: TRANSFORMING A TABLE SCHEMA INTO DOCUMENT FORMAT AND UPLOADING IT TO CLOUD STORAGE

Solution

1. Create a directory called **Activity10.01** in the **Chapter10** directory to store the files for this activity.

2. Open your Terminal (macOS or Linux) or Command Prompt (Windows), navigate to the **Chapter10** directory, and type in **jupyter notebook**.

3. In the Jupyter Notebook, click the **Activity10.01** directory, create a notebook file with a Python 3 kernel, and add the following code:

```
import os
import json
import boto3
import shutil
import pandas as pd


# set your bucket name here
# 'ch10-data' is NOT your bucket. It's just an example here
# you should replace your bucket below
BUCKET_NAME = 'ch10-data'


# 1. download data from S3 bucket
s3_resource = boto3.resource('s3')
try:
    s3_resource.Bucket(BUCKET_NAME).download_file(
        'New_York_City_Leading_Causes_of_Death.csv',
        './tmp/New_York_City_Leading_Causes_of_Death.csv')
except FileNotFoundError:
    os.mkdir('tmp/')
    s3_resource.Bucket(BUCKET_NAME).download_file(
        'New_York_City_Leading_Causes_of_Death.csv',
        './tmp/New_York_City_Leading_Causes_of_Death.csv')


# read data
```

```
df_data =
  pd.read_csv('tmp/New_York_City_Leading_Causes_of_Death.csv')
df_data.head()
```

You should get the following output:

	Year	Leading Cause	Sex	Race Ethnicity	Deaths	Death Rate	Age Adjusted Death Rate
0	2010	Influenza (Flu) and Pneumonia (J09-J18)	F	Hispanic	228	18.7	23.1
1	2008	Accidents Except Drug Posioning (V01-X39, X43,...	F	Hispanic	68	5.8	6.6
2	2013	Accidents Except Drug Posioning (V01-X39, X43,...	M	White Non-Hispanic	271	20.1	17.9
3	2010	Cerebrovascular Disease (Stroke: I60-I69)	M	Hispanic	140	12.3	21.4
4	2009	Assault (Homicide: Y87.1, X85-Y09)	M	Black Non-Hispanic	255	30	30
5	2012	Mental and Behavioral Disorders due to Acciden...	F	Other Race/ Ethnicity	.	.	.
6	2012	Cerebrovascular Disease (Stroke: I60-I69)	F	Asian and Pacific Islander	102	17.5	20.7

Figure 10.43: View of the New_York_City_Leading_Causes_of_Death.csv file

We downloaded a file from the S3 bucket first. Once the data was downloaded into the local directory, we used **pandas** to read the **New_York_City_Leading_Causes_of_Death.csv** file. Based on your first impression of the data, you can see that the data needs some preprocessing. For example, the value at row five of the **Deaths** column is **.**, whose data type is different compared to the rest of the values in the same column. There are two approaches we can use: we can either replace **.** with **0** or remove the row containing the **.** value.

4. Preprocess the data using the following code:

```
# 2. replace "." with value 0 & and convert to float type
df_data_cleaned = df_data.replace('.', 0).astype({\
                                'Deaths': float})


# check dtypes
df_data_cleaned.dtypes
```

You should get the following output:

```
Year                        int64
Leading Cause               object
Sex                         object
Race Ethnicity              object
Deaths                      float64
Death Rate                  object
Age Adjusted Death Rate     object
dtype: object
```

Figure 10.44: Data type of each column

5. Iterate over the different race-ethnicities and aggregate the death counts for each death that was caused, as shown in the following code:

```
# 3. get top 3 death causes for each ethnicity
top_causes = {}

for ethnicity, df_g in df_data_cleaned.groupby([\
                          'Race Ethnicity']):
    df_top_3_causes = df_g.groupby('Leading
       Cause')[['Deaths']].sum().sort_values(\
                'Deaths', ascending=False).head(3)
    top_3_causes = df_top_3_causes.index.values.tolist()
    top_causes.update({ethnicity: top_3_causes})


top_causes
```

You should get the following output:

```
{'Asian and Pacific Islander': ['Malignant Neoplasms (Cancer: C00-C97)',
 'Diseases of Heart (I00-I09, I11, I13, I20-I51)',
 'All Other Causes'],
 'Black Non-Hispanic': ['Diseases of Heart (I00-I09, I11, I13, I20-I51)',
 'Malignant Neoplasms (Cancer: C00-C97)',
 'All Other Causes'],
 'Hispanic': ['Diseases of Heart (I00-I09, I11, I13, I20-I51)',
 'Malignant Neoplasms (Cancer: C00-C97)',
 'All Other Causes'],
 'Not Stated/Unknown': ['Diseases of Heart (I00-I09, I11, I13, I20-I51)',
 'All Other Causes',
 'Malignant Neoplasms (Cancer: C00-C97)'],
 'Other Race/ Ethnicity': ['Diseases of Heart (I00-I09, I11, I13, I20-I51)',
 'Malignant Neoplasms (Cancer: C00-C97)',
 'All Other Causes'],
 'White Non-Hispanic': ['Diseases of Heart (I00-I09, I11, I13, I20-I51)',
 'Malignant Neoplasms (Cancer: C00-C97)',
 'All Other Causes']}
```

Figure 10.45: Output of group by aggregate

We group the death count based on each death cause per ethnicity. Then, we sort death causes based on their death counts and keep the top three death causes for each ethnicity.

> **NOTE**
>
> The **groupby** clause in **pandas** can be hard to comprehend if this is your first time using **pandas**. It might be useful if you print the output of the data when you use the **groupby** clause so that you can understand what **groupby** is doing.

6. As we can see, the last piece of output is a dictionary object. We need to dump this data into JSON format and write it to a JSON file using the following code:

```
with open('tmp/top_causes_per_ethnicity.json', 'w') as fout:
    json.dump(top_causes, fout)
```

7. The last step in this **Extract**, **Transform**, and **Load** (**ETL**) pipeline is to upload this output JSON file to the S3 bucket and remove the **tmp/** directory, as shown in the following code:

```
# 5. upload data to S3
s3_resource.Bucket('ch10-data').upload_file(
    'tmp/top_causes_per_ethnicity.json',
    'top_causes_per_ethnicity.json')

# clean up tmp
shutil.rmtree('./tmp')
```

8. Create a Python script named **ETL_pipeline.py** in the **Activity10.01** directory and put all of the code snippets together in the Python script file. The file should contain the following code:

ETL_pipeline.py

```
14    # 1. download data from S3 bucket
15    s3_resource = boto3.resource('s3')
16    try:
17        s3_resource.Bucket(BUCKET_NAME).download_file(
18            'New_York_City_Loading_Causes_of Death.csv',
19            './tmp/New_York_City_Leading_Causes_of_Death.csv')
20    except FileNotFoundError:
21        os.mkdir('tmp/')
22        s3_resource.Bucket(BUCKET_NAME).download_file(
23            'New_York_City_Leading_Causes_of_Death.csv',
24            './tmp/New_York_City_Leading_Causes_of_Death.csv')
```

The full code is available at https://packt.live/2Wh3e6q.

9. Create a Python script named **ETL_pipeline.py** in the **Activity10.01** directory. Add the preceding code snippet inside this Python script and save it.

10. To run this pipeline, we will run the following command in a Terminal:

```
python ETL_pipeline.py
```

The script will create a **top_causes_per_ethnicity.json** file and upload it to the cloud.

11. After the program has finished executing, we can verify whether the new data has been generated successfully or not by running the AWS CLI:

```
aws s3 ls s3://${BUCKET_NAME}/
```

You should get the following output:

```
2020-01-12 16:15:42    91294 New_York_City_Leading_Causes_of_Death.csv
2020-01-15 23:22:04      513 New_York_City_Top10_Causes.csv
2020-01-19 18:08:17      811 top_causes_per_ethnicity.json
```

Figure 10.46: Final output data has been successfully uploaded to the S3 bucket

NOTE

To access the source code for this specific section, please refer to https://packt.live/38TFoCt.

CHAPTER 11: BUILDING AN ARTIFICIAL INTELLIGENCE ALGORITHM

ACTIVITY 11.01: IMPLEMENTING A DOUBLE DEEP Q-LEARNING ALGORITHM TO SOLVE THE CART POLE PROBLEM

Solution

1. In the **Chapter11** directory, launch a Jupyter Notebook in your Terminal (macOS or Linux) or Command Prompt window (Windows).

2. After the Jupyter Notebook is launched, create a new directory named **Activity11.01**. Inside the **Activity11.01** directory, create a Python 3 notebook.

3. Inside the Python 3 notebook, import all necessary modules and seed the environment as shown in the following code:

```
# import module
import random
import numpy as np
from itertools import count
from collections import deque

import torch
import torch.nn as nn
import torch.nn.functional as F
import torch.optim as optim

import gym

# make game
env = gym.make('CartPole-v1')

# seed the experiment
env.seed(9)
np.random.seed(9)
random.seed(9)
torch.manual_seed(9)
```

4. Let's define our DQN as shown in the following code:

```
# define our policy
class DQN(nn.Module):
```

```
def __init__(self, observation_space, action_space):
    super(DQN, self).__init__()
    self.observation_space = observation_space
    self.action_space = action_space
    self.fc1 = nn.Linear(self.observation_space, 32)
    self.fc2 = nn.Linear(32, 16)
    self.fc3 = nn.Linear(16, self.action_space)

def forward(self, x):
    x = F.relu(self.fc1(x))
    x = F.relu(self.fc2(x))
    x = self.fc3(x)
    return x
```

We will use the three-layer feedforward neural network model from *Exercise 11.04, Implementing a Deep Q-Learning Algorithm in PyTorch to Solve the Classic Cart Pole Problem.*

5. Next, we will create the agent to play the game as shown in the following code:

act01-notebook.ipynb

```
# define our agent
class Agent:
    def __init__(self, policy_net, target_net):
        MEMORY_SIZE = 10000
        GAMMA = 0.6
        BATCH_SIZE = 128
        EXPLORATION_MAX = 0.9
        EXPLORATION_MIN = 0.05
        EXPLORATION_DECAY = 0.95
        # 1, 2, 3
        TARGET_UPDATE = 1

        self.policy_net = policy_net
        self.target_net = target_net
        self.target_net.load_state_dict(policy_net.state_dict())
        self.target_net.eval()

        self.optimizer = optim.RMSprop(\
                        policy_net.parameters(), lr=1e-3)
        self.memory = deque(maxlen=MEMORY_SIZE)
        self.gamma = GAMMA
        self.batch_size = BATCH_SIZE
        self.exploration_rate = EXPLORATION_MAX
        self.exploration_min = EXPLORATION_MIN
        self.exploration_decay = EXPLORATION_DECAY
        self.target_update = TARGET_UPDATE
```

The full code is available at https://packt.live/309071q.

Notice that we implement one more method: **update_target_net()**. This method is for updating the weights for the parameters in the target network. During the training time, we keep the weights in the target network frozen for a while to add stability to the expected Q value. We only update the weights for the target network every once in a while. How often we update the weights for the target network is controlled using the **TARGET_UPDATE** parameter. **TARGET_UPDATE = 1** means we update the weights for the target network after every 1 episode.

In the **experience_replay()** method, we train our policy network for every timestep. Recall the following loss function:

$$Loss = Q^{\pi}(s, a) - (r + \gamma \max_{a} Q^{\pi}(s', \pi(s')))$$

Figure 11.34: Loss function of the Q-learning algorithm

We use a policy network to calculate the Q values (the first term in the equation) for every previous state and chosen action. We then use the target network to calculate the expected Q values (the second term in the previous equation) using the next state, and the optimal action for every timestep.

6. Finally, we will implement the double deep Q-learning training loop as shown in the following code:

act01-notebook.ipynb

```
# create policy
observation_space = env.observation_space.shape[0]
action_space = env.action_space.n
policy_net = DQN(observation_space, action_space)
target_net = DQN(observation_space, action_space)

# create agent
agent = Agent(policy_net, target_net)

# play game
game_durations = []
for i_episode in count(1):
    state = env.reset()
    state = torch.tensor([state]).float()
    print("[ episode {} ] state={}".format(i_episode, state))
    for t in range(1, 10000):
        action = agent.select_action(state)
        state_next, reward, done, _ = env.step(action.item())
        if done:
            state_next = None
```

The full code is available at https://packt.live/309071q.

Notice that we created two deep Q neural network instances, **policy_net** and **target_net**, using the same model architecture with the **DQN** class to successfully create the double deep-Q learning algorithm.

After you run the training loop, you should get the following output:

```
[ episode 133 ][ timestamp 476 ] state=tensor([[-2.2825, -0.4745, -0.0851, -0.6469]]), action=tensor([[1]]), reward=
1.0, next_state=tensor([[-2.2920, -0.2783, -0.0980, -0.9651]])
[ Experience replay ] starts
[ episode 133 ][ timestamp 477 ] state=tensor([[-2.2920, -0.2783, -0.0980, -0.9651]]), action=tensor([[1]]), reward=
1.0, next_state=tensor([[-2.2976, -0.0821, -0.1174, -1.2869]])
[ Experience replay ] starts
[ episode 133 ][ timestamp 478 ] state=tensor([[-2.2976, -0.0821, -0.1174, -1.2869]]), action=tensor([[1]]), reward=
1.0, next_state=tensor([[-2.2992,  0.1144, -0.1431, -1.6139]])
[ Experience replay ] starts
[ episode 133 ][ timestamp 479 ] state=tensor([[-2.2992,  0.1144, -0.1431, -1.6139]]), action=tensor([[1]]), reward=
1.0, next_state=tensor([[-2.2970,  0.3108, -0.1754, -1.9476]])
[ Experience replay ] starts
[ episode 133 ][ timestamp 480 ] state=tensor([[-2.2970,  0.3108, -0.1754, -1.9476]]), action=tensor([[1]]), reward=
1.0, next_state=None
[ Experience replay ] starts
[ Ended! ] Episode 133: Exploration_rate=0.05. Score=480.
[ Solved! ] Score is now 480
```

Figure 11.35: Output logs of the main training loop

> **NOTE**
>
> We will have loads of output data with 133 episode results. We have only shown the last section of it in the preceding figure.

7. Lastly, let's visualize how our double deep Q-learning algorithm improves over time by running the following code:

```
import matplotlib.pyplot as plt
plt.style.use('ggplot')

plt.scatter(range(i_episode), game_durations)
plt.title('Game Duration Over Time')
plt.xlabel('game episode')
plt.ylabel('game duration')
plt.tight_layout()
```

You should get the following output:

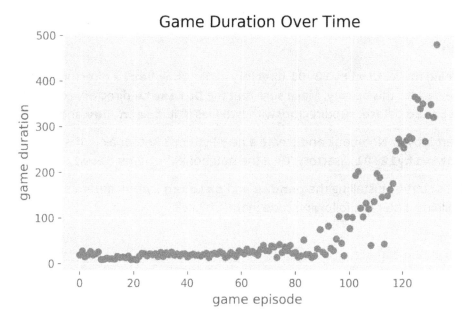

Figure 11.36: Game duration increases to 500 after 133 episodes

With the double Q-learning algorithm, the algorithm can converge faster in a more consistent fashion.

NOTE

To access the source code for this specific section, please refer to https://packt.live/309071q.

CHAPTER 12: PRODUCTIONIZING YOUR AI APPLICATIONS

ACTIVITY 12.01: PREDICTING THE CLASS OF A PASSENGER ON THE TITANIC

Solution

1. Create the **Activity12.01** directory in the **Chapter12** directory to store the files for this activity. Make sure that the **Datasets** directory contains the **Titanic** dataset subdirectory with two files in it: **train.csv** and **test.csv**.

2. Start Jupyter Notebook and create a new Python 3 notebook in the **Activity12.01** directory. Give the notebook the name **development**.

3. Let's start by installing the **pandas** and **sklearn** Python libraries for model building. Enter the following code in the first cell:

```
!pip install pandas
!pip install sklearn
```

It should give the following output:

```
WARNING: pip is being invoked by an old script wrapper. This will fail in a future version of pip.
Please see https://github.com/pypa/pip/issues/5599 for advice on fixing the underlying issue.
To avoid this problem you can invoke Python with '-m pip' instead of running pip directly.
Collecting pandas
  Downloading pandas-1.0.1-cp37-cp37m-manylinux1_x86_64.whl (10.1 MB)
    |████████████████████████████████| 10.1 MB 4.1 MB/s eta 0:00:01
Requirement already satisfied: python-dateutil>=2.6.1 in /home/bas/anaconda3/lib/python3.7/site-packages (from pan
das) (2.8.0)
Requirement already satisfied: pytz>=2017.2 in /home/bas/anaconda3/lib/python3.7/site-packages (from pandas) (201
9.3)
Requirement already satisfied: numpy>=1.13.3 in /home/bas/anaconda3/lib/python3.7/site-packages (from pandas) (1.1
7.2)
Requirement already satisfied: six>=1.5 in /home/bas/anaconda3/lib/python3.7/site-packages (from python-dateutil>=
2.6.1->pandas) (1.12.0)
Installing collected packages: pandas
Successfully installed pandas-1.0.1
WARNING: pip is being invoked by an old script wrapper. This will fail in a future version of pip.
Please see https://github.com/pypa/pip/issues/5599 for advice on fixing the underlying issue.
To avoid this problem you can invoke Python with '-m pip' instead of running pip directly.
Collecting sklearn
  Downloading sklearn-0.0.tar.gz (1.1 kB)
Requirement already satisfied: scikit-learn in /home/bas/anaconda3/lib/python3.7/site-packages (from sklearn) (0.2
1.3)
Requirement already satisfied: scipy>=0.17.0 in /home/bas/anaconda3/lib/python3.7/site-packages (from scikit-learn
->sklearn) (1.3.1)
Requirement already satisfied: joblib>=0.11 in /home/bas/anaconda3/lib/python3.7/site-packages (from scikit-learn-
>sklearn) (0.13.2)
Requirement already satisfied: numpy>=1.11.0 in /home/bas/anaconda3/lib/python3.7/site-packages (from scikit-learn
->sklearn) (1.17.2)
Building wheels for collected packages: sklearn
  Building wheel for sklearn (setup.py) ... done
  Created wheel for sklearn: filename=sklearn-0.0-py2.py3-none-any.whl size=1316 sha256=b98efd69f9eecee888121087ba
29933b9ee2d9a94ba4dce7f3013ae945cca958
  Stored in directory: /home/bas/.cache/pip/wheels/46/ef/c3/157e41f5ee1372d1be90b09f74f82b10e391eaacca8f22d33e
Successfully built sklearn
Installing collected packages: sklearn
Successfully installed sklearn-0.0
```

Figure 12.42: Installing pandas and sklearn

It will download the libraries and install them within your active Anaconda environment. There is a good chance that both frameworks are already available in your system, as part of Anaconda or from previous installations. If they are already installed, you will have the following output:

```
Requirement already satisfied: pandas in /usr/local/anaconda3/lib/python3.7/site-packages (0.25.1)
Requirement already satisfied: pytz>=2017.2 in /usr/local/anaconda3/lib/python3.7/site-packages (from pand
as) (2019.3)
Requirement already satisfied: numpy>=1.13.3 in /usr/local/anaconda3/lib/python3.7/site-packages (from pan
das) (1.17.2)
Requirement already satisfied: python-dateutil>=2.6.1 in /usr/local/anaconda3/lib/python3.7/site-packages
(from pandas) (2.8.0)
Requirement already satisfied: six>=1.5 in /usr/local/anaconda3/lib/python3.7/site-packages (from python-d
ateutil>=2.6.1->pandas) (1.12.0)
Requirement already satisfied: sklearn in /usr/local/anaconda3/lib/python3.7/site-packages (0.0)
Requirement already satisfied: scikit-learn in /usr/local/anaconda3/lib/python3.7/site-packages (from skle
arn) (0.21.3)
Requirement already satisfied: scipy>=0.17.0 in /usr/local/anaconda3/lib/python3.7/site-packages (from sci
kit-learn->sklearn) (1.3.1)
Requirement already satisfied: numpy>=1.11.0 in /usr/local/anaconda3/lib/python3.7/site-packages (from sci
kit-learn->sklearn) (1.17.2)
Requirement already satisfied: joblib>=0.11 in /usr/local/anaconda3/lib/python3.7/site-packages (from scik
it-learn->sklearn) (0.13.2)
```

Figure 12.43: pandas and sklearn already installed

4. Next, we import the **pickle**, **pandas**, and **sklearn** libraries:

```
import pickle
import pandas as pd
from sklearn.linear_model import LogisticRegression
```

5. Load the training data into a pandas DataFrame object called **train** with the following statement:

```
# load the datasets
train = pd.read_csv('../../Datasets/Titanic/train.csv')
```

6. After loading the training dataset, load the testing dataset with the following command:

```
test = pd.read_csv('../../Datasets/Titanic/test.csv')
```

7. We'll now continue to prepare and clean the training dataset as mentioned in *Chapter 3, Data Preparation*. Machine learning models work best with numerical datatypes for all columns, so let's convert the **Sex** column to zeros and ones:

```
train.Sex = train.Sex.map({'male':0, 'female':1})
```

We have now transformed the values in the **Sex** column to either 0 (for **male**) or 1 (for **female**).

8. Create the training set, **X**, which doesn't contain the **Pclass** output column, using the following code:

```
y = train.Pclass.copy()
X = train.drop(['Pclass'], axis=1)
```

Since the **Pclass** column contains our output value on which we have to train our model (the target values), we have to extract that from the dataset. We create a new dataset for it called **y** and then remove the column from the training dataset. We call the new training set **X**.

Now, let's do some feature engineering. We can be quite certain that a lot of the columns will not hold any predictive value as to whether a person survived. For example, the name of someone and their passenger ID are interchangeable and will not contribute much to the predictive power of the machine learning model.

9. Remove the following set of columns that are not needed to predict the surviving passengers:

```
X.drop(['Name'], axis=1, inplace=True)
X.drop(['Embarked'], axis=1, inplace=True)
X.drop(['PassengerId'], axis=1, inplace=True)
X.drop(['Cabin'], axis=1, inplace=True)
X.drop(['Ticket'], axis=1, inplace=True)
# X.drop(['Fare'], axis=1, inplace=True)
```

10. Replace the empty values with the **mean** age of all the passengers in the dataset:

```
X.Age.fillna(X.Age.mean(), inplace=True)
X.info()
```

11. Let's now train the actual model. We'll create a logistic regression model, which is essentially a mathematical algorithm to separate the survivors from the people who died in the accident:

```
# create and train a simple model
from sklearn.linear_model import LogisticRegression
model = LogisticRegression()
model.fit(X, y)
```

12. To see what the accuracy of the model is, we can run the full set of test data into the model and see in how many cases the algorithm produced a correct outcome using the following code:

```
# evaluate the model
model.score(X, y)
```

13. Finally, we have to serialize (export) the model to a file with the **pickle** framework:

```
# export the model to pickle file
file = open('model.pkl', 'wb') # write in bytes
pickle.dump(model, file)
file.close()
```

The **pickle.dump** method serializes the model to a **model.pkl** file in the activity directory. The file has been opened as **wb**, which means that it will write bytes to disk. You can check the file exists in the **Activity12.01** folder.

The second part of this activity is to load the model from disk in an API and expose the API.

14. Create a new Python 3 notebook in the **Activity12.01** directory and give it the name **production**.

> **NOTE**
>
> The **production.ipynb** file can be found here:
> https://packt.live/2ZqDc2n.

15. In the first cell, enter the following line to install Flask:

```
!pip install flask
```

You'll get the following output:

```
WARNING: pip is being invoked by an old script wrapper. This will fail in a future version of pip.
Please see https://github.com/pypa/pip/issues/5599 for advice on fixing the underlying issue.
To avoid this problem you can invoke Python with '-m pip' instead of running pip directly.
Collecting flask
  Downloading Flask-1.1.1-py2.py3-none-any.whl (94 kB)
     |████████████████████████████████| 94 kB 2.2 MB/s eta 0:00:011
Requirement already satisfied: Werkzeug>=0.15 in /home/bas/anaconda3/lib/python3.7/site-packages (from flask) (0.1
6.0)
Requirement already satisfied: click>=5.1 in /home/bas/anaconda3/lib/python3.7/site-packages (from flask) (7.0)
Requirement already satisfied: itsdangerous>=0.24 in /home/bas/anaconda3/lib/python3.7/site-packages (from flask)
(1.1.0)
Requirement already satisfied: Jinja2>=2.10.1 in /home/bas/.local/lib/python3.7/site-packages (from flask) (2.11.
1)
Requirement already satisfied: MarkupSafe>=0.23 in /home/bas/anaconda3/lib/python3.7/site-packages (from Jinja2>=
2.10.1->flask) (1.1.1)
Installing collected packages: flask
Successfully installed flask-1.1.1
```

Figure 12.44: Installing flask

If Flask is already installed, you will have the following output:

```
Requirement already satisfied: flask in /usr/local/anaconda3/lib/python3.7/site-packages (1.1.1)
Requirement already satisfied: itsdangerous>=0.24 in /usr/local/anaconda3/lib/python3.7/site-packages (fro
m flask) (1.1.0)
Requirement already satisfied: click>=5.1 in /usr/local/anaconda3/lib/python3.7/site-packages (from flask)
(7.0)
Requirement already satisfied: Jinja2>=2.10.1 in /usr/local/anaconda3/lib/python3.7/site-packages (from fl
ask) (2.10.3)
Requirement already satisfied: Werkzeug>=0.15 in /usr/local/anaconda3/lib/python3.7/site-packages (from fl
ask) (0.16.0)
Requirement already satisfied: MarkupSafe>=0.23 in /usr/local/anaconda3/lib/python3.7/site-packages (from
Jinja2>=2.10.1->flask) (1.1.1)
```

Figure 12.45: Flask already installed

16. Now import the required libraries to deserialize the model and to create an API:

```
from flask import Flask, jsonify, request
import pickle
```

17. The model that we trained in the first part was stored in the **model.pkl** file. Let's get that file from disk and deserialize the model into memory:

```
# load the model from pickle file
file = open('model.pkl', 'rb')  # read bytes
model = pickle.load(file)
file.close()
```

We now have the same model running in our **production** environment (the **production** Jupyter Notebook) as in our **model training** environment (the **development** notebook from part 1 of this activity). To test the model, we can make a few predictions.

18. Call the **predict** function of the model and give it an array of values that represent a person; the values are in the same order as the columns of the training set, as shown in the following code:

```
# get predictions from the model
print(model.predict([[1,0,36,2,0,14.67]]))
print(model.predict([[0,1,42,1,1,96.61]]))
```

19. Let's continue with productionizing our model. We first must set up a Flask app, which we'll call **ClassPredictor**:

```
# create an API with Flask
app = Flask('ClassPredictor')
```

20. The **app** is now an empty API with no HTTP endpoint exposed. To make things a bit more interesting and test whether it works, add a **GET** method that returns a simple string:

```
# call this: curl -X GET http://127.0.0.1:5000/foo
@app.route('/hi', methods=['GET'])
def bar():
    result = 'hello!'
    return result
```

21. We have to add a new method to the API that executes the model and returns the result, as shown in the following code:

```
@app.route('/class', methods=['POST'])
def predict_class():
    payload = request.get_json()
    person = [payload['Survived'], payload['Sex'], \
              payload['Age'], payload['SibSb'], \
              payload['Parch'], payload['Fare']]
    result = model.predict([person])
    print(f'{person} -> {str(result)}')
    return f'I predict that person {person} was in \
    class {result} of the Titanic\n'


app.run()
```

These lines define an HTTP **POST** method under the URL **'/class'**. When called, a **person** object is generated from the JSON payload. The **person** object, which is an array of input parameters for the model, is then passed to the model in the **predict** statement that we've seen before. The result is finally wrapped up in a string and returned to the caller.

Run the app again in the **production** Jupyter Notebook by executing the cell with the **app.run()** statement (stop it by clicking on the interrupt icon (■) or by typing *Ctrl + C* first, if needed). This time, we will test the API with a **curl** statement. cURL is a program that allows you to make an HTTP request across a network in a similar way to how a web browser makes requests, only the result will be just text instead of a graphical interface.

22. Open a new Terminal window or Anaconda prompt, and enter the following command:

```
curl -X POST -H "Content-Type: application/json" -d '{"Survived":
1, "Sex": 0, "Age": 72, "SibSb": 2, "Parch": 0, "Fare": 28.35}'
http://127.0.0.1:5000/class
```

> **NOTE**
>
> There is a Jupyter Notebook called **validation** in GitHub that contains the same command in a cell. It can be found here: https://packt.live/3erHRp4

The string after the **-d** parameter in the **curl** script contains a JSON object with passenger data. The fields for a person have to be named explicitly.

After running this, you'll see the result of your API call that executed the model as follows in your Terminal or Anaconda prompt:

```
I predict that person [1, 0, 72, 2, 0,28.35] was in class [2] of
the Titanic
```

> **NOTE**
>
> To access the source code for this specific section, please refer to https://packt.live/38T32iF.

ACTIVITY 12.02: DEPLOYING A MACHINE LEARNING MODEL TO A KUBERNETES CLUSTER TO PREDICT THE CLASS OF TITANIC PASSENGERS

Solution

1. Create a new directory, **Activity12.02**, in the **Chapter12** directory to store the files for this activity.

2. Copy the **model.pkl** file from the **Chapter12/Activity12.01** directory into the **Activity12.02** directory.

3. Let's create a new API first. Since we'll work outside of Jupyter on this one, we have to make sure that it runs on our local machine. Create a new file in the **Activity12.02** folder called **api.py**. Open it with a text editor or an IDE and enter the following Python code:

```python
from flask import Flask, request
import pickle

# load the model from pickle file
file = open('model.pkl', 'rb')  # read bytes
model = pickle.load(file)
file.close()

# create an API with Flask
app = Flask('Titanic')

@app.route('/class', methods=['POST'])
def predict_class():
    payload = request.get_json()
    person = [payload['Survived'], payload['Sex'],
        payload['Age'], payload['SibSb'], payload['Parch'],
        payload['Fare']]
    result = model.predict([person])
    print(f'{person} -> {str(result)}')
    return f'I predict that person {person} was in passenger
        class {result} of the Titanic\n'

app.run()
```

4. Open a new Terminal or Anaconda Prompt and navigate to the **Activity12.02** folder using the **cd** command. Enter the following command to check whether the API is working:

```
python api.py
```

If all is OK, you should get the message that your API is running on localhost.

5. Test the local API by opening a new Terminal or Anaconda Prompt and enter the following code:

```
curl -X POST -H "Content-Type: application/json" -d '{"Survived":
    0, "Sex": 1, "Age": 52, "SibSb": 1, "Parch": 0, "Fare": 82.35}'
    http://127.0.0.1:5000/class
```

6. Add a new file in the **Activity12.02** folder with the name **requirements. txt**. If we will deploy the API to another environment, it's good practice to indicate which libraries were used. To do so, Python offers the **requirements. txt** file, where you can write the dependencies of an application.

7. Open the **requirements.txt** file in an editor and enter the following lines:

```
Flask
sklearn
pandas
```

> **NOTE**
>
> It's possible to specify the exact version, for example, by entering **Flask==1.1.1**. To see which version you have in your development environment, enter **pip freeze**. To export the current list of dependencies and store them as **requirements.txt**, enter **pip freeze > requirements.txt**.

Now, let's continue with containerizing the API to make it ready to deploy to a production environment. We need to create a Docker *image*, which is a template for creating the actual Docker *containers* that will be deployed.

8. First, let's verify that Docker is installed correctly on your system. If you have followed the installation instructions, enter the following command in a Terminal or Anaconda Prompt:

```
sudo docker run hello-world
```

9. Instead of working with a pre-defined Docker image, we want to set up our own. Docker images that are defined in strict files are named **Dockerfile**. So, let's add a new text file in the same directory and call it **Dockerfile** (without an extension). Open **Dockerfile** in a text editor or IDE, enter the following code and save the file:

```
FROM python:3.7

RUN mkdir /api
WORKDIR /api
ADD . /api/

RUN pip install -r ./requirements.txt

EXPOSE 5000

ENV PYTHONPATH="$PYTHONPATH:/api"
CMD ["python", "/api/api.py"]
```

10. Open a new Terminal window (or Anaconda Prompt) in the **Activity12.02** directory and enter the following command to create an image:

```
sudo docker build -t titanic .
```

In the output, it becomes clear that all steps in our **Dockerfile** have been followed. The Flask library is loaded, port **5000** is exposed, and the API is running.

11. Let's check our local image repository again:

```
sudo docker image ls
```

It's great that we have a Docker image now, but that image has to be published to a Docker **registry** when we want to deploy it. Docker Hub is the central repository, but we don't want our Titanic API to end up there.

12. Create a local **registry** and publish our image there, ready to be deployed to Kubernetes in the next part of this activity. Enter the following command in your Terminal:

```
docker run -d -p 6000:5000 --restart=always --name registry
    registry:2
```

This will download the registry libraries and will run a local registry. The output is as follows:

```
--restart=always --name registry registry:2
Unable to find image 'registry:2' locally
2: Pulling from library/registry
486039affc0a: Pull complete
ba51a3b098e6: Pull complete
8bb4c43d6c8e: Pull complete
6f5f453e5f2d: Pull complete
42bc10b72f42: Pull complete
Digest: sha256:7d081088e4bfd632a88e3f3bcd9e007ef44a796fddfe3261407a3f9f04abe1e7
Status: Downloaded newer image for registry:2
98869ab76c12c518e80e86fd8a2c6d19333bc85c2d2fe20d4671577a4f7f1428
```

Figure 12.46: Creating a local registry

13. Tag your Docker image with the following command to give it a suitable name:

```
docker tag titanic localhost:6000/titanic
```

14. Now, push the **titanic** image to the running local registry with the following command:

```
docker push localhost:6000/titanic
```

This generates the following output:

```
The push refers to repository [localhost:5000/titanic]
0a14e0dbee45: Pushed
6559eb7a04fd: Pushed
f83dca05cf96: Pushed
e77942735734: Pushed
69f872f2a048: Pushed
2de70d7dbfa9: Pushed
3dffd131f01f: Pushed
271910c4c150: Pushed
6670e930ed33: Pushed
c7f27a4eb870: Pushed
e70dfb4c3a48: Pushed
1c76bd0dc325: Pushed
latest: digest: sha256:3c0d9793acccdebb021848cf2ce8f4bd2d3ab2ab6e214ccf479ff68ac82be82a size: 2844
```

Figure 12.47: Pushing a Docker image to a local registry

15. To verify that your image is pushed to the local registry, enter the following command:

```
curl -X GET http://localhost:6000/v2/titanic/tags/list
```

If all is well, you'll see the image with the **latest** tag:

```
{"name":"titanic","tags":["latest"]}
```

We have just successfully published our **titanic** image to that registry.

In the second part of this activity, we'll use the Docker image to host a container on a Kubernetes cluster.

16. We first have to make sure that Minikube, the local version of Kubernetes, is running. Minikube is, in fact, a local virtual machine. Type the following command in a new Terminal or Anaconda Prompt:

```
minikube version
```

If all is OK, this will produce an output like **minikube version: v1.8.1**, along with a **commit** hashtag as follows:

```
minikube version: v1.8.1
commit:   cbda04cf6bbe65e987ae52bb393c10099ab62014
```

17. Start **minikube** by entering the following command in a new Terminal window or Anaconda Prompt:

```
minikube start --insecure-registry="localhost:6000"
```

If you're running within a virtual machine like VirtualBox, the command will be **minikube start --driver=none**. If all goes well, you'll get the following output:

Figure 12.48: Starting Minikube

18. Confirm that Minikube is running by using the following command:

```
minikube status
```

You'll get a status update like the following:

```
host:  Running
kubelet:  Running
apiserver:  Running
kubeconfig:  Configured
```

19. If you want to look even further into your Kubernetes cluster, start up a dashboard with the following command:

```
minikube dashboard
```

This will open a browser window with a lot of useful information and configuration options:

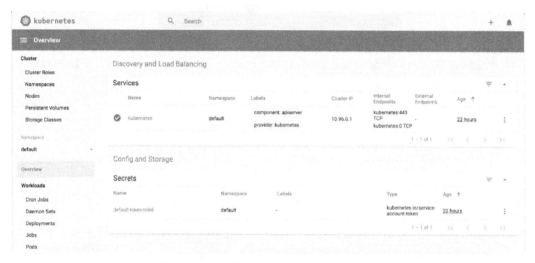

Figure 12.49: The Kubernetes dashboard

20. We have to connect our now-running Minikube Kubernetes cluster to a command-line interface, called **kubectl**. Type in the following command in your Terminal or Anaconda Prompt, as was already suggested in the previous output:

```
minikube kubectl
```

This will produce an output like the following:

```
   > kubectl.sha256: 65 B / 65 B [————————————————————] 100.00% ? p/s 0s
   > kubectl: 47.28 MiB / 47.28 MiB [————————————] 100.00% 1.25 MiB p/s 38s
kubectl controls the Kubernetes cluster manager.

Find more information at: https://kubernetes.io/docs/reference/kubectl/overview/

Basic Commands (Beginner):
  create       Create a resource from a file or from stdin.
  expose       Take a replication controller, service, deployment or pod and expose it as a new Kubernetes Service
  run          Run a particular image on the cluster
  set          Set specific features on objects

Basic Commands (Intermediate):
  explain      Documentation of resources
  get          Display one or many resources
  edit         Edit a resource on the server
  delete       Delete resources by filenames, stdin, resources and names, or by resources and label selector
```

Figure 12.50: Connecting Minikube to your kubectl command

You can read the **kubectl controls the Kubernetes cluster manager** line, which indicates that we can now use the **kubectl** tool to give commands to our cluster. You can get more information about the cluster with the **kubectl version** and **kubectl get nodes** commands:

```
         ~  kubectl version                                                ✓  anaconda3    minikube    15:02:12 ☉
Client Version: version.Info{Major:"1", Minor:"15", GitVersion:"v1.15.5", GitCommit:"20c265fef0741dd71a66480e35bd69f18351daea", Git
TreeState:"clean", BuildDate:"2019-10-15T19:16:51Z", GoVersion:"go1.12.10", Compiler:"gc", Platform:"darwin/amd64"}
Server Version: version.Info{Major:"1", Minor:"17", GitVersion:"v1.17.3", GitCommit:"06ad960bfd03b39c8310aaf92d1e7c12ce618213", Git
TreeState:"clean", BuildDate:"2020-02-11T18:07:13Z", GoVersion:"go1.13.6", Compiler:"gc", Platform:"linux/amd64"}
         ~  kubectl get nodes                                              ✓  anaconda3    minikube    15:24:05 ☉
NAME    STATUS   ROLES    AGE    VERSION
m01     Ready    master   27m    v1.17.3
```

Figure 12.51: Getting information from Minikube

21. We have already created a Docker image and practiced deploying that to a Docker registry in part 1 of this activity. We'll rebuild the image now, but with the Docker daemon (process) of the Kubernetes cluster to make sure it lands on the Minikube cluster. Start by entering the following command:

```
eval $(minikube docker-env)
```

This command points our Terminal to use a different **docker** command, namely, the one in the Minikube environment.

22. Let's build the container again but now with the Minikube. Enter the following command:

```
docker build -t titanic .
```

You'll get the same output as in part 1 of this activity.

23. Create a Kubernetes deployment with the following command:

```
kubectl run titanic --image=titanic --image-pull-policy=Never
```

If this is successful, you'll see the following output:

```
deployment.apps/titanic created
```

24. Verify the deployment of the container with the following command:

```
kubectl get deployments
```

This will produce a list of deployed containers:

```
NAME      READY    UP-TO-DATE    AVAILABLE    AGE
titanic   1/1      1             1            3m
```

Figure 12.52: Retrieving a list of deployed containers from Kubernetes

You can also check the Kubernetes dashboard, if you have started it, and check whether the deployment and Pod have been created:

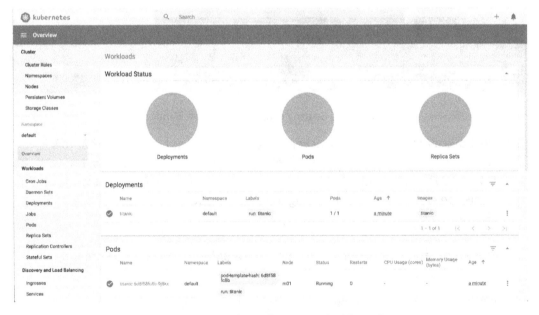

Figure 12.53: The Kubernetes dashboard

25. Now that our app is deployed, we want to connect to the **titanic** API. By default, the running containers can only be accessed by other resources (containers) in the same Kubernetes Pod. But we can instruct our cluster to create a network connection that allows communication from the outside world.:

```
kubectl port-forward titanic-6d8f58fc8b-znmx9 5000:5000
```

> **NOTE**
>
> In the preceding code, replace the name of the pod with your own; you can find it in the Kubernetes dashboard or by entering **kubectl get pods**

This will create a new port forwarding service in your Kubernetes cluster that forwards network traffic to the **titanic** Pod.

26. To test the API, enter the following command in a new terminal window (while keeping the port forwarding running):

```
curl -X POST -H "Content-Type: application/json" \
# -d '{"Survived": 1, "Sex": 0, "Age": 72, "SibSb": 2, "Parch":
  0, "Fare": 68.35}' http://127.0.0.1:5000/class
```

This will produce the following output:

```
I predict that person [1, 0, 72, 2, 0, 68.35] was in passenger
  class [1] of the Titanic
```

> **NOTE**
>
> To access the source code for this specific section, please refer to https://packt.live/32s2PC3.

ACTIVITY 12.03: PREDICTING THE CLASS OF TITANIC PASSENGERS IN REAL TIME

Solution

1. In the **Chapter12** directory, create the **Activity12.03** directory to store the files for this activity.

2. Open Jupyter Notebook and create a new Python 3 notebook called **export_pmml** in the **Activity12.03** folder.

3. Copy the **model.pkl** file from the **Activity12.01** directory to the **Activity12.03** directory. This can be done manually, or with the following statement in a notebook:

```
# copy the trained model for predicting the passenger
#class of a person
!cp ../Activity12.01/model.pkl .
```

4. Enter the following line in the first cell of the new notebook:

```
!pip install sklearn2pmml
```

This will install the required library, which can export our models to the PMML format:

```
WARNING: pip is being invoked by an old script wrapper. This will fail in a future version of pip.
Please see https://github.com/pypa/pip/issues/5599 for advice on fixing the underlying issue.
To avoid this problem you can invoke Python with '-m pip' instead of running pip directly.
Collecting sklearn2pmml
  Downloading sklearn2pmml-0.55.1.tar.gz (5.6 MB)
     |████████████████████████████████| 5.6 MB 4.1 MB/s eta 0:00:01
Requirement already satisfied: joblib>=0.13.0 in /home/bas/anaconda3/lib/python3.7/site-packages (from sklearn2pmm
l) (0.13.2)
Requirement already satisfied: scikit-learn>=0.18.0 in /home/bas/anaconda3/lib/python3.7/site-packages (from sklea
rn2pmml) (0.21.3)
Collecting sklearn-pandas>=0.0.10
  Downloading sklearn_pandas-1.8.0-py2.py3-none-any.whl (12 kB)
Requirement already satisfied: numpy>=1.11.0 in /home/bas/anaconda3/lib/python3.7/site-packages (from scikit-learn
>=0.18.0->sklearn2pmml) (1.17.2)
Requirement already satisfied: scipy>=0.17.0 in /home/bas/anaconda3/lib/python3.7/site-packages (from scikit-learn
>=0.18.0->sklearn2pmml) (1.3.1)
Requirement already satisfied: pandas>=0.11.0 in /home/bas/anaconda3/lib/python3.7/site-packages (from sklearn-pan
das>=0.0.10->sklearn2pmml) (1.0.1)
Requirement already satisfied: python-dateutil>=2.6.1 in /home/bas/anaconda3/lib/python3.7/site-packages (from pan
das>=0.11.0->sklearn-pandas>=0.0.10->sklearn2pmml) (2.8.0)
Requirement already satisfied: pytz>=2017.2 in /home/bas/anaconda3/lib/python3.7/site-packages (from pandas>=0.11.
0->sklearn-pandas>=0.0.10->sklearn2pmml) (2019.3)
Requirement already satisfied: six>=1.5 in /home/bas/anaconda3/lib/python3.7/site-packages (from python-dateutil>=
2.6.1->pandas>=0.11.0->sklearn-pandas>=0.0.10->sklearn2pmml) (1.12.0)
Building wheels for collected packages: sklearn2pmml
  Building wheel for sklearn2pmml (setup.py) ... done
  Created wheel for sklearn2pmml: filename=sklearn2pmml-0.55.1-py3-none-any.whl size=5636073 sha256=467d42831a6f06
ba4c2825110a63e7c4458c2fceffb284733e9fc88c26f7bd11
  Stored in directory: /home/bas/.cache/pip/wheels/d1/43/46/407e4a36a26f43a599e7ca185b894237be91355dbc8aa1ebde
Successfully built sklearn2pmml
Installing collected packages: sklearn-pandas, sklearn2pmml
Successfully installed sklearn-pandas-1.8.0 sklearn2pmml-0.55.1
```

Figure 12.54: Installation of sklearn2pmml

If **sklearn2pmml** is already installed, you will have the following output:

```
Requirement already satisfied: sklearn2pmml in /usr/local/anaconda3/lib/python3.7/site-packages (0.55.4)
Requirement already satisfied: joblib>=0.13.0 in /usr/local/anaconda3/lib/python3.7/site-packages (from sk
learn2pmml) (0.13.2)
Requirement already satisfied: scikit-learn>=0.18.0 in /usr/local/anaconda3/lib/python3.7/site-packages (f
rom sklearn2pmml) (0.21.3)
Requirement already satisfied: sklearn-pandas>=0.0.10 in /usr/local/anaconda3/lib/python3.7/site-packages
(from sklearn2pmml) (1.8.0)
Requirement already satisfied: numpy>=1.11.0 in /usr/local/anaconda3/lib/python3.7/site-packages (from sci
kit-learn>=0.18.0->sklearn2pmml) (1.17.2)
Requirement already satisfied: scipy>=0.17.0 in /usr/local/anaconda3/lib/python3.7/site-packages (from sci
kit-learn>=0.18.0->sklearn2pmml) (1.3.1)
Requirement already satisfied: pandas>=0.11.0 in /usr/local/anaconda3/lib/python3.7/site-packages (from sk
learn-pandas>=0.0.10->sklearn2pmml) (0.25.1)
Requirement already satisfied: python-dateutil>=2.6.1 in /usr/local/anaconda3/lib/python3.7/site-packages
(from pandas>=0.11.0->sklearn-pandas>=0.0.10->sklearn2pmml) (2.8.0)
Requirement already satisfied: pytz>=2017.2 in /usr/local/anaconda3/lib/python3.7/site-packages (from pand
as>=0.11.0->sklearn-pandas>=0.0.10->sklearn2pmml) (2019.3)
Requirement already satisfied: six>=1.5 in /usr/local/anaconda3/lib/python3.7/site-packages (from python-d
ateutil>=2.6.1->pandas>=0.11.0->sklearn-pandas>=0.0.10->sklearn2pmml) (1.12.0)
```

Figure 12.55: sklearn2pmml already installed

5. Create a new cell and enter the following lines to start using the **pickle** and **sklearn2pmml** libraries:

```
from sklearn2pmml import sklearn2pmml, make_pmml_pipeline
import pickle
```

6. Deserialize the pickled model by executing the following code in a new cell:

```
# load the model from pickle file
file = open('model.pkl', 'rb')  # read bytes
model = pickle.load(file)
file.close()
```

Now we can export the model to the PMML format.

7. Enter the following lines in a new cell to create a pipeline, export the method, and write it to the PMML file:

```
pmml_pipeline = make_pmml_pipeline(model)
sklearn2pmml(pmml_pipeline, 'titanic_class.pmml')
```

8. Check the contents of the generated **titanic_class.pmml** file by entering the following line in a new cell:

```
! cat titanic_class.pmml
```

This will produce the following output:

```xml
<?xml version="1.0" encoding="UTF-8" standalone="yes"?>
<PMML xmlns="http://www.dmg.org/PMML-4_3" xmlns:data="http://jpmml.org/jpmml-model/InlineTable" version="4.3">
    <Header>
        <Application name="JPMML-SkLearn" version="1.5.32"/>
        <Timestamp>2020-03-09T20:11:55Z</Timestamp>
    </Header>
    <DataDictionary>
        <DataField name="y" optype="categorical" dataType="integer">
            <Value value="1"/>
            <Value value="2"/>
            <Value value="3"/>
        </DataField>
        <DataField name="x1" optype="continuous" dataType="double"/>
        <DataField name="x2" optype="continuous" dataType="double"/>
        <DataField name="x3" optype="continuous" dataType="double"/>
        <DataField name="x4" optype="continuous" dataType="double"/>
        <DataField name="x5" optype="continuous" dataType="double"/>
        <DataField name="x6" optype="continuous" dataType="double"/>
    </DataDictionary>
```

Figure 12.56: Viewing a PMML model in a Jupyter Notebook

Now let's create a streaming job. We'll use Apache Flink for this. We'll write the code with an example from IntelliJ IDEA. You can also choose another IDE. It's also possible (though not recommended) to use a plain text editor and run the code from a Terminal or command line.

9. Open a new Terminal window in the **Activity12.03** directory and enter the following lines to set up Apache Flink:

```
mvn archetype:generate                                \
        -DarchetypeGroupId=org.apache.flink           \
        -DarchetypeArtifactId=flink-quickstart-java    \
        -DarchetypeVersion=1.10.0
```

10. During the installation, enter the following values:

```
groupId: com
artifactId: titanic_class
version: 0.0.1
package: packt
```

This will generate a project from a template.

11. Open your favorite IDE (IntelliJ IDEA, Eclipse, or another) and import the generated files by selecting the **pom.xml** file.

12. Remove the **BatchJob.java** file; we don't need it since we're only building a streaming job.
13. Copy the **titanic_class.pmml** file that we created in *Step 8* to the **titanic_class/src/resources** directory.
14. If you installed JDK and Maven, you should be able to compile the code. Type the following in your Terminal to compile the code and package it into a JAR file:

```
mvn clean package
```

In the output, you'll see **BUILD SUCCESS**, which indicates that you now have a working Java program.

Let's start by testing the job – first, the Maven template generated code that can be deployed to a Flink cluster.

15. Since we want to test locally first, let's change the configuration a bit. Open the **pom.xml** file and remove the **<provided>** tags in the dependencies of **flink-java** and **flink-streaming-java**; these tags are single lines within the **<dependency>** elements. The file will look like the following:

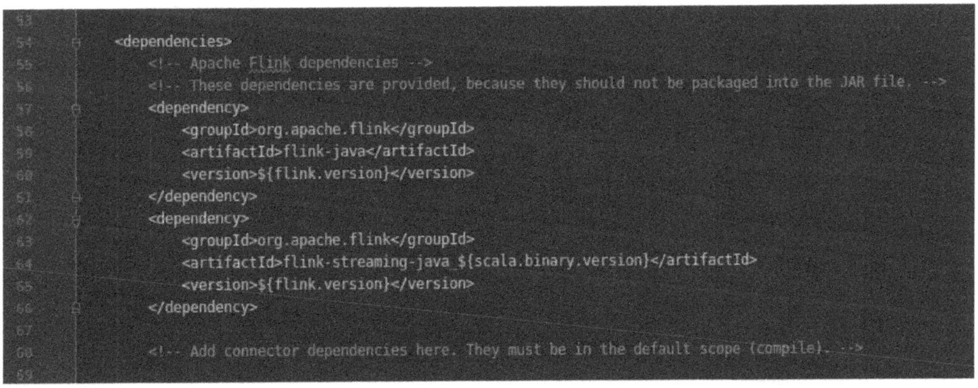

```
<dependencies>
    <!-- Apache Flink dependencies -->
    <!-- These dependencies are provided, because they should not be packaged into the JAR file. -->
    <dependency>
        <groupId>org.apache.flink</groupId>
        <artifactId>flink-java</artifactId>
        <version>${flink.version}</version>
    </dependency>
    <dependency>
        <groupId>org.apache.flink</groupId>
        <artifactId>flink-streaming-java_${scala.binary.version}</artifactId>
        <version>${flink.version}</version>
    </dependency>

    <!-- Add connector dependencies here. They must be in the default scope (compile). -->
```

Figure 12.57: The pom.xml file with the provided tags removed for Flink dependencies

Save the file, then import the Maven changes.

16. Open the **StreamingJob.java** file in your IDE and add the following lines at the top:

```
import \
org.apache.flink.api.common.serialization.SimpleStringEncoder;
import org.apache.flink.core.fs.Path;
import org.apache.flink.streaming.api.datastream.DataStream;
import org.apache.flink.api.common.functions.MapFunction;
import \
        org.apache.flink.streaming.api.datastream.\
        SingleOutputStreamOperator;
import \
        org.apache.flink.streaming.api.environment.\
        StreamExecutionEnvironment;
import \
        org.apache.flink.streaming.api.functions.\
        sink.filesystem.StreamingFileSink;
```

These lines will add the necessary Flink libraries to our **class** file.

17. In the middle part of the file, within the **main** function, add the following lines:

```
DataStream<String> dataStream = env.socketTextStream(\
                                "localhost", 1234, "\n");

StreamingFileSink<String> sink = StreamingFileSink\
    .forRowFormat(new Path("out"), \
    new SimpleStringEncoder<String>("UTF-8"))
    .build();


dataStream.addSink(sink);
```

This code sets up a data stream that listens to a local socket on port **1234**. It takes the lines and writes (sinks) the lines to a file in the **out** directory.

18. To write to a local socket, we use the Netcat tool. Test the simple code by opening a Terminal and typing the following command:

```
nc -l -p 1234
```

You get a prompt to enter lines. Leave it open for now.

19. Go back to your IDE and run the **StreamingJob.main()** method. Alternatively, type the following in a Terminal to build and run the code:

```
mvn clean package
mvn exec:java -Dexec.mainClass="packt.StreamingJob"
```

These lines are the Maven instructions to compile the code, package it into a JAR file, and run the file with an entry point in the **packt.StreamingJob** class that contains our **main** function.

20. In the Terminal that's still running Netcat, type a few lines of text, followed by *Enter* for each line, as follows:

Figure 12.58: Running a Flink job that processes input from a local socket

21. Now check the output in the **out** directory. You should see a subfolder with a date and time, and a file for each line that was entered in the input socket.

22. We have to add JPMML dependencies to our project for this. Add the following lines to **pom.xml** and import the Maven changes:

```xml
<dependency>
    <groupId>org.jpmml</groupId>
    <artifactId>pmml-evaluator</artifactId>
    <version>1.4.15</version>
</dependency>
<dependency>
    <groupId>org.jpmml</groupId>
    <artifactId>pmml-evaluator-extension</artifactId>
    <version>1.4.15</version>
</dependency>
```

23. Edit the **StreamingJob.java** file again and add the following lines at the top of the file:

```java
import org.dmg.pmml.FieldName;
import org.jpmml.evaluator.*;
import org.jpmml.evaluator.visitors.DefaultVisitorBattery;
import java.io.File;
import java.util.LinkedHashMap;
import java.util.List;
import java.util.Map;
```

These lines import the required PMML and common Java libraries for working with PMML files.

24. Enter the following lines at the beginning of the **main** method:

```java
// load PMML
Evaluator evaluator=
getPmmlEvaluator("titanic_class.pmml");

// get input fields
List<? Extends InputField> inputFields =
evaluator.getInputFields();
```

This code deserializes the model from PMML, loading the model in memory ready to be executed. The list of input fields will come in handy in the next step.

We are now ready to parse incoming messages as persons that can be evaluated according to whether they survived the Titanic disaster. The input values will be in a comma-separated string with the values **survived**, **sex**, **age**, **number of siblings on board**, **number of parents on board**, and **fare paid**. For example, "1,1,19,0,2,77.15" indicates a woman of 19 years old who entered the ship with her two parents and paid 77.15 for the ticket.

25. The PMML evaluator expects a map of **FieldName** value objects so we first have to convert the input strings. To do so, add the following lines to the class:

```
SingleOutputStreamOperator<String> mapped = dataStream.map(new
  MapFunction<String, String>() {
    @Override
    public String map(String s) throws Exception {
        System.out.println("EVENT: " + s);

        Map<FieldName, FieldValue> arguments = new
          LinkedHashMap<>();
        String[] values = s.split(",");

        // prepare model evaluation
        for (int i = 0; i < values.length; i++) {
            FieldName inputName = inputFields.get(i).getName();
            FieldValue inputValue =
              inputFields.get(i).prepare(values[i]);
            arguments.put(inputName, inputValue);
        }

        // execute the model
        Map<FieldName, ?> results =
          evaluator.evaluate(arguments);

        // Decoupling results from the JPMML-Evaluator runtime
          environment
        Map<String, ?> resultRecord =
          EvaluatorUtil.decodeAll(results);

        System.out.println(resultRecord);
        return s;
    }
});
```

This code adds a **map** method to the stream processing job that splits the input string and uses the resulting string values to build a set of arguments for the machine learning model.

26. Now, if you run this code together with Netcat for the input socket, you can get real-time predictions from the model. Enter the following set of comma-separated values in the **nc** prompt that is already running, followed by *Enter*, and see what predictions are made:

```
1,1,13,1,56.91
0,0,81,0,0,120.96
1,0,41,1,1,18.11
```

You should get the following output:

```
EVENT: 1,1,13,3,1,56.91
{y=2, probability(1)=0.05043057638902439, probability(2)=0.6529580945632668, probability(3)=0.29661132904770876}
EVENT: 0,0,81,0,0,120.96
{y=1, probability(1)=0.9528716249898004, probability(2)=0.047128334498293026, probability(3)=4.051190659187697E-8}
EVENT: 1,0,41,1,1,18.11
{y=3, probability(1)=0.02728000899651477, probability(2)=0.32434910004056755, probability(3)=0.6483708909629177}
```

Figure 12.59: Output of the Flink job that predicts the class of Titanic passengers

> **NOTE**
>
> To access the source code for this specific section, please refer to https://packt.live/38SssNr.

INDEX

A

apiserver: 550

d

databricks: 44,
    287-288, 290,
    295-296, 298-299,
    304, 306, 313-315,
    320, 322, 557
datastream:
    566-567, 570

F

flink-java: 566, 573

G

gaussian: 468
gettysburg: 313
ggplot: 468, 482,
    493, 511
groupbykey: 305, 321

J

javascript: 202-203,
    206-207, 212-214,
    216, 249-250,
    336, 452, 455
joblib: 521
json-based: 558
jsonify: 531

k

kubeconfig: 550
kubectl: 540, 551-553

kubelet: 540, 550
kubernetes:
    520, 538-541,
    548-556, 576
kwargs: 395-396

L

linearsvc: 18, 35

P

pickle: 520-522,
    524, 530-532,
    535-536, 541-543,
    557-558, 561, 572
pycharm: 115
pyplot: 152, 468, 482,
    493, 511, 514
pyspark: 83, 95-98,
    102, 117-119, 123
pytorch: 461, 477-484,
    486, 488, 491-492,
    494-495, 498, 500,
    506, 513, 516

Q

q-function: 504

S

snowflake: 61, 66,
    192, 194-195
spark-avro: 268
sparked: 72
sparkml: 557
sparks: 128
sqlite: 401, 441

U

unix-based: 83, 101

W

watermark: 114,
    121, 124

X

xml-based: 558